Lecture Notes in Mathematics

Edited by A. Dold and B. Eckmann

658

Geometric Applications of Homotopy Theory II

Proceedings, Evanston, March 21–26, 1977

Edited by
M. G. Barratt and M. E. Mahowald

Springer-Verlag
Berlin Heidelberg New York 1978

Editors
M. G. Barratt
M. E. Mahowald
Department of Mathematics
Northwestern University
Evanston, IL 60201/USA

AMS Subject Classifications (1970): 55D99, 55E99, 55G99

ISBN 3-540-08859-8 Springer-Verlag Berlin Heidelberg New York
ISBN 0-387-08859-8 Springer-Verlag New York Heidelberg Berlin

Printed in Germany

Printing and binding: Beltz Offsetdruck, Hemsbach/Bergstr.
2141/3140-543210

INTRODUCTION

A conference on geometric applications of homotopy theory jointly supported by the National Science Foundation and Northwestern University was held at Evanston March 21-26, 1977. These proceedings contain papers presented at the meeting and some related papers by mathematicians who attended the conference.

M. G. Barratt

M. E. Mahowald

Evanston December 1977

C O N T E N T S

CONTENTS OF VOLUME I *

* Volume I appeared as volume 657 of the Lecture Notes in Mathematics

---

NOT APPEARING IN THIS VOLUME:

MAPS BETWEEN CLASSIFYING SPACES

J.F. Adams

In what follows, G and G' will be compact Lie groups; BG and BG' will be their classifying spaces; and I want to study the classification of maps

$$f: BG \longrightarrow BG'.$$

What happens may be described in general terms; BG has a very rich and a very rigid structure, and the effect of this is that there are very few maps compared with what one might expect.

One can illustrate this by looking at a classical example. Take

$$G = G' = S^3 = Sp(1) = SU(2) = Spin(3).$$

Then

$$BG = BG' = HP^\infty,$$

the infinite-dimensional projective space over the quaternions. Its cohomology ring is a polynomial algebra:

$$H^*(HP^\infty; Z) = Z[x], \; x \in H^4.$$

For any map

$$f: HP^\infty \longrightarrow HP^\infty$$

we must have

$$f^*x = dx$$

for some $d \in Z$. We call d the <u>degree</u> of f.

<u>Proposition 1</u>

The integers d which arise as the degrees of maps $f: HP^\infty \longrightarrow HP^\infty$ are precisely 0 and the odd squares.

To prove that the degree is necessarily a square k^2 is not hard; there is a choice of methods. To prove that k is either zero or odd one uses symplectic K-theory. By far the most substantial part of the proof is the construction of maps with the degrees stated; this is due to Sullivan [5].

In particular, most of the maps $f: HP^\infty \longrightarrow HP^\infty$ constructed by Sullivan are not of the form $B\theta$ for any homomorphism $\theta: S^3 \longrightarrow S^3$ of Lie groups; for the integers d which arise as the degrees of maps $B\theta$ are precisely 0 and 1.

I have stated Proposition 1 in terms of ordinary cohomology. However, a theory which relies wholly on ordinary cohomology cannot be expected to work in a convenient and satisfactory way when the group G is not connected. At this point I should explain that my renewed interest in this subject was stimulated by conversations with C.B. Thomas. The direction of his work may be seen from [6]; and in his work the group G is finite. So I will try to cover the case in which G is not connected.

The appropriate measure is to classify maps $f: BG \longrightarrow BG'$ according to the induced map of K-theory

$$f^*: K(BG) \longleftarrow K(BG').$$

(When G is connected this gives the same classification as that in [1].) Here I recall that K(X) means the generalised cohomology theory of Grothendieck-Atiyah-Hirzebruch; for our purposes we should use representable K-theory,

$$K(X) = [X, X \times BU],$$

where [X,Y] means the set of homotopy classes of maps from X to Y. This is the best definition when X is an infinite complex, and BG is usually infinite.

The use of K-theory would hardly be profitable if we had no means of computing K(BG); fortunately we do. Let RG be the representation ring of the compact Lie group G. If $\theta: G \longrightarrow U(n)$ is a representation, we can form the composite

$$BG \xrightarrow{B\theta} BU(n) \subset n \times BU \subset Z \times BU,$$

and this composite gives an element $\alpha(\theta) \in K(BG)$; this construction defines a homomorphism of rings

$$\alpha: RG \longrightarrow K(BG).$$

<u>Proposition 2</u> [2,3,4].

The map α induces an isomorphism

$$\alpha^\wedge: RG^\wedge \longrightarrow K(BG).$$

Here $RG^\wedge$ means the completion of RG with respect to a topology which one has to describe. Consider the map of groups $1 \longrightarrow G$. This induces the "augmentation" map

$$\varepsilon: RG \longrightarrow R1 = Z,$$

which assigns to each representation $\theta: G \longrightarrow U(n)$ its dimension n. The augmentation ideal $I \subset RG$ is defined to be Ker ε. The topology in question is that in which the neighbourhoods of 0 are the powers I^n of the augmentation ideal.

This means that when G is a given finite group, general results expressed in terms of K(BG) can be interpreted by calculations with the character table of G; and these are calculations which algebraists prefer to homological calculations.

For example, take G = SL(2,5), the binary icosahedral group, and take G' = SU(2). We want to know the possible values for

$$f^*: K(BSL(2,5)) \longleftarrow K(BSU(2)).$$

Now the composite

$$BSU(2) \subset 2 \times BU \subset Z \times BU$$

defines an element $i_2 \in K(BSU(2))$; and it is sufficient to know f^*i_2, because this determines f^*x for every other element $x \in K(BSU(2))$. So we wish to know the composite

$$BSL(2,5) \xrightarrow{f} BSU(2) \xrightarrow{i_2} Z \times BU.$$

In order to describe it, let

$$i: SL(2,5) \longrightarrow SU(2)$$

be a fixed choice of one of the two standard embeddings. Then the general results I shall present specialise as follows.

Proposition 3.

(a) For any map

$$f: BSL(2,5) \longrightarrow BSU(2)$$

the composite

$$BSL(2,5) \xrightarrow{f} BSU(2) \xrightarrow{i_2} Z \times BU$$

is equal to

$$BSL(2,5) \xrightarrow{Bi} BSU(2) \xrightarrow{\psi^k} Z \times BU$$

for some integer k.

(b) Moreover, two composites

$$BSL(2,5) \xrightarrow{Bi} BSU(2) \xrightarrow{\psi^k} Z \times BU$$

$$BSL(2,5) \xrightarrow{Bi} BSU(2) \xrightarrow{\psi^\ell} Z \times BU$$

are equal if and only if they have the same second Chern class, that is if and only if $k^2 \equiv \ell^2 \bmod 120$.

Roughly speaking, this result says that to the eyes of K-theory, any map f: BSL(2,5) ⟶ BSU(2) looks like one of the examples constructed by Milnor. Here I recall that the examples constructed by Milnor are the composites

$$BSL(2,5) \xrightarrow{Bi} BSU(2) \xrightarrow{f'} BSU(2),$$

where f' is a map of degree k^2 (see Proposition 1). Of course one can only construct such an example when k is odd (or zero). It is likely that a "best possible" version of Proposition 3 would specify that k^2 has to be odd or zero mod 120; however, for C.B. Thomas' purposes such a result would be no more useful than the one given.

The examples of Milnor show in particular that even when G is finite, there exist maps f: BG ⟶ BG' which are not of the form Bθ for any homomorphism θ: G ⟶ G'. In fact, with the notation of Proposition 3, the maps of the form Bθ have invariants $k^2 \equiv 0, 1, 49 \bmod 120$. (There is another embedding of SL(2,5) in SU(2) besides the one which was chosen as i; this gives a map Bθ with invariant $k^2 \equiv 49 \bmod 120$.)

I mention that the most important properties of SL(2,5) which are used in proving Proposition 3 also hold for the other finite groups which can act freely on spheres. This gives grounds for hoping that the method applies well to such groups.

One may note that Proposition 3 gives a classification into a finite list of possibilities (corresponding to the residue classes $k^2 \bmod 120$). This behaviour is general; when G is finite the theorems to follow always lead to a finite list of possibilities.

We now address the problem of formulating some general theorems. Suppose given a map f: BG ⟶ BG'; then we can form the following diagram.

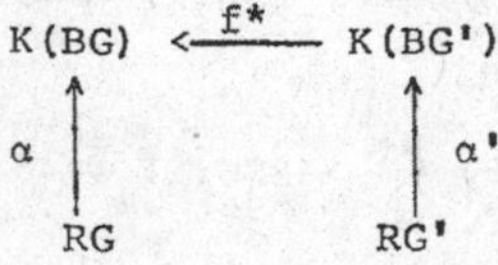

It would be very gratifying if we could prove that $f^*\mathrm{Im}\,\alpha' \subset \mathrm{Im}\,\alpha$; this would place a very substantial restriction on f, and would tend to reduce the classification to pure algebra. Unfortunately it is not true in general.

Example 4.

There is a compact Lie group G and a map

$$f: BG \longrightarrow BU(2)$$

such that the composite

$$BG \xrightarrow{f} BU(2) \subset 2 \times BU \subset Z \times BU$$

is an element $x \in K(BG)$ with $x \notin \operatorname{Im} \alpha$.

However, the example does have the property that $2x \in \operatorname{Im} \alpha$.

It is now more or less clear that we have to replace $\operatorname{Im} \alpha$ by something a bit more subtle. In general terms, we may regard the elements $x \in \operatorname{Im} \alpha \subset K(BG)$ as ones which can be constructed by finitistic, algebraic means; we may regard general elements $x \in K(BG)$ as constructed by infinitistic, topological means such as completion (see Proposition 2). There are examples, such as Example 4, of elements which can be constructed by finitistic, algebraic means although they are not in $\operatorname{Im} \alpha$. Therefore I propose to define a subset $\bar{R}G$ which we think of as "all the elements x which can be constructed by finitistic algebraic means", so that $\operatorname{Im} \alpha \subset \bar{R}G \subset K(BG)$.

At this stage I should apologise to the reader; in preparing this text I have not had time to write down all the proofs which I would like to write down. I will continue to give the statements as I made them in my lecture, because I think they are more likely to be true than false; but the reader may well treat them with caution till he sees proofs in print.

The definition which I gave in my lecture read as follows: an element $x \in K(BG)$ lies in $\bar{R}G$ if and only if there exists an integer $n \neq 0$ such that $nx \in \operatorname{Im} \alpha$. This has the effect of throwing the torsion subgroup of K(BG) into $\bar{R}G$, but I trust that this torsion subgroup is zero. So the "finitistic, algebraic means" which are allowed, in addition to those used in constructing $\operatorname{Im} \alpha$, include division by non-zero integers. I hope that this definition is good; but if I should have trouble with my proofs, I shall fall back on an earlier definition of $\bar{R}G$ which is longer and more complicated to explain.

It is now fairly clear what result I seek.

Theorem 5.

Let G and G' be compact Lie groups, and let $f: BG \longrightarrow BG'$ be a map; then

$$f^*: K(BG) \longleftarrow K(BG')$$

carries $\bar{R}G'$ into $\bar{R}G$.

The introduction of $\bar{R}G$ means that we need subsidiary results to remove it again in favourable cases.

Proposition 6.

If G is finite then $\bar{R}G = \operatorname{Im} \alpha$.

Proposition 7.

If G is a compact Lie group and its group of components, $\pi_o G$, is the union of its Sylow subgroups, then $\alpha: RG \longrightarrow K(BG)$ is mono and $\bar{R}G = \operatorname{Im} \alpha$.

Of course, neither Proposition 6 nor Proposition 7 applies to the group G used in Example 4; for that one, G is not finite and $\pi_o G$ is not the union of its Sylow subgroups.

The only reasonable way to prove a result like Theorem 5 is to characterise $\bar{R}G$ in some topological way which is preserved by induced maps f^*. For this purpose I need the exterior power operations. It is also convenient to introduce the total exterior power λ_t; this is given by

$$\lambda_t(x) = \sum_{i=o}^{\infty} \lambda^i(x)\, t^i;$$

it lies in the ring of formal power series K(BG)[[t]], where t is a new variable introduced for the purpose.

Theorem 8.

Suppose G is a compact Lie group and $x \in K(BG)$ is an element such that $\lambda_t(x)$ is a polynomial in t, i.e. $\lambda^i(x) = 0$ for i sufficiently large. Then $x \in \bar{R}G$.

Proof of Theorem 5 from Theorem 8.

Suppose $x \in \bar{R}G'$. Then there exists $n \neq 0$ such that $nx \in \operatorname{Im} \alpha'$; say $nx = \alpha'(y-z)$ for some

$$y: G' \longrightarrow U(q), \quad z: G' \longrightarrow U(r).$$

Then

$$\lambda^i y = 0 \text{ for } i > q, \quad \lambda^i z = 0 \text{ for } i > r.$$

Therfore

$$\lambda^i(f^*\alpha'y) = 0 \text{ for } i > q, \quad \lambda^i(f^*\alpha'z) = 0 \text{ for } i > r.$$

By Theorem 8,

$$f^*\alpha'y \in \bar{R}G, \quad f^*\alpha'z \in \bar{R}G.$$

So

$$f^*\alpha'(y-z) \in \bar{R}G,$$

that is

$$nf^*x \in \bar{R}G.$$

Hence

$$f^*x \in \bar{R}G.$$

This completes the proof.

If G is finite we can make Theorem 8 more precise.

Proposition 9.

Assume G is finite. In order that $x \in K(BG)$ should lie in $\operatorname{Im} \alpha$, it is necessary and sufficient that $\lambda_t(x)$ should be a rational function of t.

Here a formal power-series f(t) is called a "rational function of t" if it can be written as the quotient g(t)/h(t) of two polynomials g(t) and h(t), with h(t) invertible in K(BG)[[t]].

In Proposition 9, the "necessary" is obvious and requires no assumptions on G. The "sufficiency" does require assumptions.

Example 10.

There is a compact Lie group G and an element $x \in K(BG)$ such that $\lambda_t(x)$ is a rational function of t but $x \notin \bar{R}G$.

References.

[1] J.F. Adams and Z. Mahmud, Maps between Classifying Spaces, Inventiones Matnematicae 35 (1976) pp1-41.

[2] M.F. Atiyah, Characters and Cohomology of Finite Groups, Publ. Math. de l'Inst. des Hautes Études Scientifiques no.9 (1961) pp23-64.

[3] M.F. Atiyah and F. Hirzebruch, Vector Bundles and Homogeneous Spaces, Proc. Symposia in Pure Maths. vol.3, Amer. Math. Soc. 1961, pp.7-38.

[4] M.F. Atiyah and G. Segal, Equivariant K-theory and completion, Journal of Differential Geometry 3 (1969) pp1-18.

[5] D. Sullivan, Geometric Topology, Part I: Localisation, Periodicity and Galois Symmetry, mimeographed notes, MIT 1970 (especially Corollaries 5.10, 5.11).

[6] C.B. Thomas, in the proceedings of a conference held in Stanford, 1976; to appear in the series Proceedings of Symposia in Pure Mathematics, Amer. Math. Soc.

TWISTED LIE ALGEBRAS

M. G. Barratt
Northwestern University

Several forces have made me take up again the notion of homotopy envelopes, where the milling crowd of generalised Hopf invariants may be reduced to order or at least quieted. The first expository step is the description of twisted Lie algebras, the skeleton on which flesh and sinew will later be placed. This is the universal algebra of Whitehead products, with the permutation groups as coefficients permuting the factors of an argument-space. Here the analogue of the Poincaré-Witt theorem is proved for free twisted Lie algebras, and their structure noted. Only one step (Lemma 5) in the proof is less than obvious. The classification (in a later paper) of distributivity laws of relevance to homotopy theory involves some cohomology groups of the twisted Lie algebra analyser; it would be interesting if someone could find an elegant way of determining them all.

DEFINITION 1. A bi-ring is a graded ring $\Lambda = \oplus\Lambda_n$ with product

$$*: \Lambda_m \otimes \Lambda_n \to \Lambda_{m+n}$$

such that, for each n, Λ_n is a ring with product $\bullet$ satisfying the consistency condition C:

C: $$(a\bullet\lambda)*(b\bullet\mu) = (a*b)\bullet(\lambda*\mu)$$

when a and λ have the same degree, as do b and μ. If the Λ_n's have units it will be supposed that the $*$ product of units is a unit.

Examples. (i) If R is a commutative ring, $R \times \mathbf{Z}^+$ is a bi-ring if the products are defined by $(x,m)\cdot(y,m) = (xy,m)$, $(x,m)*(y,n) = (xy,m+n)$.

(ii) The <u>permutation bi-ring</u> has $\Lambda_n = \mathbb{Z}(\mathbf{S}_n)$, the group ring of the permutation group on n letters. The $*$ product, which will be always written as a tensor product, $\Lambda_m \otimes \Lambda_n \subset \Lambda_{m+n}$, is that induced by the obvious embedding $\mathbb{S}_m \times \mathbb{S}_n \subset \mathbf{S}_{m+n}$.

<u>DEFINITION</u> 2. A graded algebra $A = \oplus A_n$, with product $*$, is a <u>right Λ-algebra over the bi-ring</u> Λ if, for each n, A_n is a right Λ_n-module, and the right actions satisfy the consistency condition C above, with a in A_m, λ in Λ_m, b in A_n, μ in Λ_n, all m,n.

Examples. (iii) A graded algebra over a commutative ring R can be regarded as an algebra over the bi-ring $R \times \mathbb{Z}^+$.

(iv) Let B be any graded algebra, Λ any bi-ring. The <u>tensor product</u> $B \mathbin{\overline{\otimes}} \Lambda = \oplus B_n \otimes \Lambda_n$ is the algebra with product

$$(a \otimes \lambda)*(b \otimes \mu) = (a*b) \otimes (\lambda*\mu).$$

It is a right Λ-algebra.

(v) A bi-ring is an algebra over itself.

<u>DEFINITION</u> 3. A <u>twisted algebra</u> is an algebra over the permutation bi-ring of Example (ii). A <u>free associative twisted algebra</u> is the tensor product $B \mathbin{\overline{\otimes}} \Lambda$ of a free associative algebra B and the permutation bi-ring, defined in Example (iv).

Examples. (vi). Any algebra can be made into a twisted algebra in one of two trivial ways: by making the permutation groups act trivially, or by making them act through the signs of the permutations.

(vii) The permutation bi-ring is isomorphic to any free associative twisted algebra on one generator of dimension 1.

Let A be a twisted algebra: for any k, let $\tau_k \in S_k$ denote the cyclic permutation $(1,2,\ldots,k)$, and let 1_k denote the identity of S_k.

<u>DEFINITION</u> 4. A is a <u>twisted Lie algebra</u> if, for a in A_m, b in A_n, c in A_p,

(i) $a*b = -(b*a)\tau^n_{m+n}$,

(ii) $(a*b)*c = a*(b*c) + ((a*c)*b)(1_m \otimes \tau^p_{n+p})$.

Examples. (viii) If A is an ordinary graded Lie algebra, it is a twisted Lie algebra if the permutations act through their signs.

(ix) A twisted Lie algebra in which the permutations act trivially is, when the grading is ignored, a Lie algebra.

DEFINITION 5. Let A be an associative twisted algebra. The twisted Lie product [,] in A is defined, for a in A_m and b in A_n, by

$$[a,b] = (a*b) - (b*a)\tau^n_{m+n}.$$

This new product satisfies (i) and (ii) in Definition 4. Also, conjugation by τ^n_{m+n} will reverse the factors in $\Lambda_m \otimes \Lambda_n \subset \Lambda_{m+n}$, so [,] satisfies the consistency condition C. Hence

<u>LEMMA</u> 1. Let A be an associative twisted algebra. The same additive structure and twisted Lie product makes a twisted Lie algebra out of A.

It is not immediately obvious that the twisted Lie algebras are embedded in envelopes. For the topological applications it suffices to prove this for the free twisted Lie algebras:

DEFINITION 6. A free twisted Lie algebra is the quotient of the tensor product $B \overline{\otimes} \Lambda$ of a free non-associative algebra B and the permutation bi-ring Λ, obtained by imposing the relations (i) and (ii) of Definition (4).

THEOREM 2. A free twisted Lie algebra can be embedded by an additive homomorphism in a free associative twisted algebra, so that its product is carried to the twisted Lie product in the associative algebra. The obvious procedure works, despite the fact that, if L is a free twisted Lie algebra and A a free associative twisted algebra, A_n is a free Λ_n module and L_n is not free, in general. The proof reduces to the following lemmas, involving yet more definitions.

Let $L = \oplus L_n$ be any twisted Lie algebra with product [,], containing a, x.

DEFINITION 7. (i) $\sigma_1(a; x) = a$ and $\sigma_{n+1}(a; x) = [\sigma_n(a; x), x]$.

(ii) $\xi_n(x) = \sigma_n(x; x)$.

DEFINITION 8. (i) Let $\beta_n \in \Lambda_n$ be defined recursively by $\beta_1 = 1$, and

$$\beta_n = (1-\tau_n)\beta_{n-1}.$$

(ii) For any q, let $\beta_{1,q} = 1$, $\beta_{n,q} = (1-\tau_{nq}^q)\beta_{n-1,q}$.

(iii) $\theta_{n,q} = 1 + \tau_{nq}^q \beta_{n-1,q}$; $\theta_n = 1 + \tau_n \beta_{n-1}$.

(iv) $\mathrm{Ann}(\beta_{n,q})$ is the ideal in Λ_{nq} of right annihilators of $\beta_{n,q}$.

Thus $\beta_{n,q} = \xi_n(1_q)$ in the Lie algebra defined by the permutation biring, while

$$\theta_{n,q} + (n-1-\beta_{n-1,q}) = n-\beta_{n,q}.$$

LEMMA 3. Let L be any twisted Lie algebra, and $a \in L_m$, $x \in L_q$. Then

(i) $[a, \xi_n(x)] = \sigma_{n+1}(a;\ x)(1_m \otimes \beta_{n,q})$.

(ii) $\xi_n(x)\theta_{n,q} = 0 = \xi_n(x)(n-\beta_{n,q})$.

The first part is proved by induction on n. The next equality follows by taking $a = x$ and the last can be deduced since

$$\xi_n(x)(n-\beta_{n,q}) = [\xi_{n-1}(x)(n-1-\beta_{n-1,q}), x] + \xi_n(x)\theta_{n,q}.$$

LEMMA 4. (i) Let A be an associative twisted algebra, x an element of A_q. Then

$$\xi_n(x) = x^n \beta_{n,q}.$$

(ii) Hence $\beta_{n,q}\theta_{n,q} = 0 = \beta_{n,q}(n-\beta_{n,q})$.

The first part is trivial, since $\xi_n(1_q) = \beta_{n,q}$, and the second then follows from Lemma 3. The crux of the proof of Theorem 2 is to show that the right annihilators of $\xi_n(x)$, for x of dimension q, in a free twisted Lie algebra, are precisely the right annihilators of $\beta_{n,q}$, which follows from:

LEMMA 5. $\mathrm{Ann}(\beta_{n,q}) = \{\theta_{n,q}, \mathrm{Ann}(\beta_{n-1,q})\}\Lambda_{nq}$
$= \{(n-\beta_{n,q}), \mathrm{Ann}(\beta_{n-1,q})\}\Lambda_{nq}$.

The proof will be sketched for $q = 1$: the general case follows. Obviously $\mathrm{Ann}(\beta_n)$ contains $\mathrm{Ann}(\beta_{n-1})\Lambda_n$ since $\beta_n = (1-\tau_n)\beta_{n-1}$. Let

$$\varphi = \sum_0^{n-1} \tau_n^i \sigma_i = \sum_0^{n-1} \varphi_i \tau_n^i$$

where the σ_i and φ_i are in Λ_{n-1}. Then $\sigma_0 = \varphi_0$ and, if $\beta_n\varphi$ $(= (1-\tau_n)\beta_{n-1}\varphi)$ is 0, $\beta_{n-1}\varphi$ must factorise into $\zeta\lambda$, where $\zeta = \sum_1^n \tau_n^i$ and λ is in Λ_{n-1}. In fact, λ must be $\beta_{n-1}\sigma_0 = \beta_{n-1}\varphi_0$, the part of $\beta_{n-1}\varphi$ in Λ_{n-1}. Furthermore, if $\psi = \Sigma\psi_i\tau_n^i$ (where ψ_i is in Λ_{n-1}) also annihilates β_n, and if $\psi_0 = \varphi_0$, then $\beta_{n-1}\psi = \beta_{n-1}\varphi$. Thus $\varphi-\psi$ is in $\mathrm{Ann}(\beta_{n-1})\Lambda_n$. The lemma follows by taking $\psi = (1 + \tau_n\beta_{n-1})\varphi_0 = \theta_n\varphi_0$.

The additive structure of a free twisted Lie algebra L on a set X can now be described. Let L' be the ordinary free Lie algebra on the same set X, and let $\Omega' = \{W'_\alpha\}$ be an additive basis for L' whose elements are monomials in elements of X. To each monomial W'_α there corresponds a similar monomial W_α in L. Let Ω be the collection of monomials W_α in L, together with all $\xi_n(W_\alpha)$, $n \geq 2$. Let Ω_m be the subset of Ω of monomials of degree m: for each W_α in Ω_m there is a summand $W_\alpha \cdot \Lambda_m$ of L_m, isomorphic to Λ_m, and for each $\xi_n(W_\alpha)$ in Ω_m there is a summand $\xi_n(W_\alpha)\cdot\Lambda_m$ of L_m, isomorphic to the quotient $\Lambda_m/\mathrm{Ann}(\beta_{n,q})$ where $q = m/n$.

REMARKS. The twisted Lie products can be interpreted as Whitehead products, so that, if $a\colon \Sigma A \to X$ and $b\colon \Sigma B \to X$, $[a,b]\colon \Sigma A \wedge B \to X$; the element $\xi_n(a)\colon \Sigma A^{(n)} \to X$ can be acted on by Λ_n, on the right, by allowing S_n to permute the factors of $A^{(n)}$. The lemmas show that the only elements of Λ_n which annihilate all $\xi_n(a)$ are those of

$$\mathrm{Ann}(\beta_n) = \{n-\beta_n, n-1-\beta_{n-1}, \ldots, 3-\beta_3, 2-\beta_2 = 1+\tau_2\}\Lambda_n.$$

One implication of Lemma 4.1 is that the nth Hopf invariant of $\xi_n(a)\colon \Sigma A^{(n)} \to \Sigma A$ (with a = identity map) is $\beta_n\colon \Sigma A^{(n)} \to \Sigma A^{(n)}$. This was observed by D. H. Williams in his thesis; he showed that β_n on $\Sigma A^{(n)}$ is nonzero (and hence $\xi_n(a)$ is nonzero) for all n unless $H_*(A,\mathrm{point})$ has rank one for all fields, and be classified all such abnormal spaces (basically the spheres and certain $K(\pi,n)$'s).

I. Berstein had another easy argument for the non-nilpotence of $\Omega\Sigma A$. The relation $\beta_n^2 = n\beta_n$ in Λ_n is known as the Dynkin-Specht-Wever relation, P. M. Cohen tells me, although the proof of it here seems to be novel. I have not heard of a reference for Lemma 5, which may be new to the literature.

Supported in part by NSF Grant No. MCS76-07051 A01

Cobordism of sequences of manifolds

M. Bendersky

E. B. Curtis

1. Sequences of manifolds over a space. Let X be a topological space. We consider diagrams of the following type.

$$\begin{array}{llllll} \alpha: & M_s \xrightarrow{g_s} M_{s-1} \xrightarrow{g_{s-1}} & \cdots & \longrightarrow M_1 \xrightarrow{g_1} M_o \\ & \downarrow g & & \\ & X & & \end{array}$$

where (1) Each M_i is a stably almost complex manifold of dimension t.

(2) Each g_i is a smooth map which induces a complex linear map on the stable tangent bundle.

(3) the map g is continuous

Such a diagram α will be called an s-sequence of dimension t over X. If each M_i is without boundary, then α is called closed. Two closed s-sequences α and α' of dimension t are called cobordant if there is an s-sequence β of dimension $t+1$ over X, with $\partial\beta = \alpha \cup \alpha'$, the disjoint union. Just as in the usual situation of complex cobordism over X (which is the case $s=0$), cobordism becomes an equivalence relation on the set of s-sequences of dimension t over X. The set of equivalence classes will be denoted by $\Omega_t(X;s)$. Each $\Omega_t(X;s)$ is an abelian group, where the sum is represented by disjoint union.

Put

$$\Omega_*(X;s) = \bigoplus_{t\geqslant 0}\Omega_t(X;s)$$

The cartesian product of two closed stably almost complex manifolds is another such. Thus an s-sequence of dimension t over X may be multiplied by a closed stably almost complex manifold ofdimension n over a point to produce an s-sequence of dimension $t+n$ over X. In this way $\Omega_*(X;s)$ becomes an Ω_*-module, where $\Omega_* = \Omega_*(\text{point})$ is the usual complex cobordism ring.

We remark that $\Omega_*(X;0) = \Omega_*(X)$, the usual complex bordism groups of X. Conner and Floyd in Ref [3] define relative groups $\Omega_*(X,A)$, which are shown to form a homology theory for CW pairs (X,A). For $s \geqslant 1$, relative groups may be defined similarly, but the result is not a homology theory (the excision axiom is not satisfied). Instead, we shall give a homotopy interpretation of s-sequences, relating them to the unstable Adams-Novikov spectral sequence based on complex bordism.

2. The Fundamental chain complex. Let the space X and dimension t be fixed. For each $i = 0,1,\ldots,s+1$, there is a homomorphism

$$\delta^i : \Omega_t(X;s) \rightarrow \Omega_t(X;s+1)$$

defined as follows. For each s-sequence α, $\delta^0(\alpha)$ is the (s+1)-sequence

$$\delta^0(\alpha):\quad \begin{array}{ccccccccc} M_s & \xrightarrow{g_s} & M_{s-1} & \xrightarrow{g_{s-1}} & \cdots & \xrightarrow{g_1} & M_0 & \xrightarrow{g_0} & S^t \\ \downarrow g & & & & & & & & \\ X & & & & & & & & \end{array}$$

where S^t is the t-dimensional sphere, and g_0 is the constant map. For $1 \leq i \leq s+1$, $\delta^i(\alpha)$ is to be the (s+1)-sequence obtained from α by repeating the manifold M_{i-1}, with the identity map $1 = 1_{M_{i-1}}$.

$$\delta^i(\alpha):\quad \begin{array}{cccccccccc} M_s & \xrightarrow{g_s} & M_{s-1} & \xrightarrow{g_{s-1}} & \cdots & \longrightarrow M_{i-1} & \xrightarrow{1} & M_{i-1} \longrightarrow & \cdots & \xrightarrow{g_1} M_0 \\ \downarrow g & & & & & & & & & \\ X & & & & & & & & & \end{array}$$

Also, for each $i = 0,1,\ldots,s-1$, there is a homomorphism

$$\sigma^i : \Omega_t(X;s) \longrightarrow \Omega_t(X;s-1)$$

defined as follows. For each s-sequence α, $\sigma^i(\alpha)$ is the (s-1)-sequence obtained by deleting the manifold M_i, and inserting the composition $g_{i+1} \circ g_i$.

$$\sigma^i(\alpha):\quad \begin{array}{cccccccc} M_s & \xrightarrow{g_s} & M_{s-1} & \xrightarrow{g_{s-1}} & \cdots & M_{i+1} & \xrightarrow{g_{i+1} \circ g_{i-1}} & M_{i-1} \rightarrow \cdots \xrightarrow{g_1} M_0 \\ \downarrow g & & & & & & & \\ X & & & & & & & \end{array}$$

For $i = 0$, $\sigma^0(\alpha)$ is the (s-1) sequence obtained by deleting the manifold M_o and the map g_1.

It can easily be verified that the δ^i and the σ^i satisfy the cosimplicial identities (see Ref [2]). Thus, for each non-negative integer t, $\underset{\sim}{\Omega}_t(X)$ becomes a cosimplicial complex, with

$$\underset{\sim}{\Omega}_t(X)^s = \Omega_t(X;s)$$

Let

$$\partial^s = \sum_{i=0}^{i=s}(-1)^i\,\delta^i : \Omega_t(X;s-1) \longrightarrow \Omega_t(X;s)$$

The cosimplicial identities imply that $\partial^{s+1}\circ\partial^s = 0$ in the complex

$$\text{(2.1)}\qquad \Omega_t(X;0)\xrightarrow{\partial^1}\Omega_t(X;1)\xrightarrow{\partial^2}\Omega_t(X;2)\xrightarrow{\partial^3}\cdots$$

In section 4. (see also Ref[1]) we define an unstable Adams-Novikov spectral sequence $\{E_r^{s,t}(X;MU)\}$. For X simply connected (or X nilpotent), the $E_r^{*,*}(X;MU)$ converge to $E_\infty^{*,*}(X;MU)$ which is the graded group associated to a filtration on $\pi_*(X)$. We shall show that

$$E_2^{s;t}(X;MU) \approx \ker\partial^s/\operatorname{im}\partial^{s-1}$$

of the chain complex (2.1).

3. Homotopy interpretation of s-sequences. For each space X, with or without basepoint, let $X/\emptyset$ be the disjoint union of X with a point x_o, which is to be taken as the new basepoint. Let $\{MU_q\}$, $q = 0,1,2,\ldots$ be the spaces of the MU spectrum (ref [5]). We define

$$\mathcal{E}(X) = \lim_q \Lambda^q(MU_q \wedge (X/\emptyset))$$

where $\Lambda^q(.)$ stands for the q-fold loopspace. The unit in MU gives a map

$$\eta = \eta_X : X \longrightarrow \mathcal{E}(X)$$

The multiplication in MU gives a map

$$\mu = \mu_X : \mathcal{E}\mathcal{E}(X) \longrightarrow \mathcal{E}(X)$$

Then $(\mathcal{E}, \eta, \mu)$ forms a triple on the category of topological spaces (see also Ref [1] and Ref [2]).

The main theorem of cobordism theory asserts that

$$\Omega_t(X;0) \approx \pi_t \mathcal{E}(X)$$

In our situation, this generalizes to the following.

Theorem (3.1) For each $s \geqslant 0$, there is an isomorphism:

$$\Omega_t(X;s) \approx \pi_t \mathcal{E}^{s+1}(X)$$

The proof of this theorem results from the following lemma, together with the main theorem of cobordism theory.

Lemma (3.2) For each $s \geqslant 1$, there is an isomorphism

$$\Theta : \Omega_t(X;s) \approx \Omega_t(\mathcal{E}(X);s-1))$$

Proof. Suppose given a closed s-sequence α, of dimension t over X:

$$\begin{array}{cccccccc} \alpha : & M_s & \xrightarrow{g_s} & M_{s-1} & \longrightarrow \ \ldots \ \longrightarrow & M_1 & \xrightarrow{g_1} & M_0 \\ & {\scriptstyle g}\downarrow & & & & & & \\ & X & & & & & & \end{array}$$

Let $e : M_s \longrightarrow R^q$ be an embedding into an even-dimensional Euclidean space such that the normal bundle of M_s in R^q has a stably complex structure. Then consider the embedding

$$e \times g_s : M_s \longrightarrow R^q \times M_{s-1}$$

Let ν be the normal buncle of M_s in $R^q \times M_{s-1}$. Apply the Pontrjagin-Thom construction to obtain a map

$$S^q \times M_{s-1} \longrightarrow T(\nu) \times X/\emptyset \longrightarrow MU_q \wedge X/\emptyset$$

The adjoint of this is a map

$$M_{s-1} \longrightarrow \Lambda^q(MU_q \wedge X/\emptyset)$$

Passing to the limit over q, there is defined a map

$$f : M_{s-1} \longrightarrow \mathcal{E}(X)$$

Thus we obtain an (s-1)-sequence $\theta(\alpha)$ over $\mathcal{E}(X)$:

$$\begin{array}{cccccccc} \theta(\alpha) : & M_{s-1} & \xrightarrow{g_{s-1}} & M_{s-2} & \longrightarrow \ \ldots \ \longrightarrow & M_1 & \xrightarrow{g_1} & M_0 \\ & {\scriptstyle f}\downarrow & & & & & & \\ & \mathcal{E}(X) & & & & & & \end{array}$$

An argument similar to that of Ref[7] shows that the cobordism class of $\theta(\alpha)$ is independent of the choices.

The inverse ψ to θ is obtained as follows. Let β be an (s-1)-sequence over $\mathcal{E}(X)$:

$$\begin{array}{cl} \beta : & M_{s-1} \xrightarrow{g_{s-1}} M_{s-2} \longrightarrow \cdots \longrightarrow M_1 \xrightarrow{g_1} M_o \\ & f \downarrow \\ & \mathcal{E}(X) \end{array}$$

As M_{s-1} is compact, there is a map

$$h : M_{s-1} \longrightarrow \Lambda^q(MU_q \wedge X/\emptyset)$$

for sufficiently large q (also take q even). The adjoint of h is a map

$$\tilde{h} : M_{s-1} \times S^q \longrightarrow MU_q \wedge X/\emptyset$$

Then the composite, $F = \pi_1 \circ \tilde{h}$ (where π_1 is the projection on the first factor):

$$M_{s-1} \times S^q \xrightarrow{\tilde{h}} MU_q \wedge X/\emptyset \xrightarrow{\pi_1} MU_q$$

sends $M_{s-1} \times (\infty)$ to the basepoint. By a homotopy, F may be taken to be transverse regular on $BU(q/2) \subset MU_q$. Let

$$M_s = F^{-1}(BU(q/2))$$

which is a closed submanifold of $M_{s-1} \times S^q$. The composite

$$M_s \xrightarrow{\text{incl.}} M_{s-1} \times S^q \xrightarrow{\pi_1} M_{s-1}$$

is a smooth map which we call g_s. We also have a map

$$g : M_s \longrightarrow BU(q/2) \times X \xrightarrow{\pi_1} X$$

where $g = \pi_2 \circ (F \times \text{constant})$. Finally, $\psi(\beta)$ is taken to be the s-sequence

$$\begin{array}{l}\psi(\beta) : \quad M_s \xrightarrow{g_s} M_{s-1} \xrightarrow{g_{s-1}} \cdots \longrightarrow M_1 \xrightarrow{g_1} M_0 \\ \qquad\qquad\; g \downarrow \\ \qquad\qquad\;\; X\end{array}$$

As in Ref [7], ψ is well-defined on $\Omega_t(X;s-1)$, independent of the choices, and ψ is an inverse to θ on the cobordism classes. This establishes the lemma, and hence the theorem.

Furthermore, tracing through the above constructions shows that the maps

$$\delta^i : \Omega_t(X;s) \longrightarrow \Omega_t(X;s+1)$$

are induced by η as follows. For $i = 0$, δ^0 is the composite

$$\begin{array}{c}\Omega_t(X;s) \approx \pi_t(\mathcal{E}^{s+1}(X)) \\ \downarrow \pi_t(\eta\,\mathcal{E}^{s+1}(X)) \\ \pi_t(\mathcal{E}^{s+2}(X)) \approx \Omega_t(X;s+1)\end{array}$$

For $i > 0$, δ^i is the composite

$$\begin{array}{c}\Omega_t(X;s) \approx \Omega_t(\mathcal{E}^{s-i}(X),i) \\ \downarrow \Omega_t(\eta_{\mathcal{E}^{s-i}(X)}) \\ \Omega_t(\mathcal{E}^{s-i+1}(X),i) \approx \Omega_t(x,s+1)\end{array}$$

In a similar way, the maps σ^i are induced by μ.

4. The unstable Adams-Novikov spectral sequence. The ring spectrum $MU = \{MU_q\}$ defines a triple $(MU(.), \eta, \mu)$ on the category of topological spaces with basepoint as follows. For each space X, with basepoint, let

$$MU(X) = \lim_q \Lambda^q(MU_q \wedge X)$$

where $\Lambda^q(.)$ is the q-fold loopspace. The unit in MU gives a map

$$\eta = \eta_X : X \longrightarrow MU(X)$$

The multiplication in MU gives a map

$$\mu = \mu_X : MU(MU(X)) \longrightarrow MU(X)$$

The procedure of Bousfield-Kan (Ref [2]) applies, with the ring R replaced by the ring spectrum MU. Thus we have a fibration $D_1(X) \longrightarrow X$ which is the pullback via η of the pathspace fibration over $MU(X)$. $D_1(.)$ is a functor from spaces with basepoint to spaces with basepoint. Inductively, therefore, we obtain a fibration $D_s(X) \longrightarrow D_{s-1}(X)$ as the pullback via $D_{s-1}(\eta)$ of the pathspace fibration over $D_{s-1}(MU(X))$. The sequence of fibrations

$$\longrightarrow D_s(X) \longrightarrow D_{s-1}(X) \longrightarrow \ . \ . \ . \ \longrightarrow D_1(X) \longrightarrow X$$

is called the tower over X, derived from the triple (MU, η, μ).

The homotopy exact couple of this tower is called the homotopy spectral sequence of X with coefficients in MU. Specifically, the E_1-term is given by

$$E_1^{s,t}(X;MU) = \begin{cases} \pi_{t-s}D_s(MU(X)), & t > s \geqslant 0 \\ 0 & \text{otherwise} \end{cases}$$

For simply connected X, the proofs of Ref[2] and Ref [4] (see also Ref[1]) show that the spectral sequence converges to the homotopy groups of X. Also as in Ref [2], the E_2-term may be identified as the homology groups of a chain complex. That is,

$$E_2^{s,t}(X;MU) \approx \ker \partial^s / \operatorname{im} \partial^{s-1}$$

of the complex

$$(4.1) \qquad \pi_* MU(X) \xrightarrow{\partial^1} \pi_* MU(MU(X)) \xrightarrow{\partial^2} \cdots$$

An easy double complex argument shows that the chain complex (4.1) and the chain complex (2.1) have isomorphic homology groups. Thus we have the following.

Proposition (4.2) For each space X with basepoint, the homology groups of the chain complex (2.1) arising from sequences of almost complex manifolds over X is isomorphic to the E_2-term of an unstable Adams-Novikov spectral sequence. For X simply connected (nilpotent is sufficient), this spectral sequence converges to the homotopy groups of X.

For the identification of this E_2-term as an Ext group in a suitable category, we refer the reader to Ref [1]. In particular, for each odd-dimensional sphere S^{2n+1}, an unstable cobar complex is defined, and calculations of the 1-line are made.

We remark that the above methods apply equally well to any multiplicative cobordism theory with unit. For example, for unoriented real cobordism, the spectrum MO becomes a product of Eilenberg-MacLane spaces, and the spectral sequence becomes the unstable homotopy spectral sequence of Bousfield-Kan of Ref[2].

References

[1] M. Bendersky, E. B. Curtis, H. R. Miller, The unstable Adams spectral sequence for generalized homology (submitted for publication), preprints available.

[2] A. K. Bousfield and D. M. Kan, The homotopy spectral sequence of a space with coefficients in a ring, Topology, vol 11 (1972), 79-106.

[3] P. E. Conner and E. E. Floyd, Differentiable periodic maps, Springer-Verlag, Berlin (1964).

[4] E. B. Curtis, Some relations between homotopy and homology, Annals of Math. 83 (1965), 386-413.

[5] J. W. Milnor, On the cobordism ring Ω^* and a complex analogue, Amer. J. of Math. 82 (1960), 505-521.

[6] R. E. Stong, Notes on cobordism theory, Princeton University press, Princeton (1968).

[7] R. E. Stong, Cobordism of maps, Topology 5 (1966),245-258.

SPLITTINGS OF MU AND OTHER SPECTRA

J.M.Boardman*
Johns Hopkins University

Introduction and main results

An old question of J.M.Cohen [8] asks to what extent a spectrum having free abelian homology and homotopy groups has to look like the Thom spectrum MU. In order to study this and related questions we introduce the following class of spectra. Let M be any set of primes.

Definition We call a spectrum X *free-free* if it is highly connected and the homology groups $H_n(X)$ and homotopy groups $\pi_n(X)$ are free finitely generated modules over Z_M (the ring Z of integers localized at M) for all n.

As important examples we have the Thom spectrum MU (with M the set of all primes) and the Brown-Peterson spectrum BP (where M consists of just one prime p). Lemma 1.2 suggests that the homotopy theory of these spectra ought to be particularly accessible, as maps are represented faithfully by their induced homology homomorphisms. Clearly, any summand or localization of a free-free spectrum is again free-free.

We find the following notation extremely practical. Let E be any spectrum and G be a free abelian graded group. We define the spectrum $G\otimes E$ as $L(G)_\wedge E$, where L(G) denotes a graded Moore spectrum (a wedge of spheres in this case, one for each generator of G). So $G\otimes E$ is a graded sum of copies of E. However, $G\otimes E$ is quite independent of the choice of free generators. We can describe it better as the spectrum that represents the homology theory

*Research partially supported by the National Science Foundation under grants MCS 70-01647 and MCS 76-23466. This is an updated version of both [7] and the talk, except that part of the substance of the talk will appear separately.

$(G\otimes E)_*(X) = G\otimes E_*(X)$. (In cohomology we have to be more careful: $(G\otimes E)^*(X) = G\otimes E^*(X)$ is not in general valid without some kind of completion.) If E is M-local free-free and G is <u>of finite type</u> (that is: G_n finitely generated for all n and zero for large negative n) we see that $G\otimes E$ is again M-local free-free. Also we may write $G_M\otimes E$ for $G\otimes E$, if we wish. If G is a commutative ring and E is a ring spectrum, then $G\otimes E$ is again a ring spectrum.

We study the class of M-local free-free spectra by applying our localization theory [6] which we review in section 4. To synthesize a M-local free-free spectrum X, we need to choose for each prime p in M a p-local free-free spectrum X_p and rational isomorphisms between them. Accordingly, we first study p-local free-free spectra and maps between them in section 2. The main theorem here (not really new) is that BP is essentially the only example.

<u>THEOREM A</u> <u>Every p-local free-free spectrum is isomorphic to</u> $G\otimes BP$ <u>for some free abelian graded group</u> G <u>of finite type. The group</u> G <u>is uniquely determined up to isomorphism.</u>

In section 3, we consider multiplicative structures on these spectra. This requires some general theory of Chern classes, which we review in section 1. We write (in section 2) $\pi_*(BP)=V_p$.

<u>THEOREM B</u> <u>Every p-local free-free ring spectrum is isomorphic as a ring spectrum to some BP-algebra</u> $R\otimes_V BP$, <u>where</u> R <u>is a commutative algebra over</u> V_p <u>which as a</u> V_p<u>- module is free of finite type.</u> (We give the detailed definitions in section 3.)

In section 5, we recover the standard multiplicative splitting of MU_p due to Quillen. Our approach sheds some light on the image of the Hurewicz homomorphism $\pi_*(MU) \to H_*(MU)$. However, our theory provides in section 6 a much more natural way to synthesize free-free spectra: we construct for any set M of primes a M-local free-free ring spectrum we call BP(M), which reduces to BP for the prime p when M consists of one prime p, but which is obviously very much smaller than MU when M consists of all primes.

<u>THEOREM C</u> <u>There is a multiplicative splitting</u> $MU = E \otimes BP(U)$ <u>where</u> U <u>denotes the set of all primes, even before localization</u>.

We also find in Theorem 6.6 that BP(M) does not split further multiplicatively, by showing that BP(M) admits no summand that is a subring spectrum.

However, if we ignore the multiplicative structure, the situation becomes enormously more complicated, as we discover in section 7.

<u>THEOREM D</u> <u>Whenever</u> M <u>consists of more than one prime</u>, BP(M) <u>splits additively as a graded sum of copies of a quotient ring spectrum of</u> BP(M).

This, together with Theorem C, yields (many) additive splittings of MU finer than that of Theorem C. How far one can go in this direction is extremely unclear. In section 8 we take a different approach.

<u>THEOREM E</u> <u>Let</u> M <u>be any finite set of primes</u>. <u>Then two M-local free-free spectra are isomorphic if and only if they are rationally isomorphic</u>.

This, with Theorem A and our localization theory, immediately yields a classification (Theorem 8.4) of such spectra. Nevertheless, it is still not clear which of them are additively indecomposable. In section 9 we study in some detail the simplest case.

THEOREM F *Let* M *consist of two primes* p *and* q. *Then*

EITHER

(*a*) *There is exactly one indecomposable* M-*local free-free spectrum* X, *and any other* M-*local free-free spectrum has the form* $G \otimes X$;

OR

(*b*) *There are infinitely many distinct indecomposable* M-*local free-free spectra*.

Given particular primes p and q, it is not at all trivial to decide which of the two cases applies. All that is clear is that both cases do occur.

The moral appears to be that we should work with one prime at a time if we expect to find reasonable results, avoid mixing primes as much as possible, and put the various primes together right at the end, if ever. Not all is lost, however: in section 8 we prove the following.

THEOREM G *Let* X *be a spectrum whose homotopy and homology groups are additively isomorphic to those of the Thom spectrum* MU. *Then* X *is isomorphic to* MU.

1. Chern classes and logarithmic series

We use the idea of the logarithmic series mainly for descriptive purposes in order to organize some of the otherwise complicated formulae. This depends on the theory of Chern classes. We adapt what we need from Adams [1, part II], with minor changes.

Suppose E is a ring spectrum (by which we shall always mean one equipped with commutative associative multiplication with unit, in our stable homotopy category $\underline{S}_h$).

Definition 1.1 A first Chern class c^E for E associates to each complex line bundle ξ over each space X a class $c^E(\xi) \in E^2(X,\emptyset)$ in such a way that:

(a) $c^E(\xi)$ is natural for maps of line bundles;

(b) for the canonical line bundle γ over $P_1(C)$, $c^E(\gamma)$ is the image of $1 \in E^0(S^0) = E^2(S^2) \subset E^2(S^2,\emptyset)$.

Remark All our cohomology and homology theories are taken in the reduced sense. For a space X, as distinct from a spectrum, we have also the absolute groups, which we shall write as $E^*(X,\emptyset)$, etc.

The existence of Chern classes in all the cases that concern us is immediate from the following lemma. We do not need its generality here, but we shall later.

LEMMA 1.2 *Suppose given a (k-1)-connected spectrum* X *for which* $H_*(X)$ *is* Z_M- *free for some set* M *of primes, and a spectrum* Y *for which* $\pi_*(Y)$ *is a* Z_M- *module. Then*

(a) Any homomorphism $\pi_k(X) \dashrightarrow \pi_{k+n}(Y)$ *can be induced by some map* $X \dashrightarrow Y$ *of degree* n.

(b) Maps are represented faithfully by their effect on homology - two maps $f,g: X \dashrightarrow Y$ *coincide if and only if their induced homomorphisms* $f_*, g_*: H_*(X) \dashrightarrow H_*(Y)$ *are equal.*

Proof This result can be obtained from obstruction theory.

Alternatively (but not really so differently), in the Atiyah-Hirzebruch spectral sequence for $\{X,Y\}^*$ we have

$$E_2^{p,q} = H^p(X;\pi_{-q}(Y)) = \mathrm{Hom}(H_p(X),\pi_{-q}(Y)),$$

with no Ext term (by working in the category of Z_M-modules). Composition with $\alpha: Y \dashrightarrow Y_\emptyset$ induces a monomorphism of spectral sequences. All the differentials must therefore vanish for Y, since they do for

$Y_\emptyset$. We are therefore in the good convergence situation described in Theorem 10 of [5], in which the filtration of $\{X,Y\}^*$ is complete Hausdorff.

First, the edge homomorphism

$$\{X,Y\}_n \dashrightarrow E_2^{k,-k-n} = \mathrm{Hom}(H_k(X),\pi_{k+n}(Y)) = \mathrm{Hom}(\pi_k(X),\pi_{k+n}(Y))$$

is epimorphic, which proves (a). Second, α induces a monomorphism $\alpha_*:\{X,Y\}_* \dashrightarrow \{X,Y_\emptyset\}_*$, from which (b) follows.]]

To classify Chern classes, it is only necessary to consider the universal case of the canonical line bundle γ over $P_\infty(C)$. Following Dold (again see [1]) we have

$$E^*(P_\infty(C),\emptyset) = \pi_*(E)[[c^E(\gamma)]],$$

the ring of formal power series on one generator $c^E(\gamma)$ over the graded ring $\pi_*(E)$. From this we read off that <u>any</u> first Chern class c'^E of E has the form

$$c'^E = c^E + \Sigma_{n=1}^{\infty} a_n(c^E)^{n+1} \tag{1.3}$$

where the elements a_n in $\pi_{2n}(E)$ may be chosen arbitrarily.

For tensor products of line bundles ξ and η, consideration of universal examples shows that there exists some formula

$$c^E(\xi\otimes\eta) = \Sigma_{i,j}\, a_{i,j}c^E(\xi)^i c^E(\eta)^j = F(c^E(\xi),c^E(\eta)),$$

where $F(x,y)$ is a well-defined formal power series in two variables with coefficients $a_{i,j}$ in $\pi_{2i+2j-2}(E)$, called the <u>formal group multiplication law</u> of the Chern class. Properties of the tensor product of bundles yield immediately the identities $F(x,y) = F(y,x)$, $F(x,0) = x$, and $F(x,F(y,z)) = F(F(x,y),z)$. In the classical case of ordinary cohomology, where $H = K(Z)$ is the Eilenberg-MacLane spectrum, the only possibility, for dimensional reasons, is the simple <u>additive</u> formula

$$c^H(\xi\otimes\eta) = c^H(\xi) + c^H(\eta).$$

We are interested in the possibility of simplifying the product formula by changing the choice of Chern class. In case E is a rational spectrum it is an algebraic exercise to see that there is a unique Chern class whose formal group law is additive.

However, it is not necessary to pass to $Q\otimes E$ to achieve this. Consider the ring spectrum $E\wedge H$, which inherits two Chern classes, c^E from E and c^H from H, of which c^H is automatically additive. We know from general theory that $E\wedge H$ is a GEM (or graded Eilenberg-MacLane) spectrum, and that its homotopy groups are by definition $H_*(E) = \pi_*(E\wedge H)$, so that $E\wedge H$ is isomorphic to $K(H_*(E))$; moreover, this isomorphism can be chosen as a canonical ring spectrum isomorphism in favorable cases (some care is needed), such as when $H_*(E)$ is Z_M-free for some set of primes M. Therefore for $E\wedge H$ we can write

$$c^H = c^E + \Sigma_i\, m_i\, (c^E)^{i+1} = \log(c^E) \qquad (1.4)$$

for suitable coefficients m_i in $H_{2i}(E)$, which defines log as a formal power series; this is the <u>formal logarithmic series</u> of the Chern class c^E. In other words, we have

$$\log F(x,y) = \log x + \log y \qquad (1.5)$$

as formal power series over $H_*(E)$; hence the name `logarithm'. So our problem can be solved if we pass from $\pi_*(E)$ to $H_*(E)$ via the Hurewicz homomorphism $\pi_*(E) \dashrightarrow H_*(E)$. If $\pi_*(E)$ is torsion-free, this is a monomorphism and no information from $\pi_*(E)$ is lost.

We shall need to compare the logarithmic series log for c^E with the logarithmic series log′ (or mog in Adams [1]) for any other Chern class c'^E of E.

LEMMA 1.6 *The logarithmic series for any other Chern class* c'^E *of* E *has the form*

$$\log' z = \log z + \Sigma_{i>0} \log b_i z^{i+1},$$

for uniquely defined elements b_i *in* $H_{2i}(E)$ *lying in the image of the Hurewicz homomorphism. Conversely, if* $\pi_*(E)$ *is torsion-free, any such series is the logarithmic series of a unique Chern class for* E.

Proof A formal consequence of the formulae (1.3), (1.5), and (from (1.4))

$$c^H = \log c^E = \log' c'^E.]]]$$

The most important example of a Chern class is the Conner-Floyd Chern class c^{MU} of the Thom spectrum MU. Its logarithmic series is

$$\log z = z + m_1 z^2 + m_2 z^3 + m_3 z^4 + \dots, \qquad (1.7)$$

where the m_i form a system of polynomial generators of $H_*(MU)$. Indeed, this is the *universal* Chern class in the sense that there is a 1-1 correspondence between maps of ring spectra f:MU-->E and Chern classes c^E of any ring spectrum E, given by $c^E = f_* c^{MU}$. (See Lemma 4.6 on page 52 of [1], for example.)

2. p-local free-free spectra

In this section we prove Theorem A, that all such spectra have the form G⊗BP. We also develop information about maps between such spectra.

The structure of BP We assemble here the standard information we need about BP [1,10] and establish our notation. We shall quote this as needed, without further comment.

(a) BP is a p-local ring spectrum, with commutative associative multiplication map $BP \wedge BP \dashrightarrow BP$ and unit map $S^0 \dashrightarrow BP$.

(b) $H_*(BP) = W_p$, the localization at p of the polynomial ring $W = Z[w_1, w_2, w_3 \dots]$ having generators w_i (customarily written

m_{p^i-1}) in degree $2(p^i-1)$.

(c) BP has a Chern class c^{BP} whose logarithmic series (1.4) is

$$\log z = z + w_1 z^p + w_2 z^{p^2} + \dots \qquad (2.1)$$

(d) $\pi_*(BP) = V_p$, the localization at p of the polynomial ring $V = Z[v_1, v_2, v_3 \dots]$ on the Hazewinkel generators v_i (see [9]) in degree $2(p^i-1)$. The Hurewicz homomorphism embeds it as a subring of $H_*(BP) = W_p$; and the generators v_i are conveniently defined as elements of W by the formal identity

$$p \log z = pz + \log v_1 z^p + \log v_2 z^{p^2} + \dots$$

Equivalently, by equating coefficients, we have the inductive formula

$$v_n = pw_n - \Sigma_{i=1}^{i=n-1} w_i v_{n-1}^{p^i}. \qquad (2.2)$$

We deduce immediately by induction on n that:

(i) v_n is divisible in W by p for all n, and

(ii) $p^n w_n$ lies in V.

From (ii) it follows easily that we can make any element x of W (or W_p) lie in V (or V_p) by multiplying by enough powers of p: indeed, $p^m x$ lies in V for any element x in W of degree $2m(p-1)$.

(e) $H^*(BP;Z/p)$, as a module over the Steenrod algebra A, is isomorphic to $A/(\beta)$ where β is the Bockstein element, and is generated by the Thom class u in $H^0(BP;Z/p)$.

(f) We have the scalar multiplication map $V_p \otimes BP \dashrightarrow BP$, induced by the multiplication map of BP.

(g) We have the map $r: BP \dashrightarrow T \otimes BP$ of ring spectra described in effect by Quillen [10], where $T = Z[t_1, t_2, t_3 \dots]$ is another graded polynomial ring on generators t_i in degree $2(p^i-1)$. It is defined by its effect on the logarithmic series

$$r_*\log z = \log z + \log t_1z^p + \log t_2z^{p^2} + \dots,$$

or equivalently, by equating coefficients, $r_*:W \dashrightarrow T\otimes W$ (or $W_p \dashrightarrow T\otimes W_p$) is given by

$$r_*w_n = t_n\otimes 1 + \Sigma_{i=1}^{i=n-1} t_{n-i}^{p^i}\otimes w_i + 1\otimes w_n. \qquad (2.3)$$

Lemma 1.2 shows that this determines the map r completely.

Lemma 1.2 provides the information we need about maps to and from a general free-free spectrum. We also need the following lemma about BP.

<u>LEMMA 2.4</u> <u>Suppose</u> G <u>is a graded finitely generated free</u> $Z_{(p)}$-<u>module concentrated in degree</u> n, <u>and that</u> $f:G\otimes BP \dashrightarrow G\otimes BP$ <u>is any map that induces the identity</u> $f_* = 1:\pi_n(G\otimes BP) \dashrightarrow \pi_n(G\otimes BP)$. <u>Then</u> f <u>is an isomorphism</u>.

<u>Proof</u> Since $G\otimes BP$ is (n-1)-connected, f induces the identity homomorphism of $H_n(G\otimes BP)$, therefore of $H^n(G\otimes BP;Z/p)$ by the universal coefficient theorem, and also of $H^*(G\otimes BP;Z/p) = G\otimes H^*(BP;Z/p)$ from the known structure of $H^*(BP;Z/p)$ as A-module. Consider $f_*:H_*(G\otimes BP) \dashrightarrow H_*(G\otimes BP)$; by duality $f_*\otimes 1$ on $H_*(G\otimes BP)\otimes Z/p$ is the identity homomorphism, so that by Lemma 1.9(b) of [6], f_* is an isomorphism. Hence f is an isomorphism.]]]

<u>Proof of Theorem A</u> We split off copies of BP from X until there is nothing left.

Suppose we have constructed a (n-1)-connected free-free spectrum X_n, with a map $f_n:X_n \dashrightarrow X$. Let G_n be the graded group consisting of $\pi_n(X_n)$ concentrated in degree n and put $Y_n = G_n\otimes BP$, so that we may identify $\pi_n(Y_n)$ with $\pi_n(X_n)$. By Lemma 1.2 there exist maps $Y_n \dashrightarrow X_n$ and $X_n \dashrightarrow Y_n$ that induce the identity homomorphism of $\pi_n(X_n)$. According to Lemma 2.4, the composite $Y_n \dashrightarrow X_n \dashrightarrow Y_n$ is an isomorphism, which we may assume is the identity (after modifying the second map as necessary). We deduce a splitting $X_n = Y_n \vee X_{n+1}$, where by inspection X_{n+1} is a n-connected free-free spectrum. We equip it with the map

$$f_{n+1}: X_{n+1} \dashrightarrow X_n \xrightarrow{f_n} X$$

and define a map g_n as the composite

$$Y_n \dashrightarrow X_n \xrightarrow{f_n} X$$

This completes our induction step.

We start with $f_k : X_k = X$ for a suitable value of k and construct by induction for all $n \geq k$ spectra X_n and $Y_n = G_n \otimes BP$ with maps f_n and g_n. By induction on n these maps induce a splitting

$$X = Y_k \vee Y_{k+1} \vee Y_{k+2} \vee \ldots \vee Y_{n-2} \vee Y_{n-1} \vee X_n$$

We take the maps $g_n : Y_n \dashrightarrow X$ for all n as the coordinates of a map $g: Y \dashrightarrow X$, where $Y = \bigvee_n Y_n = G \otimes BP$ and $G = \bigoplus_n G_n$, and observe (by considering homotopy groups) that g is an isomorphism.

Finally, the rank of G in each degree is clearly determined by $H_*(X) = G \otimes W_p$.]]]

We have an obvious corollary of Theorem A.

<u>COROLLARY 2.5</u> <u>Suppose</u> X <u>is</u> <u>an</u> <u>indecomposable</u> <u>p-local</u> <u>free-free</u> <u>spectrum</u>. <u>Then</u> X <u>is</u> <u>isomorphic</u> <u>to</u> <u>some</u> <u>suspension</u> <u>of</u> BP.]]]

Theorem A and this corollary demonstrate quite convincingly the central role of the spectrum BP.

<u>Maps of p-local free-free spectra</u> Now that we have classified p-local free-free spectra, we can ask about maps between such spectra. By Lemma 1.2, maps are represented faithfully by their effect on homology. This is not good enough; we need to know exactly which homomorphisms in homology, $G \otimes W_p \dashrightarrow H \otimes W_p$, or in homotopy,

$G\otimes V_p \dashrightarrow H\otimes V_p$, are induced by maps $G\otimes BP \dashrightarrow H\otimes BP$, where we are assuming that G and H are free of finite type. (These assumptions are convenient, but quite unnecessary for most of our results.)

It is clear that any V_p-module homomorphism $G\otimes V_p \dashrightarrow H\otimes V_p$ is induced by a map of spectra, simply by using the multiplicative structure of BP. We develop this idea, but find it more convenient to work in homology instead, considering $G\otimes W_p \dashrightarrow H\otimes W_p$. We first measure how far an arbitrary additive homomorphism $G\otimes W_p \dashrightarrow H\otimes W_p$ departs from being a homomorphism of W_p-modules.

<u>LEMMA 2.6</u> <u>Given any additive homomorphism</u> $h:G\otimes W_p \dashrightarrow H\otimes W_p$, <u>there is a unique homomorphism</u> $\delta(h):G\otimes T \dashrightarrow H\otimes W_p$, <u>which we call the nonlinearity of</u> h, <u>such that</u> h <u>is the composite</u>

$$G\otimes W_p \xrightarrow{1\otimes r_*} G\otimes T\otimes W_p \xrightarrow{\delta(h)\otimes 1} H\otimes W_p\otimes W_p \xrightarrow{1\otimes \phi} H\otimes W_p,$$

<u>where the last factor uses the multiplication in</u> W_p. <u>This correspondence induces an isomorphism</u>

$$\mathrm{Hom}_*(G\otimes W_p, H\otimes W_p) = \mathrm{Hom}_*(G\otimes T, H\otimes W_p).$$

<u>Proof</u> The existence and uniqueness of $\delta(h)$ follow from the form of $r_*:W_p \dashrightarrow T\otimes W_p$ given in (2.3). More elegantly, Hopf algebra techniques yield a ring homomorphism $s:T \dashrightarrow W\otimes W$ such that the composite

$$W \xrightarrow{r_*} T\otimes W \xrightarrow{s\otimes 1} W\otimes W\otimes W \xrightarrow{1\otimes \phi} W\otimes W$$

takes x to $x\otimes 1$ for all x in W; and then $\delta(h)$ is constructed directly as the composite

$$G\otimes T \xrightarrow{1\otimes s} G\otimes W\otimes W \xrightarrow{h\otimes 1} H\otimes W_p\otimes W_p \xrightarrow{1\otimes \phi} H\otimes W_p.]]$$

With this we can state and prove the main theorem.

<u>THEOREM 2.7</u> <u>Every map</u> $G\otimes BP \dashrightarrow H\otimes BP$ <u>has the form</u>

$$G\otimes BP \xrightarrow{1\otimes r} G\otimes T\otimes BP \xrightarrow{e\otimes 1} H\otimes V_p\otimes BP \xrightarrow{1\otimes \phi} H\otimes BP$$

<u>for a unique additive homomorphism</u> $e:G\otimes T \dashrightarrow H\otimes V_p$. <u>The homomorphism</u> $h:G\otimes W \dashrightarrow H\otimes W_p$ <u>is induced by a map if and only if its nonlinearity</u> $\delta(h):G\otimes T \dashrightarrow H\otimes W_p$ <u>factors through the subgroup</u> $H\otimes V_p$ <u>of</u> $H\otimes W_p$.

<u>Proof</u> We are comparing the two graded groups $\{G\otimes BP, H\otimes BP\}^*$ and $Hom_*(G\otimes T, H\otimes V_p)$. Both are clearly additive functors of G in the sense that categorical sums are taken into products. Also, since H is assumed to be of finite type, both functors preserve products, and $H\otimes BP$ is a categorical product of copies of BP, with various suspensions. We therefore need only consider the case G = H = Z.

But this case is nothing more than a restatement of Quillen's result, Theorem 5 of [10]. This states that if for every monomial t^α in T we choose an element $e(t^\alpha)$ in $V_p = \pi_*(BP)$ of appropriate degree, the corresponding map BP-->BP is just $\Sigma_\alpha e(t^\alpha)r_\alpha$, where r_α:BP-->BP is the coefficient of t^α in r.]]]

<u>COROLLARY 2.8</u> <u>An additive homomorphism</u> $h:G\otimes W_p \dashrightarrow H\otimes W_p$ <u>is realizable by a map</u> $G\otimes BP \dashrightarrow H\otimes BP$ <u>if its nonlinearity</u> $\delta(h)$ <u>is divisible by enough powers of</u> p <u>in each degree</u>.]]]

However, this does not translate directly into a similar statement for h itself, because the divisibility conditions on h mix elements of different degrees and also require arbitrarily large powers of p.

<u>COROLLARY 2.9</u> <u>There is a canonical odd-even splitting of any</u> <u>p-local free-free spectrum</u> X <u>as</u> $X'\vee X''$ <u>where</u> X' <u>has nonzero homology and homotopy groups only in odd degrees, and</u> X'' <u>only in even degrees. This splitting is preserved by all maps of spectra</u>.

Proof Write $X = G \otimes BP$ and decompose $G = G' \oplus G''$, where G' is zero in even degrees and G'' is zero in odd degrees. Take $X' = G' \otimes BP$ and $X'' = G'' \otimes BP$.]]].

The description in Theorem 2.7 of maps is not very useful for composing maps. In particular, it is most important for the applications to be able to recognize isomorphisms readily. We generalize Lemma 2.4.

LEMMA 2.10 *The map* $f: G \otimes BP \dashrightarrow H \otimes BP$ *is an isomorphism if and only if any of the following homomorphisms is an isomorphism*:

(a) $G_p \dashrightarrow H_p$, *defined from*

$$f_*: G_p \otimes W_p = H_*(G \otimes BP) \dashrightarrow H_*(H \otimes BP) = H_p \otimes W_p$$

using $1 \in W_p$ *and* $W_p \dashrightarrow Z_{(p)}$;

(b) $G_p \dashrightarrow H_p$, *defined from* $\delta(f_*): G_p \otimes T_p \dashrightarrow H_p \otimes V_p$, *using* $1 \in T_p$ *and* $V_p \dashrightarrow Z_{(p)}$;

(c) $G_p \dashrightarrow H_p$, *defined from*

$$f_*: G_p \otimes V_p = \pi_*(G \otimes BP) \dashrightarrow \pi_*(H \otimes BP) = H_p \otimes V_p,$$

using $1 \in V_p$ *and* $V_p \dashrightarrow Z_{(p)}$;

(d) *Any of the above homomorphisms tensored with* Z/p.

Proof Because we are dealing with free $Z_{(p)}$-modules, tensoring with Z/p does not affect the property of a homomorphism being an isomorphism, which yields (d). In particular, we need only consider f_* in homology with coefficients Z/p, which in terms of the nonlinearity of f_* is the composite

$$G \otimes W \otimes Z/p \xrightarrow{1 \otimes r_*} G \otimes T \otimes W \otimes Z/p \xrightarrow{\delta(f_*)}$$

$$H \otimes V \otimes Z/p \otimes W \otimes Z/p \xrightarrow{1 \otimes \phi} H \otimes W \otimes Z/p$$

However, the augmentation ideal of V is divisible in W by p and therefore annihilated by the last factor, so that we may rewrite f_*

as

$$G\otimes W\otimes Z/p \xrightarrow{1\otimes r_*} G\otimes T\otimes W\otimes Z/p \xrightarrow{e\otimes 1} H\otimes Z/p\otimes W\otimes Z/p = H\otimes W\otimes Z/p$$

where e denotes the composite

$$G\otimes T \xrightarrow{\delta(f_*)} H\otimes V\otimes Z/p \dashrightarrow H\otimes Z/p$$

and we use the augmentation $V\otimes Z/p \dashrightarrow Z/p$. If we now filter W by degree, considered as an increasing filtration, the form (2.3) of r_* shows that f_* preserves filtration and takes the form

$$f_*(g\otimes x) = u(g)\otimes x + \text{terms of lower filtration},$$

where $u: G\otimes Z/p \dashrightarrow H\otimes Z/p$ is given by $u(g) = e(g\otimes 1)$. From this it is clear that f_* is an isomorphism if and only if u is, which yields (a). But we can construct u even more directly from $\delta(f_*)$, which gives (b). Naturality of the Hurewicz homomorphism shows that the homomorphisms (a) and (c) also coincide after tensoring with Z/p, which gives (c). (In fact, the homomorphisms (a) and (c) are identical, although (b) may differ before tensoring with Z/p.)]]]

3. p-local free-free ring spectra

Although Theorems A and 2.7 describe the category of p-local free-free spectra completely, they do not end the theory. The important example $BP_\wedge BP$ is clearly free-free and therefore expressible in the form $G\otimes BP$; indeed

$$BP_\wedge BP \xrightarrow{r_\wedge 1} T\otimes BP_\wedge BP \xrightarrow{1\otimes \phi} T\otimes BP$$

is such an isomorphism. In principle this is sufficient to classify p-local free-free ring spectra, but we find it preferable to proceed somewhat differently to prove Theorem B.

Obviously $A\otimes BP$ is a ring spectrum if A is a $Z_{(p)}$-free commutative graded ring. (We assume throughout that all our rings and algebras are commutative and of finite type.) Unfortunately, this is not general enough. Suppose R is any V_p-algebra; then we can define the ring spectrum $E = R\otimes_V BP$ as giving rise to the homology theory $E_*(Y) = R\otimes_V BP_*(Y)$, provided only that R is a flat V_p-module, so that the exactness axiom holds. (By Lemma 1.4 of [6], it makes no difference here or later whether we tensor over V or V_p; but the former is more convenient typographically.) The coefficient ring of E is just R. The spectrum E is equipped with an obvious ring map $BP \dashrightarrow R\otimes_V BP$ determined by the unit element 1 in R. We therefore call E a <u>BP-algebra</u>. Theorem B states that this is the general p-local free-free ring spectrum.

<u>Remark</u> The ring R is obviously unique up to isomorphism, being the coefficient ring. However, the V_p-algebra structure need not be unique. We shall see why later, in Lemma 3.2 and Theorem 3.4.

Suppose E is a p-local free-free ring spectrum. The first step is to find a ring map $BP \dashrightarrow E$. Let us write $R = \pi_*(E)$ for the coefficient ring.

<u>LEMMA 3.1</u> <u>There is a map of ring spectra</u> $\eta : BP \dashrightarrow E$.

<u>Proof</u> We see from section 1 that E has a Chern class c^E. Suppose its logarithmic series is

$$\log z = z + \Sigma_i \, d_i z^{i+1}.$$

Because E is p-local, there is no obstruction to replacing c^E by a so-called <u>typical</u> Chern class for which $d_i = 0$ whenever i+1 is not a power of p; there is a standard technique due to Cartier (see [10, section 4]). Then the universal property of BP (again see [10]) yields a ring map $\eta : BP \dashrightarrow E$ such that $\eta_* c^{BP} = c^E$.]]]

<u>LEMMA 3.2</u> <u>For any choice of</u> η, R <u>becomes a free</u> V_p- <u>module via</u> $\eta_*:V_p \dashrightarrow R$.

We know what E looks like if we ignore the ring structure; by Theorem A there is an isomorphism $f:G\otimes BP = E$. On homotopy groups we deduce the additive homomorphism

$$f_*|G:G = G\otimes 1 \subset G\otimes V_p \xrightarrow{\quad f_* \quad} R$$

which we use with a ring map η to construct a map of spectra

$$h:G\otimes BP \xrightarrow{(f_*|G)\otimes\eta} R\otimes E \dashrightarrow E,$$

and hence the additive homomorphism

$$h_*:G\otimes V_p \dashrightarrow R\otimes R \dashrightarrow R.$$

<u>LEMMA 3.3</u> <u>For any choices of</u> η <u>and</u> f, h <u>and</u> h_* <u>are isomorphisms</u>.

<u>Proofs of Lemmas 3.2 and 3.3</u> $f^{-1}\circ h:G\otimes BP \dashrightarrow G\otimes BP$ is an isomorphism by Lemma 2.10(c), because on homotopy groups it takes $g\otimes 1$ to $g\otimes 1$ for any g in G.]]]

The theorem is now immediate.

<u>Proof of Theorem B</u> We have by Lemma 3.1 the map of ring spectra

$$R\otimes_V BP \xrightarrow{1\otimes\eta} R\otimes_V E \dashrightarrow E$$

On homotopy groups it induces by design the isomorphism

$$R = R\otimes_V V \xrightarrow{1\otimes\eta_*} R\otimes_V R \dashrightarrow R,$$

which shows that we have an isomorphism of spectra. This is valid because R is a free V_p-module by Lemma 3.2.]]]

<u>Maps of p-local free-free ring spectra</u> We would like a description of ring maps between ring spectra similar to Theorem 2.7. First we consider arbitrary ring maps $\eta:BP \dashrightarrow E$ in order to see how the V_p-module structure of R can vary with the choice of η.

THEOREM 3.4 *Let R be a V_p- algebra which is a free V_p- module of finite type.*

(a) *The general ring map* $BP \dashrightarrow R\otimes_V BP$ *has the form*

$$BP \xrightarrow{r} T\otimes BP \xrightarrow{e\otimes 1} R\otimes BP \longrightarrow R\otimes_V BP$$

where $e:T\dashrightarrow R$ *is any ring homomorphism. On homotopy groups it induces*

$$V_p \xrightarrow{r_*} T\otimes V_p \xrightarrow{e\otimes 1} R\otimes V_p \dashrightarrow R\otimes_V V_p = R$$

(b) *Suppose the elements* z_i *in* $R = \pi_*(E)$ *form a* V_p- *base of the* V_p- *module* R *via* $\eta_*:V_p\dashrightarrow R$ *for one choice of ring map* $\eta:BP\dashrightarrow E$. *Then they form a base of* R *for any choice of* η.

Proof (a) This map clearly is a ring map; we have to show there are no others. Theorem 2.7 shows (after slight rewriting) that any map, multiplicative or not, has the stated form for some additive homomorphism $e:T\dashrightarrow R$. Inspection shows that the induced homology homomorphism

$$W_p \xrightarrow{r_*} T\otimes W_p \xrightarrow{e\otimes 1} R\otimes W_p \dashrightarrow R\otimes_V W_p$$

can only be a ring homomorphism if e is.

In (b), let G denote the $Z_{(p)}$-free submodule of R generated by the elements z_i. We use the first choice of η to construct the additive isomorphism

$$f:G\otimes BP \longrightarrow R\otimes BP \xrightarrow{1\otimes\eta} R\otimes E \dashrightarrow E$$

Lemma 3.2 applies to this choice of f and any choice η' of η to yield the isomorphism of V_p-modules

$$G\otimes V_p \xrightarrow{(f_*|G)\otimes\eta'_*} R\otimes R \dashrightarrow R$$

where we equip R with the V_p-module structure defined by η'.]]]

Remark One useful case of (b) is when the V_p-algebra R has the special form $V_p \otimes A$ for some $Z_{(p)}$-algebra A. Then $E = R \otimes_V BP = A \otimes BP$, and R is isomorphic to $V_p \otimes A$ as V_p-algebra for <u>any</u> choice of η.

The generalization to ring maps between general ring spectra is unavoidably more complicated.

<u>THEOREM 3.5</u> <u>Let</u> R <u>and</u> S <u>be</u> V_p- <u>free</u> V_p- <u>algebras of finite type</u>. <u>Then any ring map</u> $R \otimes_V BP \dashrightarrow S \otimes_V BP$ <u>has the form</u>

$$R \otimes_V BP \xrightarrow{1 \otimes r} R \otimes_V (T \otimes BP) = R \otimes_V (T \otimes V) \otimes_V BP \xrightarrow{e \otimes 1} S \otimes_V BP$$

<u>where</u> $e: R \otimes_V (T \otimes V) \dashrightarrow S$ <u>is a homomorphism of</u> V- <u>algebras</u>. (Here the right V-module structure of $T \otimes V$ is the obvious one and is used to make $R \otimes_V (T \otimes V)$ a V-algebra, while the left V-module structure of $T \otimes V$ must be that defined via $r_*: V \dashrightarrow T \otimes V$.)

<u>Proof</u> Similar to that of 3.4, with extra complications.]]]

It is clear that to carry this approach much further, we either have to be very careful with all the different V-module actions, or else find a better theory. An appropriate theory is Adams' treatment of multiplicative ring spectra (see [1] or [2, Lecture 3]), in which our $T \otimes V_p$ is canonically identified from the start with $BP_*(BP)$ equipped with V_p-bimodule structure, and r is written $BP \dashrightarrow BP_*(BP) \otimes_V BP$.

<u>4. Localization and free-free spectra</u>

In this section we recall the major results we need from our localization theory [6], with the simplifications that apply in our present situation.

The theory is conveniently summarized by saying that it mimics very closely the localization theory for abelian groups, with no real surprises. Given any set M of primes, we have functorially the

M-localization $X_M = X \wedge L(Z_M)$ of X, where $L(Z_M)$ denotes the Moore spectrum for the M-localization of the ring Z of integers.

The important particular cases are the p-localization X_p of X, when M consists of a single prime p, and the rationalization $X_\emptyset$ of X. The natural map $\alpha: X \dashrightarrow X_\emptyset$ localizes to yield maps $\alpha_p: X_p \dashrightarrow X_\emptyset$ which are rational isomorphisms.

We have $\pi_*(X_M) = \pi_*(X)_M$ and $H_*(X_M) = H_*(X)_M$; in other words, the homotopy and homology group functors commute with localization.

Synthesis The more significant part of the theory is the recovery of X from its p-localizations X_p. Suppose given a set M of primes, a Z_M-free graded module G (which is to be $H_*(X)$), and for each p in M a highly-connected M-local spectrum X_p such that $H_*(X)$ is torsion-free.

THEOREM 4.1 *Suppose given* M, G, *and the* X_p *as above. Then to obtain a* M- *local spectrum* X, *well defined up to isomorphism, having the* X_p *as its* p- *localizations and* $H_*(X) = G$, *we have only to choose arbitrarily for each prime* p *in* M *an isomorphism* $\alpha_{p*}: H_*(X_p) = G_p$. *All such spectra* X *are obtained in this manner, up to isomorphism. Further, if each* X_p *is a ring spectrum,* G *is a ring, each* α_{p*} *is a ring isomorphism, and the* X_p *are free-free, then* X *will be a ring spectrum.*

Proof For the main part see Theorem 2.20 of [6], and especially the discussion following it. According to the Remark after Theorem 2.21, X has a canonical ring spectrum structure provided there are no non-trivial locally zero maps $X \dashrightarrow X$, $X \wedge X \dashrightarrow X$, or $X \wedge X \wedge X \dashrightarrow X$; which is obviously the case when X is free-free by Lemma 1.2.]]

Of course, $H_*(X_p)$ and G_p need to be isomorphic.

Our theory synthesizes maps, too. The free-free case is particularly simple.

THEOREM 4.2 Suppose X and Y are M- local spectra. Then given any maps $f_p:X_p \dashrightarrow Y_p$ for each prime p in M whose rationalizations $(f_p)_\emptyset:X_\emptyset \dashrightarrow Y_\emptyset$ agree, there is a map $f:X \dashrightarrow Y$ whose p- localizations are the given f_p. Further, f is unique if X and Y are free-free.

Proof This is Theorem 2.17 of [6]. In the free-free case, there are obviously no non-zero locally zero maps, by Lemma 1.2.]]]

Uniqueness. Let us fix a particular spectrum X and consider another spectrum X′ obtained by changing α_{p*} to $\alpha'_{p*}=\theta(p)\circ\alpha_{p*}$ for each p, where $\theta(p)$ in $Aut(G_p)$ is an arbitrary automorphism of G_p. The new spectrum X′ may or may not be different from X.

LEMMA 4.3 The spectrum X′ obtained above is isomorphic to X if and only if we can find automorphisms ρ in Aut(G) and f(p) in $Aut(X_p)$ for each p in M such that

$$\rho_p\circ\theta(p) = f(p)_* \text{ for all p in M} \tag{4.4}$$

where f(p) in $Aut(X_p)$ induces $f(p)_*$ in $Aut(G_p)$ via the isomorphism α_{p*}.

Proof We need to produce isomorphisms f(p) and $K(\rho_\emptyset)$ which make the diagrams

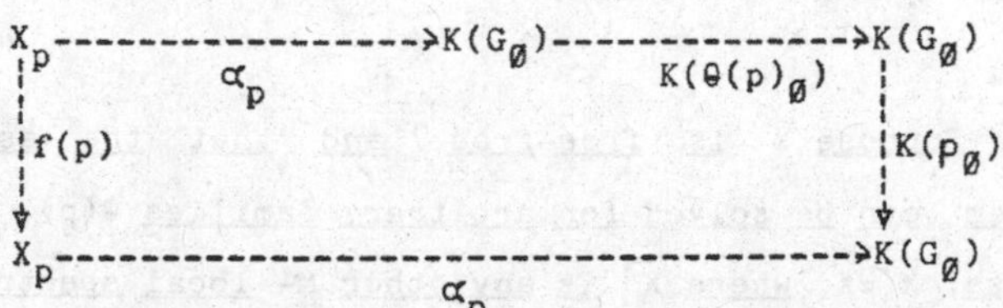

commute. Since $K(G_\emptyset)$ is a rational spectrum, it is enough to check commutativity at the homology level.]]]

It is by no means clear in practice whether the equations (4.4) admit a solution for a given family of $\theta(p)$, or for all families of $\theta(p)$. However, we can simplify the problem by filtering it. Let us assume for convenience that X is (-1)-connected. We have an obvious decreasing filtration of Aut(G) etc., if we define $F^n\text{Aut}(G)$ as the subgroup of all automorphisms of G that are the identity in all degrees $\leq$ n. These filtrations of Aut(G) and $\text{Aut}(G_p)$ are automaticallly complete Hausdorff. We write $f_1 = f_2 \bmod F^n$ to mean $f_1 \circ f_2^{-1}\ F^n$.

<u>LEMMA 4.5</u> *Suppose* X *is free-free. Then this filtration of* $\text{Aut}(G_p)$ *induces a complete filtration of* $\text{Aut}(X_p)$.

<u>Proof</u> By Theorem A we may write $X_p = C \otimes BP$. In Theorem 2.7 we gave a good algebraic description of any map $f: C\otimes BP \dashrightarrow C\otimes BP$ in terms of a homomorphism $e: C\otimes T \dashrightarrow C\otimes V_p$. For two such maps, we have $f_1 = f_2 \bmod F^n$ if and only if $e_1 = e_2$ in degrees less than n.]]]

<u>Remark</u> In this case, our filtration of $\text{Aut}(X_p)$ coincides with the skeleton filtration; however, the skeleton filtration is always available and guaranteed complete by [4], 5.17.

This suggests considering the <u>associated graded</u> problem, which is formally the same as (4.4) except that $\theta(p)$ is given in F^n, we require p and f(p) to lie in F^n, and we demand equality only modulo F^{n+1}. In other words, we consider only the effect on the nth homology groups.

<u>THEOREM 4.6</u> *Suppose* X *is free-free, and that the associated graded problem can be solved for arbitrary families* $\theta(p)$. *Then we have uniqueness,* $X'=X$, *where* X' *is any other* M- *local spectrum such that* X'_p *is isomorphic to* X_p *for all* p *and* $H_*(X')$ *is isomorphic to* $H_*(X)$.

LEMMA 4.7 Suppose given a family $\theta(p)$ and automorphisms ρ and $f(p)$ such that

$$\rho_p \circ \theta(p) = f(p)_* \bmod F^n$$

Then we can improve the agreement to

$$\rho'_p \circ \theta(p) = f(p)'_* \bmod F^{n+1}$$

by changing ρ to ρ' and $f(p)$ to $f(p)'$, where $\rho' = \rho \bmod F^n$ and $f(p)' = f(p) \bmod F^n$.

Proof We know $\sigma(p) = \rho_p \circ \theta(p) \cdot f(p)_*^{-1}$ lies in F^n. By the associated graded problem, we can write $\gamma_p \circ \sigma(p) = g(p) \bmod F^{n+1}$, with γ and $g(p)$ in F^n. We may then take $\rho' = \gamma \circ \rho$ and $f(p)' = g(p) \cdot f(p)$.]]]

Proof of Theorem 4.5 We construct by induction on n sequences of automorphisms $\rho(n)$ of G and $f(p,n)$ of X_p, starting at $\rho(0)=1$ and $f(p,0) = 1$, and a given family of automorphisms $\theta(p)$ of G_p. We construct $\rho(n+1)$ and $f(p,n+1)$ from $\rho(n)$ and $f(p,n)$ by Lemma 4.7, so that

$$\rho(n) \circ \theta(p) = f(p,n)_* \bmod F^n$$

holds for all n (trivially for n = 0), and $\rho(n+1) = \rho(n) \bmod F^n$ and $f(p,n+1) = f(p,n) \bmod F^n$. By completeness in Lemma 4.5, there exist automorphisms ρ of G and $f(p)$ of X_p such that $\rho = \rho(n) \bmod F^n$ and $f(p) = f(p,n) \bmod F^n$ for all n. Therefore these solve (4.4) mod F^n for all n, and hence exactly.]]]

Remark We are not claiming necessity for the hypotheses of 4.6. In fact, there are examples (in which X is not free-free) where the associated graded problem fails but uniqueness nevertheless holds.

5. The standard splitting of MU.

In this section we reconstitute the Thom spectrum MU from the spectra BP for each prime p by applying the localization and synthesis techniques outlined in section 4, thus reversing Quillen's route.

To synthesize a spectrum X we need a rational spectrum $X_\emptyset$, a p-local spectrum X_p for each prime p, and a rational isomorphism $\alpha_p: X_p \dashrightarrow X_\emptyset$ for each prime p. We choose $X_\emptyset$ as the GEM-spectrum $K(M_\emptyset)$, where M denotes the polynomial ring $Z[m_1, m_2, m_3, \ldots]$ on generators m_i in degree 2i for all i > 0, to make $X_\emptyset = MU_\emptyset$. We choose $X_p = E \otimes BP$, where E = E(p) is the polynomial ring $Z[e_1, e_2, e_3, \ldots]$ on generators e_i in degree 2i with e_i <u>omitted</u> whenever i+1 is a power of p; we chose the size of the padding E to make X_p rationally isomorphic to $X_\emptyset$.

The choice of $\alpha_p: X_p \dashrightarrow X_\emptyset$ is not obvious without some knowledge of Chern classes. The Chern class c^{MU} of MU has the logarithmic series (1.7)

$$\log^{MU} z = z + m_1 z^2 + m_2 z^3 + m_3 z^4 + \ldots$$

We define a ring isomorphism $E \otimes W = M$ by equating $\log^{MU}$ to the formal power series

$$\log^{BP} z + \Sigma_i \log^{BP} e_i z^{i+1}$$

where we recall from (2.1) that

$$\log^{BP} z = z + \Sigma_j w_j z^{p^j}.$$

Explicitly, on matching coefficients, we first find the two extreme cases

(i) $m_n = w_i$ if $n+1 = p^i$; (5.1)

(ii) $m_n = e_n$ if p does not divide n+1;

and then the general mixed case

$$\text{(iii)}\ m_n = e_n + w_1 e^{p}_{n_{r-1}} + w_2 e^{p^2}_{n_{r-2}} + \ldots w_r e^{p^r}_{n_0}$$

if $n+1 = p^r s$, where p does not divide s and we write $n_i = p^i s - 1$. We use this isomorphism $E \otimes W = M$ to define a rational isomorphism

$$\alpha_p : X_p = E \otimes BP \dashrightarrow K(M_\emptyset) = X_\emptyset.$$

<u>THEOREM 5.2</u> <u>The ring spectrum</u> X <u>synthesized from</u> $X_\emptyset$ <u>and the spectra</u> X_p <u>and maps</u> $\alpha_p : X_p \dashrightarrow X_\emptyset$ <u>is canonically isomorphic to</u> MU, <u>and we have</u> $MU_p = X_p = E \otimes BP$ <u>as</u> p- <u>local ring spectra</u>.

<u>Proof</u> From Lemma 1.6, what we were really doing in choosing α_p was changing the Chern class c^{BP} in $E \otimes BP$ to a different one, c' say, and forcing c' to correspond to the Chern class c^{MU} for $MU_\emptyset = X_\emptyset$. Then X becomes a ring spectrum equipped with a Chern class c^X, and the universal property of MU yields a ring map MU-->X taking c^{MU} to c^X and therefore $\log^{MU} z$ to $\log^X z$. Now $H_*(X_p) = E \otimes W_p = M_p$ for all p, from which it follows that $H_*(X) = M$. Thus MU-->X induces a homology isomorphism $H_*(MU) = H_*(X)$ and is therefore an isomorphism of ring spectra.]]]

<u>Remark</u> The resulting splitting of MU_p into copies of BP is exactly the same as Quillen's [10, Theorem 4].

If we apply the homotopy functor π_* we obtain the corresponding localization data for $\pi_*(X) = \pi_*(MU)$.

<u>COROLLARY 5.3</u> <u>The image of the Hurewicz homomorphism</u> $\pi_*(MU) \dashrightarrow H_*(MU)$ <u>is precisely the intersection of the subgroups</u> $E \otimes V_p$ <u>of</u> $M_\emptyset$, <u>taken over all primes</u> p.]]]

In other words, to decide whether or not a given polynomial in the m_i lies in the image of the Hurewicz homomorphism and so determines a homotopy class, it is sufficient to consider the problem locally at each prime p. At the prime p, we write the polynomial by means of (5.1) in terms of the e_i and w_j and decide whether or not it lies in $E\otimes V$. Of course, we need to know exactly how V lies in W, which is given by (2.2). For example, $m_5 + 9m_1m_2^2 + 8m_1^3m_2$ represents a homotopy class (only the primes 2 and 3 need to be checked) and can therefore be taken as one of the polynomial generators of $\pi_*(MU)$. This rapidly becomes very complicated if large powers of primes are involved, but one immediate consequence can be drawn.

THEOREM 5.4 (Alexander, also see [9]) *If we express the Hazewinkel generators v_i (or indeed any element of V rather than V_p) as a polynomial in the m_j by writing $w_n = m_{p^n-1}$ for all n, they lie in the image of the Hurewicz homomorphism and lift to $\pi_*(MU)$.*

Proof We apply Corollary 5.3. At primes other than p, case (ii) of 5.1 applies and there is no condition. At the prime p the condition holds by hypothesis.]]]

6. Other multiplicative splittings of MU

In this section we prove Theorem C, that MU splits multiplicatively even before localization. Whereas this can be deduced without much difficulty from section 1 and the theory of formal groups, it is quite easy to see from our present point of view.

Because we have many primes to consider, we find it necessary to append (p) to our previous notation to indicate the prime p in question; for example, BP(p) for BP, W(p) for W, and $v_i(p)$ for v_i. In fact we find it useful to extend the notation even further in certain cases to any set M of primes, by defining larger polynomial rings $W(M) = \bigotimes_{p\in M} W(p)$, and similarly V(M) and T(M).

<u>The spectra BP(M).</u> The way we assembled the spectra BP(p) in section 5 to form MU was by no means obvious to anyone not well acquainted with Chern classes and the associated algebraic theory. There is a much more direct way to put the BP(p) together, again using synthesis and Theorem 2.20 of [6].

<u>Definition 6.1</u> Given any set M of primes, we define the M-local ring spectrum BP(M) as having the localizations $BP(M)_\emptyset = K(V(M)_\emptyset)$, and $BP(M)_p = V(M-p) \otimes BP(p)$ for each prime p in M, using the obvious rational isomorphisms $\alpha_p : BP(M)_p \dashrightarrow BP(M)_\emptyset$ defined on homotopy groups by

$$\pi_*(BP(M)_p) = V(M-p) \otimes V(p)_p = V(M)_p \dashrightarrow V(M)_\emptyset = \pi_*(BP(M)_\emptyset).$$

In particular, BP(M) = BP(p) if M consists of exactly one prime p, and BP(∅) = K(Q). On the other hand, if M consists of all primes, BP(M) is obviously very much smaller than MU.

We constructed BP(M) to have the homotopy groups $\pi_*(BP(M)) = V(M)_M$. (The homotopy functor π_* preserves the localization diagram (compare 1.14 and 2.15 of [6]). However, the homology groups are equally simple.

<u>LEMMA 6.2</u> <u>We have</u> $\pi_*(BP(M)) = V(M)_M$ <u>and</u> $H_*(BP(M)) = W(M)_M$.

<u>Proof</u> We have $H_*(BP(M)_p) = V(M-p) \otimes W(p)_p$. However, V(M-p) and W(M-p) look alike at the prime p, that is $V(M-p)_p = W(M-p)_p$, because $V(q)_p = W(q)_p$ for all $q \neq p$. This enables us to write $H_*(BP(M)_p) = W(M)_p$, which we can treat the same as we did homotopy.]]]

In other words we could equally well have constructed BP(M) in terms of homology groups, by $BP(M)_p = W(M-p) \otimes BP(p)$, etc.

The result of localizing BP(M) is obvious.

<u>LEMMA 6.3</u> <u>We have multiplicative splittings</u> $BP(M)_N = V(M-N)\otimes BP(N)$ <u>whenever</u> N <u>is a subset of</u> M.]]]

<u>Maps of BP(M)</u> We would like to describe maps BP(M)-->BP(M) by generalizing Theorem 2.7. To do this we need a ring map $r(M):BP(M)\to T(M)\otimes BP(M)$. On homotopy groups it obviously ought to induce the homomorphism given by

$$V(M) = \bigotimes_{p\in M} V(p) \xrightarrow{\bigotimes_p r(p)_*} \bigotimes_{p\in M} T(p)\otimes V(p) = T(M)\otimes V(M),$$

which we prematurely call $r(M)_*$. Then we can apply Theorem 2.17 of [6] to construct the ring map r(M) as having the p-localizations $r(M)_p:BP(M)_p\to T(M)\otimes BP(M)_p$ for all p in M given by

$$BP(M)_p = V(M-p)\otimes BP(p) \xrightarrow{r(M-p)_*\otimes r(p)}$$

$$T(M-p)\otimes V(M-p)\otimes T(p)\otimes BP(p) = T(M)\otimes BP(M)_p.$$

This does indeed induce $r(M)_*$ in homotopy.

<u>THEOREM 6.4</u> <u>Every map</u> BP(M)-->BP(M) <u>has the form</u>

$$f:BP(M)\xrightarrow{r(M)} T(M)\otimes BP(M) \xrightarrow{e\otimes 1} V(M)_M\otimes BP(M)\to BP(M)$$

<u>for a unique additive homomorphism</u> $e:T(M)\to V(M)_M$. <u>It is a ring map if and only if</u> e <u>is a ring homomorphism</u>.

<u>Proof</u> For each prime p in M, Theorem 2.7 yields an additive homomorphism $e(p):V(M-p)\otimes T(p)\to V(M)_p$ such that the localization f_p is the composite

$$V(M-p)\otimes BP(p) \xrightarrow{1\otimes r(p)} V(M-p)\otimes T(p)\otimes BP(p)$$

$$\xrightarrow{e(p)\otimes 1} V(M)_p\otimes BP(p)\to BP(M)$$

Because $r(M-p)_* v_i(q) = v_i(q) + qt_i(q) +$ decomposables, and we can divide by q in $V(M)_p$ if $q\neq p$, we can construct by induction on degree (or by more elegant Hopf algebra techniques) a well defined homomorphism $e'(p):T(M)\dashrightarrow V(M)_p$ such that $e(p)$ is the composite

$$V(M-p)\otimes T(p)\xrightarrow[r(M-p)_*\otimes 1]{}T(M-p)\otimes V(M-p)\otimes T(p)$$

$$= T(M)\otimes V(M-p)\xrightarrow[e'(p)\otimes 1]{}V(M)_p\otimes V(M-p)\dashrightarrow V(M)_p$$

Then on homotopy groups, f_p induces

$$V(M)_p\xrightarrow[r(M)_p]{}T(M)\otimes V(M)_p\xrightarrow[e'(p)\otimes 1]{}V(M)_p\otimes V(M)_p\dashrightarrow V(M)_p$$

Since the rationalized homomorphisms $(f_*)_\emptyset$ must all agree, we deduce that $e'(p)_\emptyset:T(M)_\emptyset\dashrightarrow V(M)_\emptyset$ must be independent of the choice of p in M. Since it factors through $V(M)_p$ for all p in M, it must factor through $V(M)_M$ to yield the desired homomorphism e.]]]

Splittings of MU Denote by U the set of all primes; then, as we pointed out, the ring spectrum BP(U) is much smaller than MU because $H_*(BP(U)) = W(U)$ has generators $w_i(p)$ only in degrees $2(p^i-1)$ for all primes p and positive integers i. We pad it by introducing the ring $E = Z[e_5,e_9,e_{11},\dots]$ to supply the missing generators e_n in degree 2n for all positive n not of the form p^i-1.

From section 1, ring maps $MU\dashrightarrow E\otimes BP(U)$ are determined by Chern classes in $E\otimes BP(U)$. We shall look for Chern classes in BP(M) for any set M.

LEMMA 6.5 *Let c be any Chern class for BP(M), and let*

$$\log z = z + a_1z^2 + a_2z^3 + a_3z^4 + \dots$$

be its logarithmic series, with coefficients in $H_*(BP(M)) = W(M)_M$. *Then whenever* $j = p^i-1$ *with* p *in* M, *we have* $a_j = nw_i(p)$ + decomposable terms, *where* $n \equiv 1 \bmod p$. *Further, there is a Chern class for which* n = 1 *for all* p *in* M *and all* i.

Proof We know from Lemma 1.2 that BP(M) admits Chern classes. We localize at p to get $W(M-p)\otimes BP(p)$, and compare c with the standard

Chern class c^{BP} for BP(p). According to Lemma 1.6, if we ignore the image of the Hurewicz homomorphism and decomposable elements, the logarithmic series must agree with the standard one (2.1), from which the first assertion follows. From the same Lemma, if we ignore only the decomposables the logarithmic series for any other Chern class in BP(M) has the form

$$\log' z = \log z + \Sigma_i\ b_i z^{i+1} \text{ mod decomposables},$$

where the elements b_i may be chosen arbitrarily in $V(M)_M$. This yields the second assertion.]]]

Proof of Theorem C Let c be a Chern class for BP(U) provided by Lemma 6.5 with logarithmic series

$$\log z = z + \Sigma_{i,p}\ w_i(p) z^{p^i} + \text{decomposables},$$

where we sum over all primes p and all positive integers i. We may regard c as a Chern class for E⊗BP(U) having the same logarithmic series. We modify it to a Chern class c′ by involving coefficients from E, again using Lemma 1.6: we take

$$\log' z = \log z + \Sigma_j \log e_j z^{j+1},$$

summing over all applicable j. This determines a map of ring spectra f:MU-->E⊗BP(U) which induces in homology the ring homomorphism

$$f_*: M = H_*(MU) \dashrightarrow H_*(E \otimes BP(U)) = E \otimes W(U)$$

defined by the formal identity

$$\log' z = z + f_* m_1 . z^2 + f_* m_2 . z^3 + \ldots$$

We arranged for f_*, and hence f, to be isomorphisms.]]]

It follows immediately by localization and Lemma 6.3 that we have multiplicative splittings $MU_M = W(U-M) \otimes E \otimes BP(M)$ for all M.

There are no more surprises in this direction.

<u>THEOREM 6.6</u> <u>There is no multiplicative splitting of</u> BP(M), <u>for any set of primes</u> M. <u>In fact, there is no ring subspectrum of</u> BP(M) <u>that is a summand.</u>

<u>Proof</u> Suppose on the contrary that X is such a subspectrum, and consider $H_*(X)$ as a subring of $H_*(BP(M))$. By Lemma 1.6, let c be a Chern class for X, and therefore also for BP(M), with logarithmic series

$$\log z = z + a_1z^2 + a_2z^3 + a_3z^4 + \dots$$

where a_i lies in $H_*(X)$. From Lemma 6.5 the elements a_n, for n of the form $n = p^i-1$ with p in M, must form a set of polynomial generators of the rationalized ring $W(M)_\emptyset$, so that $H_*(X)_\emptyset = W(M)_\emptyset$. Since X is a summand, this can only happen if X = BP(M).]]]

<u>7.Some additive splittings</u>

In this section we prove Theorem D, that the spectrum BP(M) splits additively whenever M contains more than one prime.

First suppose that M consists of just two primes, p and q. There are many examples of rings C that can be embedded in V(p) and in V(q) in such a way that both rings become free C-modules. Such a ring C gives rise to a quotient ring spectrum Y of BP(M) for which $\pi_*(Y) = V(p)\otimes_C V(q)_M$, by using the localizations $Y_p = V(q)\otimes_C BP(p)$, $Y_q = V(p)\otimes_C BP(q)$, and $Y_\emptyset = K(V(p)\otimes_C V(q)_\emptyset)$, with the obvious rational isomorphisms $Y_p \dashrightarrow Y_\emptyset$ and $Y_q \dashrightarrow Y_\emptyset$. We plan to split BP(M) additively into copies of Y.

More generally, suppose M is any set of primes and that I is an ideal in the ring V(M). We look for a quotient ring spectrum Y of BP(M) for which $\pi_*(Y) = (V(M)/I)_M$, and a splitting of BP(M) into copies of Y. We therefore need an idempotent map f:BP(M)-->BP(M). According to Theorem 6.4, maps BP(M)-->BP(M) correspond to additive homomorphisms $e:T(M)\to V(M)_M$. We find it notationally convenient to extend e to a homomorphism of V(M)-modules, $e:T(M)\otimes V(M)\to V(M)_M$, so that on homotopy groups f_* induces the composite

$$V(M)\xrightarrow{r(M)_*}T(M)\otimes V(M)\xrightarrow{e}V(M)_M$$

<u>LEMMA 7.1</u> <u>Suppose we can find an ideal</u> I <u>in</u> V(M) <u>and a</u> V(M)-<u>module homomorphism</u> $e:T(M)\otimes V(M)\to V(M)$ <u>such that</u>

(a) e <u>takes values in the ideal</u> I <u>except that</u> $e(1\otimes x) = x$;

(b) e <u>annihilates</u> $r(M)_*I$;

(c) I^n/I^{n+1} <u>is a free</u> V(M)/I -<u>module for all</u> $n\geq 0$.

<u>Then we can find a quotient ring spectrum</u> Y <u>of</u> BP(M) <u>with</u> $\pi_*(Y) = (V(M)/I)_M$ <u>and an additive splitting</u> $BP(M) = G\otimes Y$ <u>of</u> BP(M) <u>as a graded sum of copies of</u> Y. <u>Moreover, the splitting induces</u> $V(M) = G\otimes(V(M)/I)$ <u>even before localization</u>.

<u>Proof</u> We write X = BP(M) and V = V(M) for simplicity. The homomorphism e induces (as noted above) a map $f:X\to X$, for which $f_* = 1 \bmod I$ by (a), and $f_*I = 0$ by (b). So f is idempotent, and we deduce a splitting $X = Z\vee Y$ for which $\pi_*(Z) = I_M$, $\pi_*(Y) = (V/I)_M$, and the projection map $X\to Y$ induces $V\to V/I$ and admits a splitting map $j:Y\to X$. Further, the multiplication map $X\wedge X\to X$ factors to yield $Y\wedge Y\to Y$, which makes Y a quotient ring spectrum of X as asserted.

Let G^n be the free graded abelian group generated by elements of I^n that yield a base of the free V/I-module I^n/I^{n+1}, and let G be the direct sum of all the G^n, so that G is a subgroup of V with inclusion $i:G\subset V$. Consider the composite map, which will be the desired splitting isomorphism

$$h:G\otimes Y\xrightarrow{i\otimes j}V\otimes X\xrightarrow{\phi}X.$$

On homotopy groups it induces (before localization)

$$h_*:G\otimes(V/I) \xrightarrow{i\otimes j_*} V\otimes V \xrightarrow{\phi} V$$

Our choice of G makes this an isomorphism. (In detail, if we filter V by powers of the ideal I, h_* induces isomorphisms $G^n\otimes(V/I) = I^n/I^{n+1}$ for all n.]]]

We now abandon all pretensions to generality and consider only the simplest case, when M consists of two primes p and q. Our methods here are horribly explicit and offer little prospect of generalization. As the ring C we take the polynomial ring Z[c] on one generator c in degree 2(p-1)(q-1), which we interpret either as $v_1(p)^{q-1}$ in V(p) or as $v_1(q)^{p-1}$ in V(q). We shall construct a V(M)-module homomorphism $e:T(M)\otimes V(M) \dashrightarrow V(M)$ which sends $Z[t_1(p),t_1(q)]$ into $Z[v_1(p),v_1(q)]$ and all other monomials in T(M) to 0. To simplify the notation, let us write $x = v_1(p)$, $y = v_1(q)$, $t = t_1(p)$, and $u = t_1(q)$, and ignore all the other polynomial generators as harmless. We take I as the ideal generated by the element $\alpha = x^{q-1}-y^{p-1}$, which obviously satisfies the necessary freeness condition (c) on I^n/I^{n+1}. The homomorphism $r_* = r(M)_*$, as much of it as concerns us, is given by $r_*x = x+pt$ and $r_*y = y+qu$, and we define the element

$$\beta = r_*\alpha = (x+pt)^{q-1} - (y+qu)^{p-1}.$$

To define e satisfying (a) we need to choose a value $e(t^iu^j)$ in I for all $i,j\geq 0$, except that $e(1) = 1$. Condition (b) will hold if e annihilates the ideal generated by β, that is, if $e(\beta t^iu^j) = 0$ for all $i,j\geq 0$.

We choose $e(t^iu^j)$ by induction on the degree of t^iu^j. To start, we must choose $e(1) = 1$. In degrees between zero and 2(p-1)(q-1) we may as well choose $e(t^iu^j) = 0$. In degree 2(p-1)(q-1) we have to choose $e(t^{q-1})$ and $e(u^{p-1})$ in I to make

$e(\beta) = 0$. Taking into account the choices so far, we have

$$e(\beta) = \alpha + p^{q-1}e(t^{q-1}) - q^{p-1}e(u^{p-1}).$$

It is easy enough to make this zero: choose integers a and b such that $ap^{q-1} - bq^{p-1} = 1$ and take $e(t^{q-1}) = -a\alpha$, $e(u^{p-1}) = -b\alpha$.

In the general degree above $2(p-1)(q-1)$ we need to choose a set of values $z_m = e(t^{i(m)}u^{j(m)})$ in I for $m = 0,1,2,\ldots,k$, for some $k>0$, where the exponents have the forms $i(m) = i(0) + (q-1)m$ and $j(m) = j(0) - (p-1)m$. We have to annihilate $\beta t^{i(m)}u^{j(m+1)}$ for $0 \leq m < k$. Now

$$e(\beta t^{i(m)}u^{j(m+1)}) = p^{q-1}z_{m+1} - q^{p-1}z_m + \text{terms in } I,$$

where the unstated terms depend only on previous induction steps. The following easy lemma shows that we can always satisfy the conditions.

LEMMA 7.2 *Let* p *and* q *be distinct primes,* *and* r *and* s *any positive integers. Then in any abelian group* I *there exists a solution of the system of linear equations*

$$p^r z_{m+1} - q^s z_m = c_m \quad (m=0,1,2,\ldots,k-1)$$

for the z_m $(0 \leq m \leq k)$, *whatever the elements* c_m *in* I.]]]

Proof of Theorem D. The discussion above shows that we can construct a module homomorphism e satisfying the conditions of Lemma 7.1, and obtain an additive splitting $BP(M) = G \otimes Y$ that induces $V(M) = G \otimes H$, where we write $H = V(M)/I$ for the quotient ring, and $\pi_*(Y) = H_M$.

Now suppose N is any set of primes containing M. We plan to extend the splitting of BP(M) to one of BP(N) by doing nothing at the other primes; this can be done because we were careful in Lemma 7.1 to work over Z rather than Z_M. We partition the set N of primes into the two classes M and N-M and apply Theorem 2.11 of [6] twice, also Lemma 2.10 of [6] and the discussion following it.

In detail, BP(N) has the localizations $BP(N)_M = V(N-M) \otimes BP(M)$,

$BP(N)_{N-M} = V(M) \otimes BP(N-M)$, and $BP(N)_{\emptyset} = K(V(N)_{\emptyset})$, with the obvious natural isomorphisms deduced from $V(N) = V(M) \otimes V(N-M)$. We synthesize a N-local ring spectrum Y' from the localizations $Y'_M = V(N-M) \otimes Y$, $Y'_{N-M} = H \otimes BP(N-M)$, and $Y'_{\emptyset} = K(H \otimes V(N-M)_{\emptyset})$, again using the obvious rational isomorphisms, and observe that it is indeed a quotient ring spectrum of $BP(N)$. The splitting $V(M) = G \otimes H$ yields by the uniqueness part of Theorem 2.11 of [6] an isomorphism $BP(N) = G \otimes Y'$, which is the splitting we seek.]]]

Note that there are many choices here: we can choose any two primes for M, or even combine the preceding work for several disjoint pairs of primes. And we only used one generator of $V(p)$. In particular, we make no uniqueness statement.

8. Rational isomorphisms

In this section we prove theorems E and G. Theorem E states that when the set M of primes is finite, X and X′ are isomorphic if they are rationally isomorphic; and Theorem G for MU is similar.

We apply Theorem 4.6. This means that we have to solve the equations

$$\rho \circ \theta(p) = f(p)_* \qquad (*)$$

in $H_n(X_p)$, where p runs through the set M of primes, $\theta(p)$ is any automorphism of $H_n(X_p)$, for automorphisms ρ of $H_n(X)$ and $f(p)$ in $F^n \mathrm{Aut}(G)$. That is, we work entirely with homology in degree n. Let N be the rank of the free Z_M-module $H_n(X)$, so that we may regard ρ as lying in $GL_n(Z_M)$ and $\theta(p)$ in $GL_n(Z_{(p)})$. We call an automorphism of $H_n(X)$ <u>realizable</u> if there is an automorphism $f(p)$ in $F^n \mathrm{Aut}(G)$ that induces it. From one point of view, we have to find ρ to make all the automorphisms $\rho \circ \theta(p)$ simultaneously realizable. Obviously we need a supply of realizable automorphisms.

LEMMA 8.1 *For fixed* n *and* p *there is an integer* $k>0$ *such that* $1+p^k g$ *is realizable for any endomorphism* g *of* $H_n(X_p)$.

Proof By Theorem 2.7, any map $H_*(X_p) \dashrightarrow H_*(X_p)$ is induced by some homomorphism $e: C\otimes T \dashrightarrow C\otimes W$, where we write $X_p = C\otimes BP$ by Theorem A, and is realizable if and only if its image lies in $C\otimes V_p$. If k is large enough, we have $p^k(C\otimes W_p) \subset C\otimes V_p$ in degrees $\leq n$. Let $e' = e(g)$ in degree n, and zero in other degrees; then $1+p^k e'$ corresponds to the desired map, which is automatically an automorphism by Lemma 2.10 if $k > 0$.]]]

Proof of Theorem E Lemma 8.1 shows that it is sufficient to work mod p^k for each p in M, and consider only the quotient $\overline{\theta(p)}$ in $GL_n(Z/p^k)$. It will be enough to solve the equations

$$\rho_p^{-1} = \overline{\theta(p)} \qquad \text{in } GL_n(Z/p^k).$$

This ,however , is easy. Write $s = p^k$ for the power of p, and let $M = \{p_1, p_2, \ldots, p_m\}$, say. We have to show the homomorphism

$$GL_n(Z_M) \dashrightarrow \prod_i GL_n(Z/s_i)$$

is epimorphic, where s_i denotes the apropriate power of p_i. The Chinese remainder theorem identifies the ring $\prod_i Z/s_i$ with Z/t, where $t = s_1 s_2 \ldots s_m$, and the product of groups on the right becomes $GL_n(Z/t)$. Finally, the epimorphism of rings $Z_M \dashrightarrow Z/t$ induces an epimorphism of groups $GL_n(Z_M) \dashrightarrow GL_n(Z/t)$; in fact, given a matrix in $GL_n(Z/t)$, *any* matrix A over Z_M lifting it will automatically be invertible, since det A will be a unit of Z_M (any inverse image of a unit in Z/t is a unit in Z_M).

Rational spectra are obviously classified by their Betti numbers, which are conveniently encoded in a formal power series. Merely to avoid negative exponents, we restrict attention for this purpose to (-1)-connected spectra.

Definition 8.2 Let X be a (-1)-connected spectrum, and write $b_n = \text{rank}\ \pi_n(X) = \text{rank}\ H_n(X)$ for the nth Betti number. We define the _Poincaré formal power series_ P(X) of X as

$$P(X) = b_0 + b_1 t + b_2 t^2 + b_3 t^3 + \ldots$$

Similarly we define the Poincaré series P(G) of a graded group G, by taking $b_n = \text{rank}\ G_n$.

The coefficients are obviously nonnegative integers (if finite). The function P( ) preserves sums, and takes smash and tensor products to products of series. Our most important example is P(BP(p)); since $H_*(BP(p)) = W_p = Z_{(p)}'[w_1, w_2, w_3 \ldots]$, where w_n has degree $2(p^n-1)$, we have

$$P(BP(p)) = \prod_{n=1}^{\infty}\{1+t^{2(p^n-1)}+t^{4(p^n-1)}+t^{6(p^n-1)} + \ldots\} \quad (8.3)$$

$$= \prod_{n=1}^{\infty}\{1-t^{2(p^n-1)}\}^{-1}.$$

Theorem A and the localization theory of Theorem 4.1 make it easy to characterize which series P(X) can occur for a M-local free-free spectrum X. Moreover, Theorem E yields uniqueness.

THEOREM 8.4 _A given formal power series_ y _is the Poincaré series_ P(X) _of a_ M- _local free-free_ (-1)-_connected spectrum_ X _if and only if_ y _is divisible by the series_ P(BP(p)) _for all primes_ p _in_ M, _in the sense that the coefficients of_ $y.P(BP(p))^{-1}$ _are all nonnegative. Further, if_ M _is finite, the series_ y _determines_ X _uniquely up to isomorphism_.]]

To prove Theorem G, the corresponding result for MU, we need to be more careful. In this case M is the set of all primes and we do not have a Chinese remainder theorem. We now have X = MU. The set of equations (*) is now infinite. However, all except finitely many of them may be disregarded.

LEMMA 8.5 *For fixed* n *and all sufficiently large primes* p, *every automorphism of* $H_n(X_p)$ *is realizable by some element of* $F^n\mathrm{Aut}(G)$.

Proof Referring to the proof of 8.1, we can take k=0, because in degrees $\leq$ n we have $C\otimes V_p=C\otimes W_p=C$, provided that $n < 2(p-1)$.]]]

In other words, equations (*) impose no condition on p except for finitely many primes (depending on n), say $p_1,p_2,\dots p_m$. Let us write s_i and t as before. We cannot proceed exactly as before, because the homomorphism $GL_n(Z)\dashrightarrow GL_n(Z/t)$ is obviously not epimorphic. However, we do have the following standard result.

LEMMA 8.6 *The epimorphism of rings* $Z\dashrightarrow Z/t$ *induces an epimorphism of groups*

$$SL_n(Z)\dashrightarrow SL_n(Z/t) = \prod_i SL_n(Z/s_i).$$

Proof Standard, at least in the theory of automorphic functions. One reference is Bass [3], Corollary 5.2.]]]

Proof In other words, the obvious determinantal obstruction is the only one. Let us write $\det_n(\theta)$ for the determinant of the automorphism induced on $H_n(X_p)$ by θ. Our problem is that Lemma 8.1 does not provide enough realizable automorphisms to solve (*).

LEMMA 8.7 *We assume* X = MU. *Suppose* n *is even and nonnegative, so that* $H_n(X_p) \neq 0$, *and let* u *be any unit of* $Z_{(p)}$. *Then there exists an automorphism* f *in* $F^n\mathrm{Aut}(X_p)$ *such that* $\det_n(f_*) = u$.

Proof We write $X_p=C\otimes BP$.

Case 1: C ≠ 0 in degree n. We split off a copy of BP by writing $X_p = Y\vee Y'$, where Y is a copy of S^nBP. On Y we use scalar multiplication by any unit λ of $Z_{(p)}$, and on Y′ we use the identity map. This yields an automorphism f of X for which $\det_n(f_*)=\lambda$.

<u>Case 2: p=2 and C≠0 in degree n-2.</u> We split X as $Y \vee Y'$, where $Y=S^{n-2}BP$. Given any element λ of $Z_{(2)}$, we define h as the automorphism of BP determined (see Theorem 2.7) by the homomorphism $e:T \dashrightarrow V_2$ defined by $e(1)=1$, $e(t)=\lambda v_1=2\lambda w_1$, and zero on other monomials of T. Then in $H_2(BP)$ we find $h_* w_1 = (2\lambda+1)w_1$, so that $\det_2(h_*) = 2\lambda+1$. If we define the automorphism f of X_p by using $S^{n-2}h$ on Y and the identity on Y', we have $\det_n(f_*)=2\lambda+1$. If we write u as a/b, where a and b are odd integers, we see that we have to take $\lambda=(a-b)/2b$, which does lie in $Z_{(2)}$.

Finally we have to check that these two cases cover all possibilities. If p is odd, C has a polynomial generator e_1 in degree 2, so that Case 1 always applies. If p=2, $C=Z[e_2,e_4,e_5,\ldots]$, which shows that C≠0 in degree n unless n=2 or n=6 (by using the monomials e_2^i and $e_2^i e_5$). So Case 1 applies unless n=2 or n=6, which are covered by Case 2.]]]

<u>Proof of Theorem G</u> By Lemma 8.7 we may modify each $\theta(p)$ to $\theta(p)'=\theta(p)\circ f_*$, to arrange $\det_n(\theta(p))=1$ for all p. Then Lemma 8.6 finds ρ^{-1} in $SL_n(Z)$ as required.

The proof for BP(U) is exactly the same; we only have to be sure that our proof of Lemma 8.7 still applies.]]]

<u>THEOREM 8.8</u> <u>Suppose</u> X <u>and</u> Y <u>are</u> <u>rationally</u> <u>isomorphic</u> (-<u>1</u>)-<u>connected</u> M- <u>local</u> <u>free</u>-<u>free</u> <u>spectra</u>. <u>Then</u> $X \vee MU_M$ <u>and</u> $Y \vee MU_M$ <u>are</u> <u>isomorphic</u>.

<u>Proof</u> Same as Theorem G.]]]

<u>Remark 1</u> This result, with Theorem 8.4, implies immediately that the Grothendieck group of (-1)-connected free-free spectra is just the additive group of all formal power series with integer coefficients. (Given any series y, choose a graded group G large enough that

$y.P(MU)^{-1} + P(G)$ has all its coefficients positive; then by Theorem 8.4, $y + P(G \otimes MU)$ is the Poincaré series of some free-free spectrum.)

Remark 2 The hypotheses in Theorem E are not the best possible. There are other sets of primes M for which it holds with no other assumption on X. We can prove it (but do not give the proof here) for any set M of primes for which the series $\Sigma_{p\ M} p^{-1}$ converges, without needing anything beyond Case 1 of Lemma 8.7. And of course, we have proved it in two special cases when M is the set of all primes.

The most extreme conjecture one could make is the following.

Conjecture Theorem E holds with no restriction on the set M of primes?

We still have no counterexamples.

9. Indecomposable free-free spectra

When M is a finite set of primes, Theorem 8.4 provides a purely algebraic classification of (-1)-connected M-local free-free spectra X in terms of the Poincaré series P(X). Precisely: in the set of formal power series in t with non-negative integer coefficients, P(X) can be any series that is divisible in this set by P(p) for all primes p in M, where from now on P(p) denotes the particular power series (8.3),

$$P(p) = P(BP(p)) = \Pi_{n=1}^{\infty}\{1-t^{2(p^n-1)}\}^{-1} \qquad (*)$$

Unfortunately, the factorization properties of this set of power series appear to be atrocious; for instance, there is in general no least common multiple of the set of power series P(p), as p runs through M. In a completely satisfactory classification, one would expect to be able to read off the additively indecomposable spectra. In this section we attempt to do this in the simplest case, with only limited success. Our main result is Theorem F, that we either have one indecomposable or infinitely many.

We need some terminology. Let us write x,y,z,... for power series with integer coefficients, and x_n for the coefficient of t^n in x. We call the series x <u>monic</u> if $x_0 = 1$. It is convenient to call x <u>positive</u> and to write $x \geq 0$, if $x_n \geq 0$ for all n, and more generally $x \leq y$ to mean $y-x \geq 0$. If x and y are positive power series and y is monic, y^{-1} and xy^{-1} certainly have integer coefficients; we shall say that y <u>divides</u> x if xy^{-1} is positive, in other words, if we can write x = yz with z positive.

From now on we shall confine attention to the case when M consists of just two primes p and q, with p < q, which is quite difficult enough.

If P(q) divides P(p), which obviously can happen only if p-1 divides q-1 (since $P(q) = 1 + t^{2(q-1)} + \dots$ and all exponents in P(p) are multiples of 2(p-1)), the situation is obviously very simple: the only condition on the power series P(X) is divisibility by P(p), and we might as well consider instead the arbitrary positive series $P(X).P(p)^{-1}$. The unique indecomposable (apart from suspension, or multiplication by t^n) is just P(p) itself, and X decomposes trivially as a graded sum of indecomposables, uniquely up to isomorphism.

If, however, P(q) does not divide P(p), we shall find that there are always at least two indecomposables, say x and y. In this case there is obviously no uniqueness of decomposition because we can equally well write xy as a graded sum of copies of x or of y.

Rather than use Theorem 8.4 directly, we find it convenient to return to the pair of power series $x = P(G(p)) = P(X).P(p)^{-1}$ and $y = P(G(q)) = P(X).P(q)^{-1}$. (If instead, M contained n primes, we would consider n-tuples of series.) We first restate the Theorem from this point of view.

LEMMA 9.1 *There is a 1-1 correspondence between isomorphism classes of (-1)-connected M- local free-free spectra and pairs (x,y) of positive power series that satisfy* $x.P(p) = y.P(q)$.]]]

Addition of spectra clearly corresponds in this description to addition of pairs of power series. We are looking for indecomposable pairs.

One obvious pair of series is the pair $(P(q),\ P(p))$, which corresponds to the spectrum BP(M). Every other pair of series corresponding to a M-local free-free spectrum can be written in the form $(P(q)z,\ P(p)z)$ for some series z, where z is not necessarily positive. This is our starting point in the search for indecomposables. We need a method of recognizing them.

Let us call the pair (x,y) of monic positive power series *irreducible* if $z=1$ is the only monic power series for which $0 \leq xz \leq x$ and $0 \leq yz \leq y$. (The series z will not itself be positive in any interesting case. We are not discounting the possibility of indecomposable spectra X for which P(X) is not monic.)

LEMMA 9.2 *Suppose the series* P(X) *is monic. Then the spectrum* X *is indecomposable if and only if its associated pair of series* (x,y) *is irreducible, as defined above.*

Proof If z exists, distinct from 1, the pairs (xz, yz) and (x(1-z), y(1-z)) represent spectra whose sum is X. And conversely.]]]

Irreducible pairs We therefore study pairs of positive power series from a purely algebraic point of view. Some pairs are obviously irreducible. We shall call the pair of series (x, y) disjoint if $x_n = 0$ or $y_n = 0$ for all $n > 0$, so that in positive degrees each series has gaps wherever the other has terms.

LEMMA 9.3 *Any disjoint pair of monic positive power series is irreducible.*

Proof Let (x,y) be such a pair, but suppose that on the contrary we have z as above. Since $z \neq 1$, we may write $z = 1 + et^n + \dots$ for some n, with $e \neq 0$. In degree n the conditions $0 \leq xz \leq x$ and $0 \leq yz \leq y$ yield $0 \leq x_n + e \leq x_n$ and $0 \leq y_n + e \leq y_n$. Since at least one of x_n and y_n is zero, we deduce e=0, a contradiction.]]]

There is a canonical way to start from any pair of monic power series and produce an irreducible pair.

LEMMA 9.4 *Given any pair (x,y) of (not necessarily positive) monic power series, there is a unique monic power series z such that (xz, yz) is a disjoint monic positive pair and therefore (by Lemma 9.3) irreducible.*

Proof We choose the coefficients z_n of z by induction, starting of course with $z_0 = 1$. For n>0, we choose z_n as the minimum integer (possibly negative) that makes both the expressions

$$(xz)_n = x_n + \Sigma_{i=1}^{i=n-1} x_i z_{n-i} + z_n = a + z_n, \text{ say,}$$

and similarly

$$(yz)_n = b + z_n,$$

both nonnegative, by taking $z_n = \max(-a,-b)$. This is the only possible choice for z_n to make at least one of these expressions zero, since a and b depend only on previous induction steps.]]]

Remark It is sometimes useful to modify this reduction process. If for any reason we do not wish to disturb x_i and y_i for $i \leq m$, we may take $z_i = 0$ for $0 < i \leq m$ and apply the procedure to find z_n only for $n > m$, to make $(xz)_n$ or $(yz)_n$ zero for all $n > m$.

On the other hand, irreducible pairs need not be disjoint. The point is that even for a disjoint positive monic pair, it may be possible to make xz and yz both positive without having z itself positive.

<u>LEMMA 9.5</u> <u>Suppose we have disjoint monic positive power series</u> x <u>and</u> y, <u>with</u> $x_r \neq 0$, $y_s \neq 0$, $r > 0$, $s > 0$, <u>and write</u> $k = |r-s|$. <u>Then there is exists a monic power series</u> $z = 1 + t^k + \ldots$ <u>such that</u> (xz,yz) <u>is an irreducible positive pair.</u>

<u>Proof</u> For a fixed value of k, we may assume that r and s are the <u>least</u> such numbers, and for definiteness that $r<s$, so that $s = r+k$. We construct the rest of the power series z according to the Remark above to make xz and yz positive and $(xz)_n$ or $(yz)_n$ zero for all $n > k$.

Let us look for the next nontrivial term in the series z. We assert that the series z has the form $z = 1 + t^k - et^s + \ldots$, with $e \neq 0$. Suppose that $z = 1 + t^k - et^n + \ldots$, where we do <u>not</u> initially assume that $e \neq 0$. Then $(xz)_n = x_n + x_{n-k} - e$ and $(yz)_n = y_n + y_{n-k} - e$. For these to be both nonnegative and one zero, we must have $e = \min(x_n+x_{n-k}, y_n+y_{n-k})$. Since (x,y) is a disjoint pair, $e \neq 0$ would imply that $x_n \neq 0$ and $y_{n-k} \neq 0$, or that $x_{n-k} \neq 0$ and $y_n \neq 0$. We chose r and s so that this cannot happen for $n < s$. On the other hand, if we take $n = s$, we find $e = \min(x_r,y_s) \neq 0$, which establishes our assertion.

We have to verify that the pair (xz,yz) is irreducible. Suppose on the contrary that $w = 1 + at^m + \ldots$, where $a \neq 0$, is a power series such that $0 \leq xzw \leq xz$ and $0 \leq yzw \leq yz$. In particular, $0 \leq (xz)_m + a \leq (xz)_m$ and $0 \leq (yz)_m + a \leq (yz)_m$. Since $(xz)_m$ or $(yz)_m$ is zero for all $m \neq k$, the only possibility is $m = k$. Moreover, since $(xz)_k = x_k + 1$ and $(yz)_k = y_k + 1$ and one of x_k and y_k is zero, we must have $a = -1$.

Therefore unless $zw = 1$, we can write $zw = 1 + bt^h + \ldots$, where $b \neq 0$ and $h > k$. To make xzw and yzw both positive we must have $b > 0$ (since $(xzw)_h = x_h + b$ and $(yzw)_h = y_h + b$). There are three cases:

(i) If $h < s$, $(xzw)_h \leq (xz)_h$ gives $x_h + b \leq x_h + x_{h-k}$, and hence $b \leq x_{h-k}$. Similarly, $b \leq y_{h-k}$. These yield a contradiction because one of x_{h-k} and y_{h-k} is zero.

(ii) If $h = s$, $(yzw)_s \leq (yz)_s$ gives $y_s + b \leq y_s + y_r - e = y_s - e$ and hence $b \leq -e$, a contradiction.

(iii) If $zw = 1$ or if $h > s$, $(yzw)_s \leq (yz)_s$ gives $y_s \leq y_s + y_r - e = y_s - e$, a contradiction.]]]

Proof of Theorem F We apply the preceding algebra to M-local free-free spectra, when M consists of the two primes p and q, with $p < q$. We recall that these spectra correspond to pairs of positive series $(x, y) = (P(q)z, P(p)z)$, where z is any series that makes both members of the pair positive. We shall need only monic z. According to Lemma 9.2 we are interested in choosing z to make the pair of series irreducible.

Lemma 9.4 gives a canonical way to choose z to make the pair (x,y) disjoint and hence (by Lemma 9.3) irreducible. If P(q) divides P(p), the pair (x,y) takes the form $(1, P(p).P(q)^{-1})$ and this case has already been dismissed. This is case (a) of the Theorem.

If on the other hand P(q) does not divide P(p), neither x nor y is 1, and we are able to apply Lemma 9.5 to find irreducible pairs of the form (xz, yz) which correspond to other indecomposable spectra. Moreover, unless x and y are both polynomials, Lemma 9.5 can be applied with infinitely many distinct values of k, to yield infinitely many indecomposables. The following Lemma shows that x and y are never both polynomials, since $P(p).P(q)^{-1} = yx^{-1}$.]]]

<u>LEMMA 9.6</u> $P(p).P(q)^{-1}$ <u>is never a rational function of</u> t.
<u>Proof</u> We have from (*)

$$F(t) = P(p).P(q)^{-1} = \prod_{n=1}^{\infty}\{1-t^{2(q^n-1)}\}/\{1-t^{2(p^n-1)}\}. \qquad (9.7)$$

If we regard t as a complex variable, F(t) is clearly holomorphic in $|t| < 1$. The unit circle $|t| = 1$ is its natural boundary, since one can show that for any point of the form $\tau = e^{i\pi a/p^k}$ on the unit circle, where a and k are integers (k > 0),

$$\lim_{r\to 1-} F(r\tau) = 0. \ ]]]$$

Note that in case (b) of the Theorem we have failed to give an explicit description of a single indecomposable, even the canonical one. This is due to the inscrutability of the series resulting from Lemma 9.4.

To summarize, there is a sharp division in the theory according as to whether P(q) divides P(p) or not. Obviously, we would like to know for which pairs of primes {p,q} we have divisibility; in other words, for which p and q the series (9.7) is positive. There is no simple answer.

<u>Numerical evidence</u> We consider the question of divisibility of P(p) by P(q) for various pairs of primes p and q, in other words, whether the formal power series (9.7) has all its coefficients nonnegative. As already pointed out, divisibility of q-1 by p-1 is an obvious necessary condition. Of course, as soon as one negative coefficient

is found, the question is completely settled. Otherwise, we have a certain amount of general theory that covers a few cases, and a small mountain of computer-generated data for the rest. (We spare the reader the Fortran listing, which is an easy exercise.) The computer seems to be the only practical way to find negative coefficients in many cases. In most of the others, the computer data is highly suggestive that there are no negative coefficients.

This is not a very mathematical approach. In fact, if we consider (9.7) from another point of view, regarding F(t) as a holomorphic function of the complex variable t in the unit disk, one is led to the opposite conclusion. If we take a finite subproduct with more factors from the numerator than the denominator, the resulting series has a zero at t = 1 and is therefore certainly not positive. The behavior of this function near the unit circle appears to be related to such questions as: given a number m not divisible by p, what can be said about the smallest k such that m divides p^k-1 and about $\prod_{j=1}^{j=k-1}(1-\omega^{p^j})$, where ω is a primitive mth root of unity ?

We need more practical notation. Let us write G(n) for the geometric series $1 + t^n + t^{2n} + t^{3n} + \ldots$, and $G(n_1,n_2,n_3\ldots)$ for the product $\prod_i G(n_i)$, so that

$$P(p) = G(2(p-1),\ 2(p^2-1),\ 2(p^3-1),\ldots)$$

<u>Case p=2, q=3</u> By explicit hand computation, $P(2).P(3)^{-1}$ is <u>not</u> positive, so that Case (b) of Theorem F applies. There is a negative term $-t^{52}$. After the obvious cancellation $G(2).G(4)^{-1} = 1 + t^2$, we have, retaining only terms up to degree 60,

$$P(2).P(3)^{-1} = (1+t^2).G(6,14,30).(1-t^{16})(1-t^{52}).$$

<u>Case p=2, q=5</u> Any reasonable computation fails to find a negative coefficient. It appears to be more profitable to remove as many common factors as possible, and estimate what is left. It is clear

that G(8) divides G(2), G(48) divides G(6), and that G(248) divides G(62). What to do with the next factor, G(1248), is less obvious, but the following lemma is useful.

<u>LEMMA 9.8</u> *Suppose we can write* $n = \Sigma_i \lambda_i n_i$, *with positive integer coefficients* λ_i. *Then we can write*

$$G(n_1, n_2, \ldots, n_k)(1-t^n) = \Sigma_i \; A_i \; G(n_1, \ldots, n_{i-1}, n_{i+1}, \ldots, n_k),$$

where each A_i *is a positive series.*

<u>Proof</u> (for k=2) Let us write $u = t^{n_1}$, $v = t^{n_2}$, $a = \lambda_1$, and $b = \lambda_2$. Then we have the identity

$$(1 + u + u^2 + u^3 + \ldots)(1 + v + v^2 + v^3 + \ldots)(1 - t^n) =$$
$$(1+v+v^2+v^3+\ldots+v^{b-1})(1+u+u^2+u^3+\ldots)$$
$$+ (1+u+u^2+\ldots+u^{a-1})v^b(1+v+v^2+v^3+\ldots).]]]$$

In words, each term on the right has one of the n_i removed, but we don't know which. Now return to $P(2).P(5)^{-1}$, where

P(2) = G(2, 6, 14, 30, 62, 126, 254, 510, 1022, ...)

and

P(5) = G(8, 48, 248, 1248, 6248, 31,248, ...)

We have so far used

8 = 4.2, which removes 2 from P(2),

48 = 8.6, which removes 6,

248 = 4.62, which removes 62.

For 1248, we appeal to Lemma 9.8 and write

1248 = 8.126 + 8.30, which removes 126 or 30, but we do not know which. If 126 is still available, we can next use

6248 = 1022 + 10.510 + 126, which removes 126 or 510 or 1022. Otherwise, 30 is still available and we may use instead

6248 = 4.1022 + 4.510 + 4.30, which removes 30 or 510 or 1022. For the next generator of P(5) we can use at least one of the formulae

31,248 = 248.126 = 119.254 + 1022 = 9.1022 + 735.30, which removes 30 or 126 or 254 or 1022. We continue applying Lemma 9.8 in this way with the formulae (we have saved 14)

156,248 = 4.16,382 + 6480.14,

781,248 = 73.2046 + 1239.510,

3,906,248 = 286.8190 + 382.4094,

which brings us to 19,531,248. At this point we stopped, in the absence of any general procedure. It does not seem to be too difficult in practice to find suitable linear combinations. What we have shown is that there are no negative coefficients in degrees up to 19.10^6; in fact, the first 1000 coefficients are not only positive, but increase quite rapidly. It seems highly probable that Case (a) of Theorem F applies.

<u>Case p=2, q=7</u> As in the previous case, there are no negative coefficients in degrees up to 10^6, by using the formulae

12 = 2.6
96 = 3.30 + 3.2
684 = 9.14 + 9.62,
4800 = 30.126 + 2.510,
33,612 = 84.254 + 6.2046,
235,296 = 2.32,766 + 8.8190 + 102.1022.

<u>Case p=2, q=11</u> As in the previous case, there are no negative coefficients in degrees up to 3.10^6, by use of the formulae

20 = 6 + 14,
240 = 8.30,
2660 = 13.126 + 1022,
29,280 = 75.254 + 5.2046,
322,100 = 161,050.2.

<u>Case p=2, q=13</u> Again, no negative coeffients in degrees up to 9.10^6. The formulae are becoming fewer and easier to find:

24 = 4.6,
336 = 24.14,
4392 = 26.126 + 18.62,
57,120 = 112.510,
742,584 = 371,292.2.

Case p=2, q=17 At last, we can guarantee the existence of the formulae we seek.

THEOREM 9.9 If $p-1$ divides $q-1$ and $q > p^4$, then $P(q) = P(BP(q))$ divides $P(p) = P(BP(p))$ and Case (a) of Theorem F applies.

Proof We plan to divide $G(2(q^n-1))$ into $G(2(p^{2n-2}-1), 2(p^{2n-1}-1))$ for all $n > 1$ by applying Lemma 9.8. We can always find positive integers a and b such that

$$2(q^n-1) = 2a(p^{2n-2}-1) + 2b(p^{2n-1}-1)$$

because the greatest common divisor of $p^{2n-2}-1$ and $p^{2n-1}-1$ is $p-1$ which divides q^n-1, and $q^n-1 > (p^{2n-2}-1)(p^{2n-1}-1)$.]]]

Case p=3, q=5 The first negative coefficient appears in degree 252.

Case p=3, q=7 Lemma 9.8 does not take us very far. We start out with 12 = 3.4, 96 = 6.16, but cannot express 672 in the required form. However, the coefficients in degrees less than 60,000 are all quite definitely positive and fairly steadily increasing, so that further searching seems unpromising.

Cases p=3, q=11 and q=13 No negative coefficients in degrees less than 30,000, and in fact, the coefficients increase quite rapidly in this range.

Case p=5, q=13 There is a negative coefficient, but not until degree 4648, which bears no obvious relation to the degrees of the generators of P(13).

Case p=5, q=17 The coefficients stay quite small for a long time: the largest in degrees up to 6000 is only 11, and the coefficient in degree 4962 is 4. We therefore computed the first 20,000 coefficients, which takes us up to degree 160,000, without finding any negative coefficient. But the last thousand of these coefficients all lie between 700 and 1100, so that there appears to be no point in further searching. Nevertheless, we did continue a little further, up to degree 264,000, and found no negative coefficients.

Cases p=5, q=29, 37, 41 No negative coefficients in degrees less than 264,000, and prospects beyond 264,000 seem poor.

Case p=7, q=13 There is a negative coefficient in degree 336, corresponding to one of the generators of P(13).

Case p=7, q=19 There is a negative coefficient in degree 720, again corresponding to a generator of P(19).

Cases p=7, q=31 and q=37 No negative coefficients in degrees less than 396,000.

Case p=7, q=43 No negative coefficient in degrees less than 396,000, with prospects even worse than the previous case.

Case p=11, q=31 There is a negative coefficient in degree 59,580, corresponding to the third generator of P(31).

Case p=11, q=41 We have to go all the way to degree 169,760 to find a negative coefficient, and this degree does not correspond in any obvious way to the generators of P(41).

Cases p=11, q=61 and q=71 No negative coefficient in degrees less than 660,000. In degrees between 300,000 and 400,000, all the (non-trivial) coefficients are at least 10 and less than 50.

Case p=11, q=101 No negative coefficient in degrees less than 660,000, with prospects poor beyond that; in degrees between 12,000 and 16,000 all coefficients lie in the range from 3 to 7.

Cases p=13, q=37, 61, 73 The first negative coefficient occurs in degree 2760, 7536, and 10,656* respectively, related in obvious ways

*This number was misprinted in [7].

to P(q).

Cases p=13, q=97 and q=109 No negative coeffient in degrees less than 792,000.

Cases p=17, q=97 and q=113 Negative coefficient in degrees 18,816 and 25,632 respectively.

Cases p=19, q=37 and q=73 Negative coefficients appear quite early, in degrees 2736 and 10,656 respectively.

Cases p=19, q=109 No negative coefficient in degrees less than 2,000,000, by applying Lemma 9.8 with the formulae 216 = 6.36 and 23,760 = 33.720.

Case p=19, q=127 There is a negative coefficient in degree 32,256, corresponding to the second generator of P(127).

August 1977
Department of Mathematics
Johns Hopkins University
Baltimore, Md. 21218

REFERENCES

[1] J.F.Adams, Stable homotopy and generalised homology, Chicago Lectures in Mathematics, Univ. of Chicago Press, 1974.

[2] J.F.Adams, Lectures on generalized cohomology, Lecture Notes in Mathematics, 99 (1969), 1-138 (Springer Verlag).

[3] H.Bass, K-theory and stable algebra, Publ. Math. I.H.E.S. 22 (1964), 5-60.

[4] J.M.Boardman, Stable homotopy theory, Chapter II (mimeograph), Johns Hopkins Univ., July 1970.

[5] ---, ---, Appendix B, Spectral sequences and images, November 1970.

[6] ---, ---, Appendix C, Localization theory, August 1975.

[7] ---, ---, Appendix D, Localization and splittings of MU, February 1976.

[8] Joel M. Cohen, The Hurewicz homomorphism on MU, Invent. Math. 10 (1970) 177-186.

[9] M.Hazewinkel, A universal formal group and complex cobordism, Bull. Amer. Math.Soc. 81 (1975) 930-933.

[10] D.Quillen, On the formal group laws of unoriented and complex cobordism theory, Bull. Amer. Math. Soc. 75 (1969) 1293-98.

HOMOTOPY THEORY OF Γ-SPACES, SPECTRA, AND BISIMPLICIAL SETS

A. K. Bousfield and E. M. Friedlander

In [Segal 1], Graeme Segal introduced the concept of a Γ-space and proved that a certain homotopy category of Γ-spaces is equivalent to the usual homotopy category of connective spectra. Our main purpose is to show that there is a full-fledged homotopy theory of Γ-spaces underlying Segal's homotopy category. We do this by giving Γ-spaces the structure of a closed model category, i.e. defining "fibrations," "cofibrations," and "weak equivalences" for Γ-spaces so that Quillen's theory of homotopical algebra can be applied. Actually, we give two such structures (3.5, 5.2) leading to a "strict" and a "stable" homotopy theory of Γ-spaces. The former has had applications, cf. [Friedlander], but the latter is more closely related to the usual homotopy theory of spectra.

In our work on Γ-spaces, we have adopted the "chain functor" viewpoint of [Anderson]. However, we do not require our Γ-spaces to be "special," cf. §4, because "special" Γ-spaces are not closed under direct limit constructions. We have included in §§4,5 an exposition, and slight generalization, of the Anderson-Segal results on the construction of homology theories from Γ-spaces, and on the equivalence of the homotopy categories of Γ-spaces and connective spectra.

To set the stage for our work on Γ-spaces, we have given in §2 an exposition of spectra from the standpoint of homotopical algebra. We have also included an appendix (§B) on bisimplicial sets, where we outline some well-known basic results needed in this paper and prove a rather strong fibration theorem (B.4) for diagonals of bisimplicial sets. We apply B.4 to prove a generalization of

Supported in part by NSF Grants

Quillen's spectral sequence for a bisimplicial group. In another appendix (§A), we develop some homotopical algebra which we use to construct our "stable" model categories.

The paper is organized as follows:

We work "simplicially" and refer the reader to [May 1] for the basic facts of simplicial theory.

§1. A brief review of homotopical algebra

For convenience we recall some basic notions of homotopical algebra ([Quillen 1,2]) used repeatedly in this paper.

Definition 1.1 ([Quillen 2, p. 233]). A closed model category consists of a category $\mathcal{C}$ together with three classes of maps in $\mathcal{C}$ called fibrations, cofibrations, and weak equivalences, satisfying CM1 - CM5 below. A map f in $\mathcal{C}$ is called a trivial cofibration if f is a cofibration and weak equivalence, and called a trivial fibration if f is a fibration and weak equivalence.

CM1. $\mathcal{C}$ is closed under finite limits and colimits.

CM2. For $W \xrightarrow{f} X \xrightarrow{g} Y$ in $\mathcal{C}$, if any two of f,g, and gf are weak equivalences, then so is the third.

CM3. If f is a retract of g and g is a weak equivalence, fibration, or cofibration, then so is f.

CM4. Given a solid arrow diagram

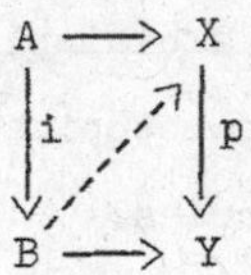

where i is a cofibration and p is a fibration, then the filler exists if either i or j is a weak equivalence.

CM5. Any map f can be factored as $f = pi$ and $f = qi$ with i a trivial cofibration, p a fibration, j a cofibration, and q a trivial fibration.

The above axioms are equivalent to the earlier more complicated ones in [Quillen 1] and are motivated in part by Example 1.3 below. They allow one to "do homotopy theory" in $\mathcal{C}$. The homotopy category $\mathrm{Ho}\,\mathcal{C}$ can be obtain from $\mathcal{C}$ by giving formal inverses to the weak equivalences. More explicitly, the objects of $\mathrm{Ho}\,\mathcal{C}$ are those of $\mathcal{C}$ and the set of morphisms, $\mathrm{Ho}\,\mathcal{C}(X,Y) = [X,Y]$, can be obtained as follows: first choose weak equivalences $X' \to X$ and $Y \to Y'$ where X' is cofibrant (i.e. $\emptyset \to X'$ is a cofibration where $\emptyset \in \mathcal{C}$ is initial) and Y' is fibrant (i.e. $Y' \to e$ is a fibration where $e \in \mathcal{C}$ is terminal); then $[X,Y] \approx [X',Y']$ and $[X',Y'] = \mathcal{C}(X',Y')/\sim$ where $\sim$ is the "homotopy relation" ([Quillen 1, I.1]). Thus $\mathrm{Ho}\,\mathcal{C}$ is equivalent to the category $\mathrm{ho}\,\mathcal{C}$ whose objects are the fibrant-cofibrant objects of $\mathcal{C}$ and whose maps are homotopy classes of maps in $\mathcal{C}$. The homotopy relation is especially manageable when $\mathcal{C}$ is a closed simplicial model category ([Quillen 1, II.2]), i.e. for objects $V, W \in \mathcal{C}$ there is a natural simplicial set $\mathrm{HOM}(V,W)$ ($= \mathrm{HOM}_{\mathcal{C}}(V,W)$) which has the properties of a function complex with vertices corresponding to the maps $V \to W$ in $\mathcal{C}$. For V cofibrant and W fibrant, one then has $[V,W] \approx \pi_0 \mathrm{HOM}(V,W)$.

It will be convenient to have

Definition 1.2. A closed model category $\mathcal{C}$ is proper if whenever a square

$$\begin{array}{ccc} A & \xrightarrow{f} & C \\ \downarrow{\scriptstyle i} & & \downarrow{\scriptstyle j} \\ B & \xrightarrow{g} & D \end{array}$$

is a pushout with i a cofibration and f a weak equivalence, then g is a weak equivalence; and whenever the square is a pullback with j a fibration and g a weak equivalence, then f is a weak equivalence.

Some needed results on proper closed model categories are proved in Appendix A, and we conclude this review with

Example 1.3. Let (s.sets) and (s.sets$_*$) denote the categories of unpointed and pointed simplicial sets respectively. These are proper closed simplicial model categories, where the cofibrations are the injections, the fibrations are the Kan fibrations, the weak equivalences are the maps whose geometric realizations are homotopy equivalences, $\mathrm{HOM}_{(\mathrm{s.sets})}(X,Y)_n$ consists of the maps $X \times \Delta[n] \to Y$ in (s.sets), and $\mathrm{HOM}_{(\mathrm{s.sets}_*)}(X,Y)_n$ consists of the maps $X \wedge (\Delta[n] \cup *) \to Y$ in (s.sets$_*$). Note that the Kan complexes are the fibrant objects and all objects are cofibrant. The associated homotopy categories Ho(s.sets) and Ho(s.sets$_*$) are equivalent to the unpointed and pointed homotopy categories of CW complexes respectively. For $X \in$ (s.sets$_*$) we will let $\pi_i X$ denote $\pi_i |X|$ where $|X|$ is the geometric realization of X.

§2. Closed model category structures for spectra

To set the stage for our study of Γ-spaces, we now discuss spectra from the standpoint of homotopical algebra. Although spectra in the sense of [Kan] admit a closed model category structure (cf. [Brown]), these spectra are not very closely related to Γ-spaces and don't seem to form a closed simplicial model category. For our purposes the appropriate spectra are old-fashioned ones equipped with a suitable model category structure. After developing that structure, we show that it gives a stable homotopy theory equivalent to the usual one.

Definition 2.1. A spectrum $\underset{\sim}{X}$ consists of a sequence $\underset{\sim}{X}^n \in (\text{s.sets}_*)$ for $n \geq 0$ and maps $\sigma^n: S^1 \wedge \underset{\sim}{X}^n \to \underset{\sim}{X}^{n+1}$ in (s.sets_*), where $S^1 = \Delta[1]/\dot{\Delta}[1] \in (\text{s.sets}_*)$. A map $f: \underset{\sim}{X} \to \underset{\sim}{Y}$ of spectra consists of maps $f^n: \underset{\sim}{X}^n \to \underset{\sim}{Y}^n$ in (s.sets_*) for $n \geq 0$ such that $\sigma^n(1 \wedge f^n) = f^{n+1}\sigma^n$; and (spectra) denotes the category of spectra.

The sphere spectrum $\underset{\sim}{S}$ is the obvious spectrum with $\underset{\sim}{S}^0 = S^0 = \Delta[0] \cup *$, $\underset{\sim}{S}^1 = S^1$, $\underset{\sim}{S}^2 = S^1 \wedge S^1$, $\underset{\sim}{S}^3 = S^1 \wedge S^1 \wedge S^1, \ldots$.

For $K \in (\text{s.sets})$ and $\underset{\sim}{X} \in (\text{spectra})$, $\underset{\sim}{X} \wedge K$ is the obvious spectrum with $(\underset{\sim}{X} \wedge K)^n = \underset{\sim}{X}^n \wedge K$ for $n \geq 0$; and for $\underset{\sim}{X}, \underset{\sim}{Y} \in (\text{spectra})$, $\text{HOM}(\underset{\sim}{X}, \underset{\sim}{Y})$ is the obvious simplicial set whose n-simplices are maps $\underset{\sim}{X} \wedge (\Delta[n] \cup *) \to \underset{\sim}{Y}$ in (spectra).

A map $f: \underset{\sim}{X} \to \underset{\sim}{Y}$ in (spectra) is a strict weak equivalence (resp. strict fibration) if $f^n: \underset{\sim}{X}^n \to \underset{\sim}{Y}^n$ is a weak equivalence (resp. fibration) in (s.sets_*) for $n \geq 0$; and f is a strict cofibration if the induced maps

$$\underset{\sim}{X}^0 \to \underset{\sim}{Y}^0 \qquad \underset{\sim}{X}^{n+1} \coprod_{S^1 \wedge \underset{\sim}{X}^n} S^1 \wedge \underset{\sim}{Y}^n \longrightarrow \underset{\sim}{Y}^{n+1}$$

are cofibrations in (s.sets_*) for $n \geq 0$. (This implies that each $f^n: \underset{\sim}{X}^n \to \underset{\sim}{Y}^n$ is a cofibration.) We let $(\text{spectra})^{\text{strict}}$ denote the category (spectra) equipped with these "strict" classes of maps.

Proposition 2.2. $(\text{spectra})^{\text{strict}}$ is a proper closed simplicial model category.

The proof is straightforward. Of course the associated homotopy category $\text{Ho}(\text{spectra})^{\text{strict}}$ is not equivalent to the usual stable homotopy category because it has too many homotopy types.

To obtain the usual stable theory, we call a map $f\colon \underset{\sim}{X} \to \underset{\sim}{Y}$ in (spectra) a <u>stable weak equivalence</u> if $f_*\colon \pi_*\underset{\sim}{X} \cong \pi_*\underset{\sim}{Y}$ where $\pi_*\underset{\sim}{X} = \varinjlim_{n} \pi_{*+n} \underset{\sim}{X}^n$; and call f a <u>stable cofibration</u> if f is a strict cofibration. Call $\underset{\sim}{X} \in$ (spectra) an <u>Ω-spectrum</u> if for each $n \geq 0$ the geometric realization $|S^1| \wedge |\underset{\sim}{X}^n| \cong |S^1 \wedge \underset{\sim}{X}^n| \xrightarrow{|\sigma^n|} |\underset{\sim}{X}^{n+1}|$ induces a weak homotopy equivalence $|\underset{\sim}{X}^n| \to |\underset{\sim}{X}^{n+1}|^{|S^1|}$. Then choose a functor $Q\colon$ (spectra) $\to$ (spectra) and a natural transformation $\eta\colon 1 \to Q$ such that $\eta\colon \underset{\sim}{X} \to Q\underset{\sim}{X}$ is a stable weak equivalence and $Q\underset{\sim}{X}$ is an Ω-spectrum for each $\underset{\sim}{X} \in$ (spectra). For instance one can let $Q\underset{\sim}{X}$ be the obvious spectrum with

$$(Q\underset{\sim}{X})^n = \lim_{i\to\infty} \operatorname{Sing} \Omega^i |\underset{\sim}{X}^{n+i}|$$

where Sing is the singular functor. Now call $f\colon \underset{\sim}{X} \to \underset{\sim}{Y}$ a <u>stable fibration</u> if f is a strict fibration and for $n \geq 0$

$$\begin{array}{ccc} \underset{\sim}{X}^n & \xrightarrow{\eta} & (Q\underset{\sim}{X})^n \\ \downarrow f^n & & \downarrow (Qf)^n \\ \underset{\sim}{Y}^n & \xrightarrow{\eta} & (Q\underset{\sim}{Y})^n \end{array}$$

is a homotopy fibre square in (s.sets_*), cf. A.2. When all the $\underset{\sim}{Y}^n$ are connected this is actually equivalent to saying that f is a strict fibration with fibre on Ω-spectrum. Let $(\text{spectra})^{\text{stable}}$ denote the category (spectra) equipped with stable weak equivalences, stable fibrations, and stable cofibrations.

<u>Theorem 2.3</u>. $(\text{spectra})^{\text{stable}}$ is a proper closed simplicial model category.

<u>Proof</u>. The usual arguments of stable homotopy theory show that if

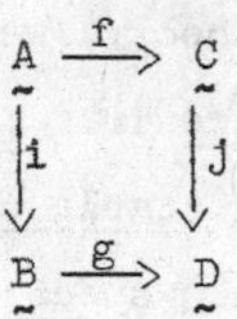

is a pushout in (spectra) with $f_*: \pi_* \underset{\sim}{A} \cong \pi_* \underset{\sim}{C}$ and with each $i^n: \underset{\sim}{A}^n \to \underset{\sim}{B}^n$ a cofibration in (s.sets$_*$), then $g_*: \pi_* \underset{\sim}{B} \cong \pi_* \underset{\sim}{D}$; and if the square is a pullback with $g_*: \pi_* \underset{\sim}{B} \cong \pi_* \underset{\sim}{D}$ and with each $j^n: \underset{\sim}{C}^n \to \underset{\sim}{D}^n$ a fibration in (s.sets$_*$) then $f_*: \pi_* \underset{\sim}{A} \cong \pi_* \underset{\sim}{C}$. Moreover, a map f: $\underset{\sim}{X} \to \underset{\sim}{Y}$ in (spectra) is a stable weak equivalence iff Qf: $Q\underset{\sim}{X} \to Q\underset{\sim}{Y}$ is a strict weak equivalence. The result now follows by using Theorem A.7 and the simpliciality criterion SM7(b) of [Quillen 1, II.2].

Note that our definition of "stable fibration" does not actually depend on the choice of Q, because the fibrations in a closed model category are determined by the trivial cofibrations.

2.4. The stable homotopy category. By 2.5 below, Ho(spectra)^{stable} is the usual stable homotopy category; and by model category theory, it is equivalent to the "concrete" category ho(spectra)^{stable} of fibrant-cofibrant spectra in (spectra)^{stable} and homotopy classes of maps. Note that a spectrum $\underset{\sim}{X} \epsilon$ (spectra)^{stable} is fibrant iff $\underset{\sim}{X}$ is an Ω-spectrum with each $\underset{\sim}{X}^n$ a Kan complex, and $\underset{\sim}{X}$ is cofibrant iff each $\sigma: S^1 \wedge \underset{\sim}{X}^n \to \underset{\sim}{X}^{n+1}$ is an injection. Also, it is easy to show that Q induces an equivalence

$$\text{Ho(spectra)}^{\text{stable}} \xrightarrow{\cong} \text{Ho}(\Omega\text{-spectra})^{\text{strict}}$$

where Ho(Ω-spectra)^{strict} is the full subcategory of Ω-spectra in Ho(spectra)^{strict}.

2.5. Equivalence of various stable homotopy theories

We wish to show that our model category (spectra)^{stable} gives a homotopy theory equivalent to that for (Kan's spectra) developed in [Kan] and [Brown]. Recall that Kan's spectra are like pointed simplicial sets, except that they have simplices in both positive and negative degrees, and have operators d_i and s_i for all $i \geq 0$. They arise as "direct limits" of Kan's prespectra, which are sequences $K^0, K^1, K^2, \ldots$ in (s.sets$_*$) together with maps $SK^n \to K^{n+1}$ for $n \geq 0$. Here, S(-) is the "small" suspension functor given in [Kan,2.2]; so for $K \epsilon$ (s.sets$_*$), the non-basepoint non-degenerate simplices of $(SK)_i$ correspond to those of K_{i-1} but have trivial i^{th} faces.

It is difficult to relate our spectra to Kan's in a purely simplicial way, because the suspension functors S(-) and $S^1 \wedge (-)$ are very different. Thus we will need the intermediate category (top. spectra) defined as in 2.1, but using pointed topological spaces and the topological suspension. We will also need the category (Kan's prespectra) defined as in 2.1, but using the "small" suspension functor S(-) as indicated above. Our categories (top. spectra) and (Kan's presepctra) differ from those discussed in [Kan], because we put no injectivity conditions on the structural maps; but there are still adjoint functors

$$\text{(spectra)} \underset{\text{Sing}}{\overset{|\,|}{\rightleftarrows}} \text{(top. spectra)}$$

$$\underset{\text{Sing}}{\overset{|\,|}{\leftrightarrows}} \text{(Kan's prespectra)} \underset{\text{Ps}}{\overset{\text{Sp}}{\rightleftarrows}} \text{(Kan's spectra)}$$

defined as in [Kan, §§3,4], where the upper arrows are the left adjoints. In particular, the realization and singular functors induce adjoint functors between (spectra) and (top. spectra), where the structural maps are handled using the natural homeomorphism $|S^1 \wedge K| \cong |S^1| \wedge |K|$ for $K \epsilon$ (s.sets$_*$). We define closed model category structures on (top. spectra) and (Kan's prespectra) by mimicing the construction of (spectra)^{stable}; in the construction for (top. spectra), we use the standard model category structure on pointed topological spaces, c.f. [Quillen 1, II.3]. The above pairs of adjoint functors all satisfy the hypotheses of [Quillen 1, I.4, Th. 3], and thus induce "equivalences of homotopy theories;" in particular, the four stable homotopy categories are equivalent. We remark that, unlike (spectra) and (top. spectra), the categories (Kan's prespectra) and (Kan's spectra) do not seem to have reasonable closed <u>simplicial</u> model category structures.

§3. <u>The strict homotopy theory of Γ-spaces</u>.

In this section, we introduce Γ-spaces and verify that they admit a "strict" model category structure similar to that of spectra. Not only does this "strict" model category structure admit applications (cf. [Friedlander]), but also it enables us to subsequently construct the "stable" model category structure on the category of Γ-spaces (whose homotopy category is the homotopy category of connected spectra).

We adopt D. Anderson's viewpoint in defining Γ-spaces. Let Γ^0 denote the category of finite pointed sets and pointed maps; Γ^0 is the dual of the category considered by G. Segal [Segal 1]. For $n \geq 0$, let n^+ denote the set $\{0,1,\ldots,n\}$ with basepoint $0 \in n^+$.

<u>Definition 3.1</u>. Let $\underline{C}$ be a pointed category with initial-terminal object $*$. A Γ-<u>object</u> over $\underline{C}$ is a functor $\underset{\sim}{A}: \Gamma^0 \to \underline{C}$ such that $\underset{\sim}{A}(0^+) = *$. A Γ-<u>space</u> is a Γ-object over the category $(s.sets_*)$ of pointed simplicial sets. $\Gamma^0\underline{C}$ is the category of Γ-objects over $\underline{C}$.

The reader should consult [Friedlander], [Segal 1] for interesting examples of Γ-topological spaces, Γ-spaces, and Γ-varieties.

For notational convenience, we shall sometimes view a Γ-object over $\underline{C}$ as a functor from the full subcategory of Γ^0 whose objects are the sets n^+, $n \geq 0$. Such a functor is the restriction of a functor $\Gamma^0 \to \underline{C}$ (determined up to canonical equivalence).

We begin our consideration of $\Gamma^0(\text{s.sets}_*)$, the category of Γ-spaces, by introducing some categorical constructions. For $\underset{\sim}{A}\in\Gamma^0(\text{s.sets}_*)$ and $K\in(\text{s.sets}_*)$, define $\underset{\sim}{A}\wedge K \in \Gamma^0(\text{s.sets}_*)$ by

$$(\underset{\sim}{A}\wedge K)(n^+) = \underset{\sim}{A}(n^+)\wedge K \qquad \text{for } n \geq 0$$

and define $\underset{\sim}{A}^K\in\Gamma^0(\text{s.sets}_*)$ by

$$\underset{\sim}{A}^K(n^+) = \underset{\sim}{A}(n^+)^K \qquad \text{for } n \geq 0$$

If $\underset{\sim}{A}$, $\underset{\sim}{B}\in\Gamma^0(\text{s.sets}_*)$, we define $\text{HOM}(\underset{\sim}{A},\underset{\sim}{B})\in(\text{s.sets}_*)$ by

$$\text{HOM}(\underset{\sim}{A},\underset{\sim}{B})_n = \text{Hom}_{\Gamma^0(\text{s.sets}_*)}(\underset{\sim}{A}\wedge(\Delta[n] \cup *),\underset{\sim}{B}).$$

<u>Definition 3.2</u>. Let $i_n\colon \Gamma^0_n \to \Gamma^0$ denote the inclusion of the full subcategory of all finite sets with no more than n non-basepoint elements. Let

$$T_n\colon \Gamma^0(\text{s.sets}_*) \to \Gamma^0_n(\text{s.sets}_*)$$

be the <u>n-truncation</u> functor defined by sending $\underset{\sim}{A}\colon \Gamma^0 \to (\text{s.sets}_*)$ to $\underset{\sim}{A}\cdot i_n\colon \Gamma^0_n \to (\text{s.sets}_*)$. The left adjoint of T_n

$$\text{sk}_n\colon \Gamma^0_n(\text{s.sets}_*) \to \Gamma^0(\text{s.sets}_*)$$

is called the n-<u>skeleton</u> functor and is given for $\underset{\sim}{A}\in\Gamma^0_n(\text{s.sets}_*)$ by

$$(\text{sk}_n\underset{\sim}{A})(m^+) = \underset{\substack{k^+\to m^+\\ k\leq n}}{\text{colim}}\ \underset{\sim}{A}(k^+).$$

The right adjoint of T_n

$$csk_n\colon \Gamma^0_n(s.sets_*) \to \Gamma^0(s.sets_*)$$

is called the n-coskeleton functor and is given for $\underset{\sim}{A} \in \Gamma^0_n(s.sets_*)$ by

$$(csk_n\underset{\sim}{A})(m^+) = \lim_{\substack{m^+ \to j^+ \\ j \le n}} \underset{\sim}{A}(j^+).$$

We shall frequently commit a slight abuse of notation and let $sk_n\underset{\sim}{A}$ $csk_n\underset{\sim}{A}$ denote $sk_n \circ T_n(A)$ $csk_n \circ T_n(\underset{\sim}{A})$ for $\underset{\sim}{A} \in \Gamma^0(s.sets_*)$.

Our construction of the strict model category for Γ-spaces depends on the following model category structure for G-equivariant homotopy theory for the groups $G = \Sigma_n$ (the groups of pointed automorphisms of n^+). For any group G, we let $G(s.sets_*)$ denote the category of pointed simplicial sets with left G-action (or, equivalently, of simplicial objects over pointed left G-sets). For $X, Y \in G(s.sets_*)$, HOM(X,Y) denotes the simplicial set defined by

$$HOM(X,Y)_n = Hom_{G(s.sets_*)}(X \wedge (\Delta[n] \cup *), Y)$$

where G acts trivially on $\Delta[n] \cup *$.

Proposition 3.3. For any G, the category $G(s.sets_*)$ is a proper closed simplicial model category when provided with the following additional structure: a G-weak equivalence (respectively, a G-fibration) is a map $f\colon X \to Y$ in $G(s.sets_*)$ which is a weak equivalence (resp., fibration) in $(s.sets_*)$; a G-cofibration is a map $f\colon X \to Y$ in $G(s.sets_*)$ which is injective and for which G acts freely on the simplices not in the image of f.

The proof of Proposition 3.3 is straight-forward; indeed, this model category is a case of that defined in [Quillen 1, II.4].

The role of Σ_n-equivariance is revealed by the following proposition, whose straight-forward proof we omit (the notation of the proposition has been chosen to fit the proof of Theorem 3.5).

<u>Proposition 3.4</u>. For $\underset{\sim}{B}, \underset{\sim}{X} \in \Gamma_n^0(\text{s.sets}_*)$, let $u_{n-1}: T_{n-1}\underset{\sim}{B} \to T_{n-1}\underset{\sim}{X}$ be a map in $\Gamma_{n-1}^0(\text{s.sets}_*)$. A map $u^n: \underset{\sim}{B}(n^+) \to \underset{\sim}{X}(n^+)$ in (s.sets_*) determines a prolongation of u_{n-1} to $u: \underset{\sim}{B} \to \underset{\sim}{X}$ in $\Gamma_n^0(\text{s.sets}_*)$ if and only if u^n is a Σ_n-equivariant map which fills in the following commutative diagram in $\Sigma_n(\text{s.sets}_*)$:

$$\begin{array}{ccccc} (sk_{n-1}\underset{\sim}{B})(n^+) & \to & \underset{\sim}{B}(n^+) & \to & (csk_{n-1}\underset{\sim}{B})(n^+) \\ \downarrow{\scriptstyle sk_{n-1}(u_{n-1})} & & \downarrow & & \downarrow{\scriptstyle csk_{n-1}(u_{n-1})} \\ (sk_{n-1}\underset{\sim}{X})(n^+) & \to & \underset{\sim}{X}(n^+) & \to & (csk_{n-1}\underset{\sim}{X})(n^+) \end{array} \tag{3.4.1}$$

Proposition 3.4 should motivate the following model category structure on $\Gamma^0(\text{s.sets}_*)$.

<u>Theorem 3.5</u>. The category of Γ-spaces becomes a proper closed simplicial model category (denoted $\Gamma^0(\text{s.sets}_*)^{\text{strict}}$), when provided with the following additional structure: a map $f: \underset{\sim}{A} \to \underset{\sim}{B} \in \Gamma^0(\text{s.sets}_*)$ is called a <u>strict weak equivalence</u> if $f(n^+): \underset{\sim}{A}(n^+) \to \underset{\sim}{B}(n^+)$ is a $(\Sigma_n$-) weak equivalence for $n \geq 1$; $f: \underset{\sim}{A} \to \underset{\sim}{B}$ is called a <u>strict cofibration</u> if the induced map

$$(sk_{n-1}\underset{\sim}{B})(n^+) \coprod_{(sk_{n-1}\underset{\sim}{A})(n^+)} \underset{\sim}{A}(n^+) \longrightarrow \underset{\sim}{B}(n^+) \tag{3.5.1}$$

is a Σ_n-cofibration for $n \geq 1$; and a map $f: \underset{\sim}{A} \to \underset{\sim}{B}$ is called a <u>strict fibration</u> if the induced map

$$\underset{\sim}{A}(n^+) \longrightarrow (csk_{n-1}\underset{\sim}{A})(n^+) \prod_{(csk_{n-1}\underset{\sim}{B})(n^+)} \underset{\sim}{B}(n^+) \tag{3.5.2}$$

is a $(\Sigma_n$-) fibration for $n \geq 1$.

This model category structure is similar to that obtained by

C. Reedy for simplicial objects over a closed model category, and our proof will somewhat resemble his.

Proof. Because finite limits, finite colimits, and weak equivalences in $\Gamma^O(s.sets_*)^{strict}$ are defined level-wise, CM1 and CM2 are immediately verified. Similarly, CM3 for $\Gamma^O(s.sets_*)^{strict}$ follows directly from CM3 for $\Sigma_n(s.sets_*)$ for each $n > 0$.

To prove one half of CM4 (we omit the similar proof of the other half) for $\Gamma^O(s.sets_*)^{strict}$, let

$$\begin{array}{ccc} \underset{\sim}{A} & \longrightarrow & \underset{\sim}{X} \\ {\scriptstyle i}\downarrow & \nearrow & \downarrow{\scriptstyle p} \\ \underset{\sim}{B} & \longrightarrow & \underset{\sim}{Y} \end{array} \tag{3.5.3}$$

be a diagram in $\Gamma^O(s.sets_*)^{strict}$ such that i is a strict trivial cofibration and p is a strict fibration. A filler $u: \underset{\sim}{B} \to \underset{\sim}{X}$ is constructed inductively by finding fillers $u_n: T_n\underset{\sim}{B} \to T_n\underset{\sim}{X}$ for the truncations $T_n(3.5.3)$ of diagram (3.5.3) for $n \geq 1$. These truncated fillers are obtained by applying Propositions 3.3 and 3.4 together with the facts that $(sk_{n-1}\underset{\sim}{B})(n^+) \coprod_{(sk_{n-1}\underset{\sim}{A})} \underset{\sim}{A}(n^+) \to \underset{\sim}{B}(n^+)$ is trivial Σ_n-cofibration and $\underset{\sim}{X}(n^+) \to (csk_{n-1}\underset{\sim}{X})(n^+) \times_{(csk_{n-1}\underset{\sim}{X})(n^+)} \underset{\sim}{Y}(n^+)$ is a $(\Sigma_n\text{-})$fibration. The second fact is immediate, and the first follows since $(sk_{n-1}\underset{\sim}{A})(n^+) \to (sk_{n-1}\underset{\sim}{B})(n^+)$ is a trivial cofibration as in the proof of 3.7 below.

To prove one half of CM5 (we omit the similar proof of the other half), we must factor a map $f: \underset{\sim}{A} \to \underset{\sim}{B}$ in $\Gamma^O(s.sets_*)$ as $f = p \cdot i$ where i is a strict trivial cofibration and p is a strict fibration. Suppose inductively that we have a factorization

$$T_{n-1}\underset{\sim}{A} \to T_{n-1}\underset{\sim}{C} \to T_{n-1}\underset{\sim}{B} \in \Gamma^O_{n-1}(s.sets_*)$$

for some $n \geq 1$. Using the closed model category structure on $\Sigma_n(\text{s.sets}_*)$ given by Proposition 3.3, we obtain a factorization in $\Sigma_n(\text{s.sets}_*)$

$$(sk_{n-1}\underset{\sim}{C})(n^+) \coprod_{(sk_{n-1}\underset{\sim}{A})(n^+)} \underset{\sim}{A}(n^+) \xrightarrow{\alpha} K \xrightarrow{\beta} (csk_{n-1}\underset{\sim}{C})(n^+) \times_{(csk_{n-1}\underset{\sim}{B})(n^+)} \underset{\sim}{B}(n^+)$$

of the canonical map with α a trivial Σ_n-cofibration and β a fibration. The desired factorization $\underset{\sim}{A} \to \underset{\sim}{C} \to \underset{\sim}{B}$ is now obtained by induction using 3.4 and the following lemma (whose proof is immediate); the map $\underset{\sim}{A} \to \underset{\sim}{C}$ is a strict trivial cofibration by a patching argument as in the proof of 3.7 below.

<u>Lemma 3.6</u>. For $\underset{\sim}{C} \in \Gamma^o_{n-1}(\text{s.sets}_*)$, let

$$(sk_{n-1}\underset{\sim}{C})(n^+) \to K \to (csk_{n-1}\underset{\sim}{C})(n^+)$$

be a factorization in $\Sigma_n(\text{s.sets}_*)$ of the canonical map. Then $\underset{\sim}{C}$ prolongs to an object $\underset{\sim}{C}' \in \Gamma^o_n(\text{s.sets}_*)$ with $\underset{\sim}{C}'(n^+) = K$ such that the given factorization equals the canonical one for $\underset{\sim}{C}'$.

This completes the proof of <u>CM5</u>, and thus of the fact that $\Gamma^o(\text{s.sets}_*)^{\text{strict}}$ is a closed model category. To prove that $\Gamma^o(\text{s.sets}_*)^{\text{strict}}$ is a simplicial closed model category, it suffices to prove for each fibration $p\colon \underset{\sim}{A} \to \underset{\sim}{B}$ in $\Gamma^o(\text{s.sets}_*)$ and each cofibration $i\colon K \to L$ in (s.sets) that the induced map in $\Gamma^o(\text{s.sets}_*)^{\text{strict}}$

$$\mu\colon \underset{\sim}{A}^L \to \underset{\sim}{A}^K \times_{\underset{\sim}{B}^K} \underset{\sim}{B}^L$$

is a fibration which is trivial whenever either p or i is trivial. This follows easily from the closed model category properties of (s.sets_*), because the maps of type (3.5.2) associated with μ are given by the maps $D^L \to D^K \times_{E^K} E^L$ in (s.sets_*) induced by

$$D = \underset{\sim}{A}(n^+) \to (csk_{n-1}\underset{\sim}{A})(n^+) \underset{(csk_{n-1}\underset{\sim}{B})(n^+)}{\times} \underset{\sim}{B}(n^+) = E.$$

Finally, to prove that $\Gamma^O(s.sets_*)^{strict}$ is a proper simplicial closed model category, it suffices to prove the following lemma and then employ the fact that $(s.sets_*)$ is a proper closed model category (one proceeds level-by-level, since strict weak equivalences are determined levelwise).

Lemma 3.7. If $f: \underset{\sim}{A} \to \underset{\sim}{B}$ is a cofibration (resp., fibration) in $\Gamma^O(s.sets_*)^{strict}$, then

$$(sk_m\underset{\sim}{A})(n^+) \to (sk_m\underset{\sim}{B})(n^+) \ \ (\text{resp., } (csk_m\underset{\sim}{A})(n^+) \to (csk_m\underset{\sim}{B})(n^+))$$

is a cofibration (resp., fibration) in $(s.sets_*)$ for all $m,n \geq 0$.

Proof. We treat the cofibration case and omit the similar proof of the fibration case. Assuming inductively that $(sk_{m-1}\underset{\sim}{A})(n^+) \to (sk_{m-1}\underset{\sim}{B})(n^+)$ is a cofibration, we will show that $(sk_m\underset{\sim}{A})(n^+) \to (sk_m\underset{\sim}{B})(n^+)$ is a cofibration. There is a push-out square

$$\begin{array}{ccc} \coprod_S sk_{m-1}\underset{\sim}{A}(S) & \to & sk_{m-1}\underset{\sim}{A}(n^+) \\ \downarrow & & \downarrow \\ \coprod_S sk_m\underset{\sim}{A}(S) & \to & sk_m\underset{\sim}{A}(n^+) \end{array}$$

where S runs through the pointed subsets of n^+ with exactly m non-basepoint elements. Note that for $n < m$ the sums on the left are trivial, and for $n \geq m$ the maps $sk_{m-1}\underset{\sim}{A}(S) \to sk_m\underset{\sim}{A}(S)$ are equivalent to the canonical maps $sk_{m-1}\underset{\sim}{A}(m^+) \to \underset{\sim}{A}(m^+)$. The fact that $sk_m\underset{\sim}{A}(n^+) \to sk_m\underset{\sim}{B}(n^+)$ is a cofibration now follows from the following lemma applied to the natural map from the above push-out square to the analogous push-out square for $\underset{\sim}{B}$.

3.8. Reedy's patching lemma ([Reedy]).

Let

$$\begin{array}{ccccc} A_2 & \longleftarrow & A_1 & \longrightarrow & A_3 \\ \downarrow{\scriptstyle f_2} & & \downarrow{\scriptstyle f_1} & & \downarrow{\scriptstyle f_3} \\ B_2 & \longleftarrow & B_1 & \longrightarrow & B_3 \end{array}$$

be a diagram in a closed model category, e.g. (s.sets$_*$). If f_3 and $A_2 \coprod_{A_1} B_1 \to B_2$ are cofibrations (resp. trivial cofibrations), then $A_2 \coprod_{A_1} A_3 \to B_2 \coprod_{B_1} B_3$ is a cofibration (resp. trivial cofibration).

This follows since the maps

$$A_2 \coprod_{A_1} A_3 \to A_2 \coprod_{A_1} B_3 \cong (A_2 \coprod_{A_1} B_1) \coprod_{B_1} B_3 \to B_2 \coprod_{B_1} B_3$$

are cofibrations (resp. trivial cofibrations). Of course, there is also a dual result.

We observe in passing that Theorem 3.5 is valid more generally for Γ-objects over certain other pointed model categories $\underline{C}$ besides (s.sets$_*$). To obtain such a generalization, one must be able to impose a suitable model category structure on the category $\dot{\Sigma}_n\underline{C}$ of left Σ_n-objects over $\underline{C}$ for each $n \geq 1$. In general, this may not be feasible; however, in favorable cases (e.g., when $\underline{C}$ is Quillen's model category of pointed topological spaces [Quillen 1, II.3]), $\Sigma_n\underline{C}$ has a closed model category structure such that a map f in $\Sigma_n\underline{C}$ is a Σ_n-fibration if and only if f is a fibration in $\underline{C}$, and f is a Σ_n-weak equivalence if and only if f is a weak equivalence in $\underline{C}$. (The Σ_n-cofibrations are then determined by closure, and are cofibrations in $\underline{C}$). In these favorable cases, one obtains a closed model category $\Gamma^O\underline{C}^{strict}$ as in Theorem 3.5.

Finally, we remark that $\Gamma^O(s.sets_*)$ admits a second reasonable "strict" model category structure. This is obtained by [Bousfield-Kan, p. 314] and has weak equivalences (resp. fibrations) given by the termwise weak equivalences (resp. fibrations). However, our version seems to be more useful in applications and allows the symmetric groups to play a more explicit role.

§4. The construction of homology theories from Γ-spaces

In this section we give an exposition, and slight generalization, of some results of [Anderson] and [Segal 1]. In particular, we show that a Γ-space $\underset{\sim}{A}$ induces a generalized homology theory $h_*(\ ;\underset{\sim}{A})$ which can be directly computed when $\underset{\sim}{A}$ is "(very) special" by using $\underset{\sim}{A}$ as a chain functor. The constructions and proofs in this section will be used in §5 to compare Γ-spaces with spectra and to develop the "stable" model category structure for Γ-spaces.

We begin by showing that a Γ-space $\underset{\sim}{A}\colon \Gamma^O \to (s.sets_*)$ prolongs successively to functors $\underset{\sim}{A}\colon (sets_*) \to (s.sets_*)$, $\underset{\sim}{A}\colon (s.sets_*) \to (s.sets_*)$, and $\underset{\sim}{A}\colon (spectra) \to (spectra)$. For $W \in (sets_*)$ define $\underset{\sim}{A}(W) \in (s.sets_*)$ by

$$\underset{\sim}{A}(W) = \operatorname*{colim}_{\substack{V \subset W \\ V \in \Gamma^O}} \underset{\sim}{A}(V).$$

For $K \in (s.sets_*)$ define $\underset{\sim}{A}K \in (s.sets_*)$ by $(\underset{\sim}{A}K)_n = (\underset{\sim}{A}K_n)_n$ for $n \geq 0$ with the obvious face and degeneracy operators. Thus $\underset{\sim}{A}K$ is the diagonal of the bisimplicial set $(\underset{\sim}{A}K_*)_*$, cf. Appendix B. In order to prolong $\underset{\sim}{A}$ to spectra, note that for $K, L \in (s.sets_*)$ there is a natural simplicial map $L \wedge \underset{\sim}{A}K \to \underset{\sim}{A}(L \wedge K)$ sending $x \wedge y \in L_n \wedge (\underset{\sim}{A}K_n)_n$ to the image of y under the map $\underset{\sim}{A}(x \wedge \_)_n\colon \underset{\sim}{A}(K_n)_n \to \underset{\sim}{A}(L_n \wedge K_n)$. Now for $\underset{\sim}{X} \in (spectra)$

define $\underset{\sim}{A}X \in$ (spectra) by $(\underset{\sim}{A}X)^n = \underset{\sim}{A}(X^n)$ with the obvious structural maps

$$S^1 \wedge \underset{\sim}{A}(X^n) \to \underset{\sim}{A}(S^1 \wedge X^n) \to \underset{\sim}{A}(X^{n+1}).$$

Finally, for $K, L \in (\text{s.sets}_*)$ and $X \in$ (spectra), there are pairings

$$(\underset{\sim}{A}K) \wedge L \to \underset{\sim}{A}(K \wedge L) \in (\text{s.sets}_*)$$

$$(\underset{\sim}{A}X) \wedge L \to \underset{\sim}{A}(X \wedge L) \in (\text{spectra})$$

whose definitions are now obvious. In particular, $\underset{\sim}{A}$ preserves the simplicial homotopy relation for maps in (s.sets_*) and (spectra).

A Γ-space $\underset{\sim}{A}$ determines a spectrum $\underset{\sim}{A}\underset{\sim}{S}$ where $\underset{\sim}{S}$ is the sphere spectrum, and we let $\tilde{h}_*(\ ;\ \underset{\sim}{A})$ be the associated homology theory, i.e. $\tilde{h}_*(K;\ \underset{\sim}{A}) = \pi_*(\underset{\sim}{A}\underset{\sim}{S}) \wedge K$ for $K \in (\text{s.sets}_*)$. An alternative construction of $\tilde{h}_*(K;\ \underset{\sim}{A})$ is given by

Lemma 4.1. If $\underset{\sim}{A}$ is a Γ-space and $K \in (\text{s.sets}_*)$, then the map $(\underset{\sim}{A}\underset{\sim}{S}) \wedge K \to \underset{\sim}{A}(\underset{\sim}{S} \wedge K)$ is a stable weak equivalence, cf. 2.3, and thus

$$\tilde{h}_*(K;\ \underset{\sim}{A}) \cong \operatorname*{colim}_n \pi_{*+n}\, \underset{\sim}{A}(S^n \wedge K).$$

The proof is in 4.8. To give an even more direct construction of $\tilde{h}_*(K;\ \underset{\sim}{A})$, we must put conditions on $\underset{\sim}{A}$. A Γ-space $\underset{\sim}{A}$ is special if the obvious map $\underset{\sim}{A}(V \vee W) \to \underset{\sim}{A}V \times \underset{\sim}{A}W$ is a weak equivalence for $V, W \in \Gamma^0$. This is equivalent to requiring that for $n \geq 1$ the map

$$\underset{\sim}{A}(p_1) \times \cdots \times \underset{\sim}{A}(p_n) \colon \underset{\sim}{A}(n^+) \to \underset{\sim}{A}(1^+) \times \cdots \times \underset{\sim}{A}(1^+)$$

is a weak equivalence where $p_i\colon n^+ \to 1^+$ is defined by $p_i(i) = 1$ and $p_i(j) = 0$ for $j \neq i$. For $\underset{\sim}{A}$ special, $\pi_0\underset{\sim}{A}(1^+)$ is an abelian monoid with multiplication

$$\pi_0\underset{\sim}{A}(1^+) \times \pi_0\underset{\sim}{A}(1^+) \xleftarrow[\cong]{(p_1)_* \times (p_2)_*} \pi_0\underset{\sim}{A}(2^+) \xrightarrow{\mu_*} \pi_0\underset{\sim}{A}(1^+)$$

where μ: $2^+ \to 1^+$ is defined by $\mu(0) = 0$, $\mu(1) = 1$, and $\mu(2) = 1$. A Γ-space $\underset{\sim}{A}$ is <u>very special</u> if $\underset{\sim}{A}$ is special and $\pi_0\underset{\sim}{A}(1^+)$ is an abelian group.

The following theorem shows that a very special Γ-space can be used as a chain functor.

<u>Theorem 4.2</u>. (cf. [Anderson, p. 3], [Segal, 1, 1.4]). If $\underset{\sim}{A}$ is a very special Γ-space and $K \epsilon (\text{s.sets}_*)$, then $\underset{\sim}{A}(\underset{\sim}{S} \wedge K)$ is an Ω-spectrum and $\tilde{h}_*(K; \underset{\sim}{A}) \cong \pi_*\underset{\sim}{A}K$.

This is an easy consequence of 4.1 and

<u>Lemma 4.3</u>. If $\underset{\sim}{A}$ is a very special Γ-space and $L \subset K \epsilon (\text{s.sets}_*)$, then

$$\underset{\sim}{A}L \to \underset{\sim}{A}K \to \underset{\sim}{A}(K/L)$$

is a homotopy fibration, i.e. $\underset{\sim}{A}K$ maps by a weak equivalence to the homotopy theoretic fibre of $\underset{\sim}{A}K \to \underset{\sim}{A}(K/L)$.

<u>Proof</u>. It suffices to show that the bisimplicial square

$$\begin{array}{ccc} (\underset{\sim}{A}L_*)_* & \to & (\underset{\sim}{A}K_*)_* \\ \downarrow & & \downarrow \\ * & \longrightarrow & \underset{\sim}{A}(K_*/L_*) \end{array}$$

satisfies the hypotheses of Theorem B.4. The termwise homotopy fibre square condition follows since $\underset{\sim}{A}$ is special. The remaining conditions follow by B.3.1, because the maps

$$\pi_0^v(\underset{\sim}{A}K_*)_* \to \pi_0^v\underset{\sim}{A}(K_*/L_*)_*$$

$$\pi_t^v((\underset{\sim}{A}K_*)_*)_{\text{free}} \to \pi_0^v(\underset{\sim}{A}K_*)_* \quad \text{for } t \geq 1$$

$$\pi_t^v(\underset{\sim}{A}(K_*/L_*)_*)_{\text{free}} \to \pi_0^v\underset{\sim}{A}(K_*/L_*)_* \quad \text{for } t \geq 1$$

are fibrations since they are surjective homomorphisms of simplicial groups.

We now wish to generalize Theorem 4.2 to the case of a Γ-space $\underset{\sim}{A}$ which is merely special. For such $\underset{\sim}{A}$, the map $\underset{\sim}{A}(K \vee L) \to \underset{\sim}{A}K \times \underset{\sim}{A}L$ is a weak equivalence for $K,L \in (\text{s.sets}_*)$ by B.2. Thus $\pi_0\underset{\sim}{A}K$ is an abelian monoid with multiplication given by

$$\pi_0\underset{\sim}{A}K \times \pi_0\underset{\sim}{A}K \xleftarrow{\ \approx\ } \pi_0\underset{\sim}{A}(K \vee K) \xrightarrow{\ \mu_*\ } \pi_0\underset{\sim}{A}K$$

where $\mu\colon K \vee K \to K$ is the folding map.

Theorem 4.4 (cf. [Segal, 1.4]). Let $\underset{\sim}{A}$ be a special Γ-space and $K \in (\text{s.sets}_*)$. Then $\underset{\sim}{A}(\underset{\sim}{S} \wedge K)$ is an Ω-spectrum above its 0th term and thus $\tilde{h}_*(K; \underset{\sim}{A}) \approx \pi_{*+1}\underset{\sim}{A}(S^1 \wedge K)$. If $\pi_0\underset{\sim}{A}K$ is an abelian group, then $\underset{\sim}{A}(\underset{\sim}{S} \wedge K)$ is an Ω-spectrum and thus $\tilde{h}_*(K; \underset{\sim}{A}) \approx \pi_*\underset{\sim}{A}K$.

Proof. Let $\underset{\sim}{B}$ be the Γ-space with $\underset{\sim}{B}(n^+) = \underset{\sim}{A}(n^+ \wedge S^1 \wedge K)$ for $n \geq 0$, and note that $\underset{\sim}{B}$ is very special. Hence $\underset{\sim}{B}\underset{\sim}{S}$ is an Ω-spectrum by 4.2, and the first statement follows since $\underset{\sim}{B}\underset{\sim}{S}$ gives the portion of $\underset{\sim}{A}(\underset{\sim}{S} \wedge K)$ above its 0th term. The second statement follows similarly using the Γ-space $\underset{\sim}{C}$ with $\underset{\sim}{C}(n^+) = \underset{\sim}{A}(n^+ \wedge K)$.

We now turn to the proof of Lemma 4.1 which asserts that the map $(\underset{\sim}{A}\underset{\sim}{S}) \wedge K \to \underset{\sim}{A}(\underset{\sim}{S} \wedge K)$ is a stable weak equivalence. Although our proof is somewhat indirect, it allows us to introduce some notions needed in §5. It is based on the following general criterion.

Lemma 4.5. In a closed simplicial model category $\mathcal{C}$, e.g. $(\text{spectra})^{\text{stable}}$, a map $f: A \to B$ between cofibrant objects is a weak equivalence $\iff$ $f^*: \text{HOM}(B,X) \to \text{HOM}(A,X)$ is a weak equivalence in (s.sets) for all fibrant $X \in \mathcal{C}$.

Proof. f is a weak equivalence $\iff$ $f^*: [B,X] \approx [A,X]$ for all fibrant $X \in \mathcal{C}$ $\iff$ $f^*: [B,X^K] \approx [A,X^K]$ for all $K \in$ (s.sets) and fibrant $X \in \mathcal{C}$ $\iff$ $f^*: [K,\text{HOM}(B,X)] \approx [K,\text{HOM}(A,X)]$ for all $K \in (\text{s.sets}_*)$ and fibrant $X \in \mathcal{C}$ $\iff$ $f^*: \text{HOM}(B,X) \to \text{HOM}(A,X)$ is a weak equivalence for all fibrant $X \in \mathcal{C}$.

To effectively apply 4.5 in our case we need an adjointness lemma. For $\underset{\sim}{X},\underset{\sim}{Y} \in$ (spectra) define a Γ-space $\Phi(\underset{\sim}{X},\underset{\sim}{Y})$ by

$$\Phi(\underset{\sim}{X},\underset{\sim}{Y})(V) = \text{HOM}_{(\text{spectra})}(\underset{\sim}{X}^V,\underset{\sim}{Y})$$

for $V \in \Gamma^0$ where $\underset{\sim}{X}^V = \underset{\sim}{X} \times \cdots \times \underset{\sim}{X} \in$ (spectra) is the product of copies of $\underset{\sim}{X}$ indexed by the non-basepoint elements in V.

Lemma 4.6. For $\underset{\sim}{X},\underset{\sim}{Y} \in$ (spectra) and $\underset{\sim}{A} \in \Gamma^0(\text{s.sets}_*)$, there is a natural simplicial isomorphism

$$\text{HOM}_{(\text{spectra})}(\underset{\sim}{A}\underset{\sim}{X},\underset{\sim}{Y}) \approx \text{HOM}_{\Gamma^0(\text{s.sets}_*)}(\underset{\sim}{A},\Phi(\underset{\sim}{X},\underset{\sim}{Y}))$$

Proof. For a functor $T: \Gamma^0 \to (\text{sets}_*)$ with $T(0^+) = *$ and $W \in (\text{sets}_*)$, there is a natural isomorphism

$$\Big(\coprod_{n \geq 0} W^{n^+} \wedge Tn^+\Big)/\sim \xrightarrow{\approx} \overline{T}W$$

where

$$\overline{T}(W) = \operatorname*{colim}_{\substack{V\subset W\\ V\in\Gamma^0}} T(V)$$

and where $\sim$ is the equivalence relation generated by setting $\varphi_*(x) \sim \varphi^*(x)$ for each φ: $m^+ \to n^+$ in Γ^0 and each $x \in W^{n^+} \wedge Tm^+$ using

$$W^{m^+} \wedge Tm^+ \xleftarrow{\varphi^*} W^{n^+} \wedge Tm^+ \xrightarrow{\varphi_*} W^{n^+} \wedge Tn^+.$$

Thus there is a natural isomorphism

$$\underset{\sim\sim}{AX} \cong (\coprod_{n\geq 0} \underset{\sim}{X}^{n^+} \wedge \underset{\sim}{A}(n^+))/\sim \quad \epsilon(\text{spectra})$$

and the lemma follows easily.

To prove 4.1 using 4.5, we need a final technical lemma which will also be used in §5.

<u>Lemma</u> 4.7. Let f: $\underset{\sim}{B} \to \underset{\sim}{C}$ be a map of Γ-spaces, and let $\underset{\sim}{X}$ be a spectrum. Then:

(i) If f is a strict weak equivalence, then so is f_*: $\underset{\sim\sim}{BX} \to \underset{\sim\sim}{CX}$.

(ii) If f: $B(n^+) \to \underset{\sim}{C}(n^+)$ is an injection for each $n \geq 0$, then f_*: $\underset{\sim\sim}{BX} \to \underset{\sim\sim}{CX}$ is a strict cofibration.

(iii) If $\underset{\sim}{X}$ is strictly cofibrant and g: $\underset{\sim}{Y} \to \underset{\sim}{Z}$ is a strict fibration of spectra, then g_*: $\Phi(\underset{\sim}{X},\underset{\sim}{Y}) \to \Phi(\underset{\sim}{X},\underset{\sim}{Z})$ is a strict fibration.

<u>Proof</u>. Part (i) follows from B.2, and (ii) is reasonably straightforward. For (iii), it suffices to show that g_* has the right lifting property for each strict trivial cofibration f: $\underset{\sim}{B} \to \underset{\sim}{C}$ of Γ-spaces. This follows from 4.6 using (i) and (ii).

<u>4.8. Proof of 4.1</u>. By 4.7(i) we can assume $\underset{\sim}{A}$ is a strictly

cofibrant Γ-space. To show $(\underset{\sim}{A}S) \wedge K \to \underset{\sim}{A}(\underset{\sim}{S} \wedge K)$ is a stable weak equivalence, it suffices by 4.5, 4.6, and 4.7(ii) to show that the map

$$\mathrm{HOM}(\underset{\sim}{A},\Phi(\underset{\sim}{S} \wedge K,\underset{\sim}{X})) \cong \mathrm{HOM}(\underset{\sim}{A}(\underset{\sim}{S} \wedge K),\underset{\sim}{X})$$

$$\longrightarrow \mathrm{HOM}((\underset{\sim}{A}\underset{\sim}{S}) \wedge K,\underset{\sim}{X}) \cong \mathrm{HOM}(\underset{\sim}{A},\Phi(\underset{\sim}{S},\underset{\sim}{X}^K))$$

is a weak equivalence for each stably fibrant spectrum $\underset{\sim}{X}$. Now $\Phi(\underset{\sim}{S}\wedge K,\underset{\sim}{X})$ and $\Phi(\underset{\sim}{S},\underset{\sim}{X}^K)$ are strictly fibrant by 4.7(iii), and it suffices by the dual of 4.5 to show that the map $\Phi(\underset{\sim}{S} \wedge K,\underset{\sim}{X}) \to \Phi(\underset{\sim}{S},\underset{\sim}{X}^K)$ is a strict weak equivalence. This follows by 4.5 since the maps

$$(\underset{\sim}{S} \times \cdots \times \underset{\sim}{S}) \wedge K \to (\underset{\sim}{S} \wedge K) \times \cdots \times (\underset{\sim}{S} \wedge K) \in (\text{spectra})^{\text{stable}}$$

are weak equivalence's of cofibrant objects.

We conclude this section by noting that the functor $\underset{\sim}{A}: (\text{s.sets}_*) \to (\text{s.sets}_*)$ has homotopy theoretic significance even when $\underset{\sim}{A}$ is not special.

<u>Proposition 4.9</u>. For $\underset{\sim}{A} \in \Gamma^0(\text{s.sets}_*)$, if $f\colon K \to L \in (\text{s.sets}_*)$ is a weak equivalence then so is $\underset{\sim}{A}f\colon \underset{\sim}{A}K \to \underset{\sim}{A}L$. Thus $\underset{\sim}{A}$ induces a functor

$$\mathrm{Ho}\underset{\sim}{A}\colon \mathrm{Ho}(\text{s.sets}_*) \to \mathrm{Ho}(\text{s.sets}_*).$$

The proof is very similar to that of 4.1.

<u>Corollary 4.10</u>. For $\underset{\sim}{A} \in \Gamma^0(\text{s.sets}_*)$, if $K \in (\text{s.sets}_*)$ is n-connected for some $n \geq 0$ then so is $\underset{\sim}{A}K$.

<u>Proof</u>. This is clear when $K_i = *$ for $i \leq n$, and the general case now follows by 4.9.

§5. The stable homotopy theory of Γ-spaces

Following Graeme Segal, we will show that the strict homotopy category of very special Γ-spaces is equivalent to the stable homotopy category of connective spectra. Then we will develop a "stable" model category structure for Γ-spaces such that the associated homotopy category is equivalent to that of connective spectra.

By 4.6 there are functors

$$(-)\underset{\sim}{S}: \Gamma^{O}(\text{s.sets}_*) \rightleftarrows (\text{spectra}): \Phi(\underset{\sim}{S},-)$$

with $(-)\underset{\sim}{S}$ left adjoint to $\Phi(\underset{\sim}{S},-)$; indeed, there is a natural isomorphism

$$\text{HOM}(\underset{\sim}{A}\underset{\sim}{S},\underset{\sim}{X}) \cong \text{HOM}(\underset{\sim}{A},\Phi(\underset{\sim}{S},\underset{\sim}{X}))$$

for a Γ-space $\underset{\sim}{A}$ and spectrum $\underset{\sim}{X}$. By 4.7 and the dual of 4.5, $(-)\underset{\sim}{S}$ preserves weak equivalences and cofibrations in $\Gamma^{O}(\text{s.sets}_*)^{\text{strict}}$, while $\Phi(\underset{\sim}{S},-)$ preserves weak equivalences between fibrant objects and fibrations in $(\text{spectra})^{\text{strict}}$. Thus by [Quillen 1, I.4], there are induced adjoint functors

$$L^{\text{strict}}: \text{Ho}\Gamma^{O}(\text{s.sets}_*)^{\text{strict}} \rightleftarrows \text{Ho}(\text{spectra})^{\text{strict}}: R^{\text{strict}}$$

where $L^{\text{strict}}(\underset{\sim}{A}) = \underset{\sim}{A}\underset{\sim}{S}$ for $\underset{\sim}{A} \in \text{Ho}\Gamma^{O}(\text{s.sets}_*)^{\text{strict}}$ and $R^{\text{strict}}(\underset{\sim}{X}) = \Phi(\underset{\sim}{S},\underset{\sim}{X}')$ for $\underset{\sim}{X} \in \text{Ho}(\text{spectra})^{\text{strict}}$ where $\underset{\sim}{X} \to \underset{\sim}{X}'$ is a strict weak equivalence with $\underset{\sim}{X}'$ strictly fibrant. Now let

$$\text{Ho}(\text{v.s. }\Gamma\text{-spaces})^{\text{strict}} \subset \text{Ho}\Gamma^{O}(\text{s.sets}_*)^{\text{strict}}$$

$$\text{Ho}(\text{c. }\Omega\text{-spectra})^{\text{strict}} \subset \text{Ho}(\text{spectra})^{\text{strict}}$$

denote the full subcategories given by the very special Γ-spaces and the connective Ω-spectra respectively, where a spectrum $\underset{\sim}{X}$ is called connective if $\pi_i \underset{\sim}{X} = 0$ for $i < 0$.

Theorem 5.1. (cf. [Anderson, pp. 4,5], [Segal 1, 1.4]). The adjoint functors L^{strict} and R^{strict} restrict to adjoint equivalences

$$L^{strict}\colon \mathrm{Ho}(\text{v.s. } \Gamma\text{-spaces})^{strict} \rightleftarrows \mathrm{Ho}(\text{c. } \Omega\text{-spectra})^{strict} : R^{strict}.$$

Moreover, $\mathrm{Ho}(\text{c. } \Omega\text{-spectra})^{strict}$ is equivalent to the usual homotopy category of connective spectra.

Proof. The first statement is proved by combining the four facts below, and the last follows from 2.4. If $\underset{\sim}{A}$ is a very special Γ-space, then $\underset{\sim}{A}\underset{\sim}{S}$ is a connective Ω-spectrum by 4.2 and 4.10. If $\underset{\sim}{X}$ is a strictly fibrant Ω-spectrum, then $\Phi(\underset{\sim}{S},\underset{\sim}{X})$ is a very special Γ-space by 4.5 since the maps $\underset{\sim}{S} \vee \cdots \vee \underset{\sim}{S} \to \underset{\sim}{S} \times \cdots \times \underset{\sim}{S}$ are weak equivalences in $(\text{spectra})^{stable}$. If $\underset{\sim}{A}$ is a very special Γ-space and $\underset{\sim}{A}\underset{\sim}{S} \to \underset{\sim}{X}$ is a strict weak equivalence with $\underset{\sim}{X}$ strictly fibrant, then the natural map $\underset{\sim}{A} \to \Phi(\underset{\sim}{S},\underset{\sim}{X})$ is a strict weak equivalence, because both $\underset{\sim}{A}$ and $\Phi(\underset{\sim}{S},\underset{\sim}{X})$ are very special and the map

$$(\underset{\sim}{A}\underset{\sim}{S})^0 = \underset{\sim}{A}(1^+) \to \Phi(\underset{\sim}{S},\underset{\sim}{X})(1^+) = \underset{\sim}{X}^0 \in (\text{s.sets}_*)$$

is a weak equivalence. Similarly, if $\underset{\sim}{A} \to \Phi(\underset{\sim}{S},\underset{\sim}{X})$ is a strict weak equivalence for some strictly fibrant connective Ω-spectrum $\underset{\sim}{X}$, then the natural map $\underset{\sim}{A}\underset{\sim}{S} \to \underset{\sim}{X}$ is a weak equivalence.

We now wish to use our strict homotopy theory of Γ-spaces to build a corresponding stable theory, just as we previously used our strict homotopy theory of spectra to build a stable theory in 2.3.

Theorem 5.2. The category of Γ-spaces becomes a closed simplicial model category (denoted $\Gamma^{0}(\text{s.sets}_*)^{\text{stable}}$) when provided with the following additional structure: a map $f\colon \underset{\sim}{A} \to \underset{\sim}{B} \in \Gamma^{0}(\text{s.sets}_*)$ is called a stable weak equivalence if $f_*\colon \pi_*\underset{\sim}{A}\underset{\sim}{S} \cong \pi_*\underset{\sim}{B}\underset{\sim}{S}$; $f\colon \underset{\sim}{A} \to \underset{\sim}{B}$ is called a stable cofibration if it is a strict cofibration; and $f\colon \underset{\sim}{A} \to \underset{\sim}{B}$ is called a stable fibration if it has the right lifting property for the stable trivial cofibrations.

Following the proof we will say more about stable fibrations in 5.7. Our proof will rely on the formal machinery developed in Appendix A. Let Q: (spectra) $\to$ (spectra) and $\eta\colon 1 \to Q$ be such that, for each spectrum $\underset{\sim}{X}$, $\eta_X\colon \underset{\sim}{X} \to Q\underset{\sim}{X}$ is a stable weak equivalence and $Q\underset{\sim}{X}$ is a stably fibrant spectrum, cf. §2. Now define $T\colon \Gamma^{0}(\text{s.sets}_*) \to \Gamma^{0}(\text{s.sets}_*)$ by $T\underset{\sim}{A} = \Phi(\underset{\sim}{S}, Q\underset{\sim}{A}\underset{\sim}{S})$ and let $\eta\colon 1 \to T$ be the canonical transformation. Note that for each Γ-space $\underset{\sim}{A}$, $\eta_A\colon \underset{\sim}{A} \to T\underset{\sim}{A}$ is a stable weak equivalence and $T\underset{\sim}{A}$ is strictly fibrant and very special. Using the terminology of Appendix A, the T-equivalences, T-cofibrations, and T-fibrations in $\Gamma^{0}(\text{s.sets}_*)^{\text{strict}}$ are the same as the stable weak equivalences, stable cofibrations, stable fibrations, respectively. Moreover, for $\eta\colon 1 \to T$, the conditions (A.4) and (A.5) clearly hold although (A.6) doesn't, cf. 5.7. Thus by A.8(i) all the closed model category axioms hold in $\Gamma^{0}(\text{s.sets}_*)^{\text{stable}}$ except possibly for the "trivial cofibration, fibration" part of CM5. To verify an important case of that part, we use the following substitute for (A.6).

Lemma 5.3. For a pull-back square

$$\begin{array}{ccc} \underset{\sim}{A} & \xrightarrow{h} & \underset{\sim}{X} \\ \downarrow & & \downarrow j \\ \underset{\sim}{B} & \xrightarrow{k} & \underset{\sim}{Y} \end{array}$$

in $\Gamma^{0}(\text{s.sets}_*)$, suppose j is a strict fibration with $\underset{\sim}{X}$ and $\underset{\sim}{Y}$ very special and with

$$\pi_0\underset{\sim}{X}(1^+) \approx \pi_0\underset{\sim}{X}(\underset{\sim}{S}) \xrightarrow{j_*} \pi_0\underset{\sim}{Y}(\underset{\sim}{S}) \approx \pi_0\underset{\sim}{Y}(1^+)$$

onto. If k is a stable weak equivalence, then so is h.

Proof. For $K \epsilon (s.sets_*)$ consider the induced square

$$\begin{array}{ccc} \underset{\sim}{A}(K_*)_* & \to & \underset{\sim}{X}(K_*)_* \\ \downarrow & & \downarrow \\ \underset{\sim}{B}(K_*)_* & \to & \underset{\sim}{Y}(K_*)_* \end{array}$$

of bisimplicial sets. As in the proof of 4.3, $\underset{\sim}{X}(K_*)_*$ and $\underset{\sim}{Y}(K_*)_*$ satisfy the π_*-Kan condition and $j_*\colon \pi_0^v\underset{\sim}{X}(K_*)_* \to \pi_0^v\underset{\sim}{Y}(K_*)_*$ is a fibration. Thus by B.4

$$\begin{array}{ccc} \underset{\sim}{A}K & \longrightarrow & \underset{\sim}{X}K \\ \downarrow & & \downarrow \\ \underset{\sim}{B}K & \longrightarrow & \underset{\sim}{Y}K \end{array}$$

is a homotopy fibre square in $(s.sets_*)$, and the lemma follows easily.

Now using 5.3 in place of (A.6), the argument in A.10 shows that if $f\colon \underset{\sim}{X} \to \underset{\sim}{Y}$ is a map in $\Gamma^0(s.sets_*)$ with $f_*\colon \pi_0\underset{\sim}{X}(\underset{\sim}{S}) \to \pi_0\underset{\sim}{Y}(\underset{\sim}{S})$ onto, then f can be factored as $f = pi$ where i is a stable trivial cofibration and p is a stable fibration. The following lemma will complete the proof of CM5, and Theorem 5.2 will then follow using the criterion SM7(b) of [Quillen 1, II.2].

Lemma 5.4. Each map $f\colon \underset{\sim}{A} \to \underset{\sim}{B}$ in $\Gamma^0(s.sets_*)$ can be factored as $\underset{\sim}{A} \xrightarrow{u} \underset{\sim}{C} \xrightarrow{v} \underset{\sim}{B}$ where $u_*\colon \pi_0\underset{\sim}{A}\underset{\sim}{S} \to \pi_0\underset{\sim}{C}\underset{\sim}{S}$ is onto and v is a stable fibration.

To prove this (in 5.6) we will first show that the functor

$$\pi_0(-)(\underset{\sim}{S}) : \Gamma^0(\text{s.sets}_*) \to (\text{ab. gps.})$$

has a right adjoint. For an abelian group M, let $\underset{\sim}{M}$ be the usual very special Γ-space such that $\underset{\sim}{M}(V)_n = M^V$ for $V \in \Gamma^0$ and $n \geq 0$, where M^V is the product of copies of M indexed by the non-basepoint elements of V. Clearly $\underset{\sim}{M}\underset{\sim}{S}$ is an Eilenberg-MacLane spectrum of type $(M,0)$, and we identify $\pi_0\underset{\sim}{M}\underset{\sim}{S}$ with M.

Lemma 5.5. For a Γ-space $\underset{\sim}{A}$ and an abelian group M, the obvious map

$$\text{Hom}_{\Gamma^0(\text{s.sets}_*)}(\underset{\sim}{A},\underset{\sim}{M}) \to \text{Hom}_{(\text{ab. gps.})}(\pi_0\underset{\sim}{A}\underset{\sim}{S},M)$$

is a bijection.

Proof. In $\Gamma^0(\text{s.sets}_*)$, let $\underset{\sim}{B} \to \underset{\sim}{A}$ be a strict weak equivalence with $\underset{\sim}{B}$ strictly cofibrant. In the square

$$\begin{array}{ccc} \text{Hom}_{\Gamma^0(\text{s.sets}_*)}(\underset{\sim}{A},\underset{\sim}{M}) & \to & \text{Hom}_{(\text{ab. gps.})}(\pi_0\underset{\sim}{A}\underset{\sim}{S},M) \\ \downarrow & & \downarrow \\ \text{Hom}_{\Gamma^0(\text{s.sets}_*)}(\underset{\sim}{B},\underset{\sim}{M}) & \to & \text{Hom}_{(\text{ab. gps.})}(\pi_0\underset{\sim}{B}\underset{\sim}{S},M) \end{array}$$

the right map is bijective since $\pi_0\underset{\sim}{A}\underset{\sim}{S} \cong \pi_0\underset{\sim}{B}\underset{\sim}{S}$, and the left map is bijective since

$$\text{Hom}_{(\text{s.sets}_*)}(\underset{\sim}{A}(n^+),\underset{\sim}{M}(n^+)) \cong \text{Hom}_{(\text{sets}_*)}(\pi_0\underset{\sim}{A}(n^+),M^{n^+})$$

$$\cong \text{Hom}_{(\text{sets}_*)}(\pi_0\underset{\sim}{B}(n^+),M^{n^+}) \cong \text{Hom}_{(\text{s.sets}_*)}(\underset{\sim}{B}(n^+),\underset{\sim}{M}(n^+)).$$

The lemma now follows since the bottom map of the square is a composite of bijections

$$\operatorname{Hom}(\underset{\sim}{B},\underset{\sim}{M}) \xrightarrow{\approx} \pi_0\operatorname{HOM}(\underset{\sim}{B},\underset{\sim}{M}) \xrightarrow{\approx 1} \pi_0\operatorname{HOM}(\underset{\sim}{B},\Phi(\underset{\sim}{S},\underset{\sim}{MS}))$$

$$\xrightarrow{\approx} \pi_0\operatorname{HOM}(\underset{\sim}{BS},\underset{\sim}{MS}) \xrightarrow{\approx 2} \operatorname{Hom}(\pi_0\underset{\sim}{BS},M)$$

where 1 holds by 5.1 and the dual of 4.5, and 2 holds since $\pi_0\operatorname{HOM}(\underset{\sim}{BS},\underset{\sim}{MS})$ is the set of homotopy classes from the connective spectrum $\underset{\sim}{BS}$ to the Eilenberg-MacLane spectrum $\underset{\sim}{MS}$ in $(\text{spectra})^{\text{stable}}$.

5.6. Proof of 5.4. It will suffice to inductively construct a descending sequence of Γ-spaces

$$\underset{\sim}{B} = \underset{\sim}{C}^0 \supset \underset{\sim}{C}^1 \supset \underset{\sim}{C}^2 \supset \cdots \supset \underset{\sim}{C}^\alpha \supset \cdots$$

indexed by the ordinal numbers and such that: $f(\underset{\sim}{A}) \subset \underset{\sim}{C}^\alpha$ for all α; the inclusion $\underset{\sim}{C}^\alpha \xrightarrow{\subset} \underset{\sim}{B}$ is a stable fibration for all α; and, for sufficiently large α, $\underset{\sim}{C}^\alpha = \underset{\sim}{C}^{\alpha+1}$ and $f_*\colon \pi_0\underset{\sim}{AS} \to \pi_0\underset{\sim}{C}^\alpha\underset{\sim}{S}$ is onto. Given $\underset{\sim}{C}^\alpha \subset \underset{\sim}{B}$ with $f(\underset{\sim}{A}) \subset \underset{\sim}{C}^\alpha$, define $\underset{\sim}{C}^{\alpha+1} \subset \underset{\sim}{B}$ by the pull-back

$$\begin{array}{ccc} \underset{\sim}{C}^{\alpha+1} & \longrightarrow & \underset{\sim}{M}^\alpha \\ \big\downarrow \subset & & \big\downarrow \subset \\ \underset{\sim}{C}^\alpha & \longrightarrow & (\pi_0\underset{\sim}{C}^\alpha\underset{\sim}{S})_{\sim} \end{array}$$

where M^α is the image of $f_*\colon \pi_0\underset{\sim}{AS} \to \pi_0\underset{\sim}{C}^\alpha\underset{\sim}{S}$ and where the bottom map corresponds via 5.5 to the identity on $\pi_0\underset{\sim}{C}^\alpha\underset{\sim}{S}$. Note that $\underset{\sim}{C}^{\alpha+1} \xrightarrow{\subset} \underset{\sim}{C}^\alpha$ is a stable fibration because $\underset{\sim}{M}^\alpha \xrightarrow{\subset} (\pi_0\underset{\sim}{C}^\alpha\underset{\sim}{S})_{\sim}$ is one by an argument using 5.5, and note that $f(\underset{\sim}{A}) \subset \underset{\sim}{C}^{\alpha+1}$. Given a limit ordinal λ and given $\underset{\sim}{C}^\alpha \subset \underset{\sim}{B}$ with $f(\underset{\sim}{A}) \subset C^\alpha$ for all $\alpha < \lambda$, define

$\underset{\sim}{C}^{\lambda} \subset \underset{\sim}{B}$ by $\underset{\sim}{C}^{\lambda} = \bigcap_{\alpha<\lambda} \underset{\sim}{C}^{\alpha}$, and note that $f(A) \subset \underset{\sim}{C}^{\alpha}$. This completes the inductive construction of $\{\underset{\sim}{C}^{\alpha}\}$, and the desired properties are easily verified.

This concludes the proof of Theorem 5.2 and we next discuss

5.7. Stable fibrations of Γ-spaces. By A.9, a sufficient condition for a Γ-space map $f: \underset{\sim}{A} \to \underset{\sim}{B}$ to be a stable fibration is that f be a strict fibration and that

$$\begin{array}{ccc} \underset{\sim}{A} & \xrightarrow{\eta} & T\underset{\sim}{A} \\ \downarrow f & & \downarrow Tf \\ \underset{\sim}{B} & \xrightarrow{\eta} & T\underset{\sim}{B} \end{array}$$

be a homotopy fibre square in $\Gamma^{O}(\text{s.sets}_*)^{\text{strict}}$. When $f_*: \pi_0\underset{\sim}{A}\underset{\sim}{S} \to \pi_0\underset{\sim}{B}\underset{\sim}{S}$ is onto, this condition is also necessary by the argument of A.10; but it is not always necessary. To give an example, we first note that an abelian monoid M determines a Γ-space $\underset{\sim}{M}$ with $\underset{\sim}{M}(V)_n = M^V$ for $V \in \Gamma^O$ and $n \geq 0$. Letting $\overline{M}$ denote the universal abelian group generated by M, we note that the Γ-space map $\underset{\sim}{M} \to \underset{\sim}{\overline{M}}$ is a stable weak equivalence, because $\pi_*\underset{\sim}{M}S^n \xrightarrow{\approx} \pi_*\underset{\sim}{\overline{M}}S^n$ for $n \geq 1$ by [Spanier, Corollary 5.7]. Now let M be the abelian monoid given by

$$M = \{n \in Z \mid n \geq 0\} \cup \{0'\}$$

with the usual addition for the non-negative integers and with $0' + 0' = 0$, $0' + 0 = 0'$, $0' + n = n$ for $n \geq 1$. Note that $\overline{M} = Z$, and let $D = \{0, 0'\} \subset M$. Using the pull-back square

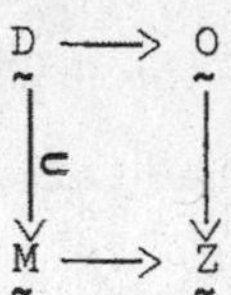

in $\Gamma_0(\text{s.sets}_*)$, one sees that $\underset{\sim}{D} \xrightarrow{\subset} \underset{\sim}{M}$ is a stable fibration although it doesn't satisfy the sufficient condition mentioned above. Since $\underset{\sim}{M} \to \underset{\sim}{Z}$ is a stable weak equivalence and $\underset{\sim}{D} \to \underset{\sim}{O}$ is not, this square also shows that (A.6) fails in our Γ-space context.

For the adjoint functors

$$(-)\underset{\sim}{S}\colon \Gamma^0(\text{s.sets}_*) \rightleftarrows (\text{spectra})\colon \Phi(\underset{\sim}{S},-)$$

it is now easy to verify that $(-)\underset{\sim}{S}$ preserves weak equivalences and cofibrations in $\Gamma^0(\text{s.sets}_*)^{\text{stable}}$, while $\Phi(\underset{\sim}{S},-)$ preserves weak equivalences between fibrant objects and fibrations in $(\text{spectra})^{\text{stable}}$. Thus by [Quillen 1, I.4] there are induced adjoint functors

$$L^{\text{stable}}\colon \text{Ho}\Gamma^0(\text{s.sets}_*)^{\text{stable}} \rightleftarrows \text{Ho}(\text{spectra})^{\text{stable}}\colon R^{\text{stable}}$$

and we let

$$\text{Ho}(\text{c.spectra})^{\text{stable}} \subset \text{Ho}(\text{spectra})^{\text{stable}}$$

denote the full subcategory given by the connective spectra. It is now easy to prove

Theorem 5.8. The adjoint functors L^{stable} and R^{stable} restrict to adjoint equivalences

$$L^{\text{stable}}\colon \text{Ho}\Gamma^0(\text{s.sets}_*)^{\text{stable}} \rightleftarrows \text{Ho}(\text{c.spectra})^{\text{stable}}\colon R^{\text{stable}}.$$

Thus the stable homotopy category of Γ-spaces is equivalent to the usual connective homotopy category of spectra. Moreover, it is easy to show that T induces an equivalence

$$\mathrm{Ho}\Gamma^0(\text{s.sets}_*)^{\text{stable}} \xrightarrow{\cong} \mathrm{Ho}(\text{v.s. }\Gamma\text{-spaces})^{\text{strict}}$$

just as Q induced an equivalence

$$\mathrm{Ho}(\text{spectra})^{\text{stable}} \xrightarrow{\cong} \mathrm{Ho}(\Omega\text{-spectra})^{\text{strict}}$$

in 2.4.

Appendix A. Proper closed model categories

In this appendix we outline some formal results on proper closed model categories (cf. 1.2) which we use in §§2,5 to pass from our "strict" to our "stable" model category structures on spectra and Γ-spaces. Some familiar examples of proper closed model categories are the (pointed) simplicial sets, (pointed) topological spaces, and simplicial groups, all equipped with the standard model structures ([Quillen 1, II.3]); however, as noted in [Quillen 2, p. 241], some closed model categories are not proper.

Our first result may be viewed as a generalization of the factorization axiom CM5 (see 1.1).

Proposition A.1. Let $\mathcal{C}$ be a proper closed model category and let $f: X \to Y$ in $\mathcal{C}$. For each factorization $[f] = vu$ in $\mathrm{Ho}\mathcal{C}$ there is a factorization $f = ji$ in $\mathcal{C}$ such that i is a cofibration, j is a fibration, and the factorization $[f] = [j][i]$ is equivalent to $[f] = vu$ in $\mathrm{Ho}\mathcal{C}$ (i.e. there exists an isomorphism w in $\mathrm{Ho}\mathcal{C}$ such that $wu = [i]$ and $[j]w = v$.)

Proof. First suppose X is cofibrant and Y is fibrant. Then choose a fibrant-cofibrant object $W \in \mathcal{C}$ and maps $X \xrightarrow{\alpha} W \xrightarrow{\beta} Y$ in $\mathcal{C}$ such that $[f] = [\beta][\alpha]$ and such that this factorization is equivalent to $[f] = vu$ in $\mathrm{Ho}\mathcal{C}$. Using CM5 and the homotopy extension theorem ([Quillen, HA, Ch. I, p. 1.7]), one then constructs the desired factorization $f = ji$. In the general case, choose weak equivalences $s: X' \to X$ and $t: Y \to Y'$ with X' cofibrant and Y' fibrant. Then apply the special case to give a factorization $tfs = \theta\omega$ where ω is a cofibration, θ is a fibration, and the factorization $[f] = ([t]^{-1}[\theta])([\omega][s]^{-1})$ is equivalent to $[f] = vu$ in $\mathrm{Ho}\mathcal{C}$. Now, using the properness of $\mathcal{C}$ and CM5, it is not hard to construct the desired factorization of f.

A.2. Homotopy fibre squares. In a proper closed model category $\mathcal{C}$, a commutative square

$$\begin{array}{ccc} A & \longrightarrow & C \\ \downarrow & & \downarrow w \\ B & \xrightarrow{v} & D \end{array}$$

is a homotopy fibre square if for some factorization $C \xrightarrow{i} W \xrightarrow{p} D$ of w with i a weak equivalence and p a fibration, the map $A \to B \times_D W$ is a weak equivalence. This easily implies that for any factorization $B \xrightarrow{j} V \xrightarrow{q} D$ of v with j a weak equivalence and q a fibration, the map $A \to V \times_D C$ is a weak equivalence. Thus in our definition we could have replaced "some" by "any" or used v in place of w. It is not hard to verify the following expected results. In a commutative diagram

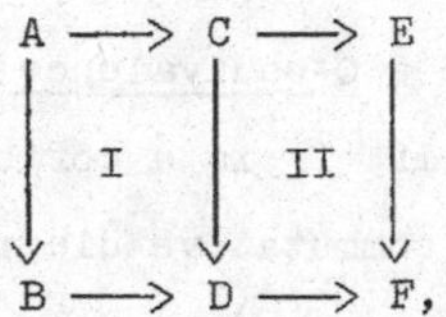

if I and II are homotopy fibre squares, so is the combined square III; and if II and III are homotopy fibre squares, so is I. If a map between homotopy fibre squares has weak equivalences at the three corners away from the upper left, then it has a weak equivalence at the upper left. A retract of a homotopy fibre square is a homotopy fibre square.

Although it does not depend on properness, we also need.

A.3. The model category $\mathcal{C}^{\text{Pairs}}$. Let $\mathcal{C}$ be a closed model category, and let $\mathcal{C}^{\text{Pairs}}$ be the category whose objects are the maps in $\mathcal{C}$

and whose maps are commutative squares in $\mathcal{C}$. A map

$$\begin{array}{ccc} A_0 & \xrightarrow{f_0} & B_0 \\ \downarrow i & & \downarrow j \\ A_1 & \xrightarrow{f_1} & B_1 \end{array}$$

from i to j in $\mathcal{C}^{\text{Pairs}}$ will be called a <u>weak equivalence</u> (resp. <u>fibration</u>) if f_0 and f_1 are weak equivalences (resp. fibrations), and a <u>cofibration</u> if $f_0\colon A_0 \to B_0$ and $(f_1, j)\colon A_1 \coprod_{A_0} B_0 \to B_1$ are cofibrations. (This implies that $f_1\colon A_1 \to B_1$ is also a cofibration.) One easily shows that $\mathcal{C}^{\text{Pairs}}$ is a closed model category which is proper if $\mathcal{C}$ is proper.

We now develop the machinery which allows us to pass from our "strict" to our "stable" model category structures on spectra and Γ-spaces. Let $\mathcal{C}$ be a proper closed model category, let $Q\colon \mathcal{C} \to \mathcal{C}$ be a functor, and let $\eta\colon 1 \to Q$ be a natural transformation. A map $f\colon X \to Y$ in $\mathcal{C}$ will be called a Q-<u>equivalence</u> if $Qf\colon QX \to QY$ is a weak equivalence, a Q-<u>cofibration</u> if f is a cofibration, and a Q-<u>fibration</u> if the filler exists in each commutative diagram

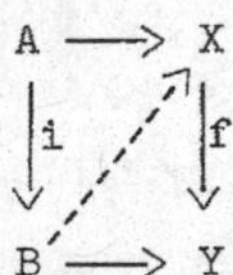

where i is a Q-cofibration and Q-equivalence. We wish to show that $\mathcal{C}^Q$ is a proper closed model category, where $\mathcal{C}^Q$ denotes $\mathcal{C}$ equipped with its Q-equivalences, Q-cofibrations, and Q-fibrations. For this we need:

(A.4) If $f\colon X \to Y$ is a weak equivalence in $\mathcal{C}$, then so is

Qf: QX → QY.

(A.5) For each $X \in \mathcal{C}$ the maps $\eta_{QX}, Q\eta_X$: QX → QQX are weak equivalences in $\mathcal{C}$.

(A.6) For a pull-back square

$$\begin{array}{ccc} A & \xrightarrow{h} & X \\ \downarrow & & \downarrow j \\ B & \xrightarrow{k} & Y \end{array}$$

in $\mathcal{C}$, if j is a Q-fibration and k is a Q-equivalence, then h is a Q-equivalence; and the dual condition holds for a push-out square.

Theorem A.7. Suppose (A.4), (A.5), and (A.6). Then $\mathcal{C}^Q$ is a proper closed model category. Moreover, a map f: X → Y in $\mathcal{C}$ is a Q-fibration <==> f is a fibration and

$$\begin{array}{ccc} X & \xrightarrow{\eta} & QX \\ \downarrow & & \downarrow Qf \\ Y & \xrightarrow{\eta} & QY \end{array}$$

is a homotopy fibre square in $\mathcal{C}$.

The proof is completed in A.10 after the following lemmas. In our Γ-space context, (A.6) does not quite hold and we use these lemmas directly.

Lemma A.8. Suppose (A.4). Then:

(i) $\mathcal{C}^Q$ satisfies CM1-CM4 and the "cofibration, trivial fibration" part of CM5.

(ii) A map f: $X \to Y$ in $\mathcal{C}$ is a trivial fibration in $\mathcal{C}^Q \Longleftrightarrow$ f is a trivial fibration in $\mathcal{C}$.

(iii) If f: $X \to Y$ is a fibration in $\mathcal{C}$ and both $\eta: X \to QX$ and $\eta: Y \to QY$ are weak equivalences, then f is a Q-fibration.

Proof. Statement (i) follows using (ii). In (ii), "$\Longleftarrow$" is clear and "$\Longrightarrow$" follows by first factoring f as f = ji with i a cofibration and j a trivial fibration, and then noting that f is a retract of j by a lifting argument using the fact that i is a Q-equivalence. For (iii), it suffices to show that the filler exists in each commutative square

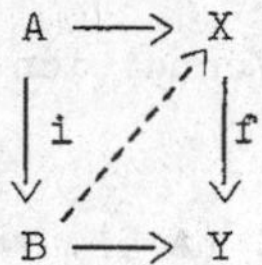

with i a trivial cofibration in $\mathcal{C}^Q$. Viewing this as a map from i to f in $\mathcal{C}^{\text{Pairs}}$, we apply A.1 and A.3 to factor it as

$$\begin{array}{ccccc} A & \longrightarrow & V & \longrightarrow & X \\ \downarrow{\scriptstyle i} & & \downarrow{\scriptstyle h} & & \downarrow{\scriptstyle f} \\ B & \longrightarrow & W & \longrightarrow & Y \end{array}$$

where h is isomorphic to Qi in Ho($\mathcal{C}^{\text{Pairs}}$). Then h is a weak equivalence, so we apply <u>CM5</u> to h and use <u>CM4</u> to obtain the desired filler.

Now, A.8(iii) easily implies

<u>Lemma A.9.</u> Suppose (A.4) and (A.5). If f: $X \to Y$ is a fibration in $\mathcal{C}$ and

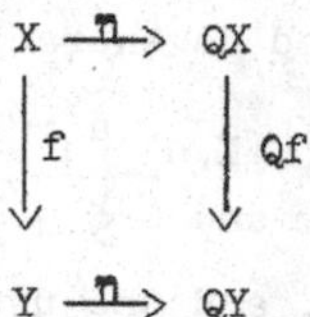

is a homotopy fibre square, then f is a Q-fibration.

A.10. Proof of A.7. We wish to factor a map $f: X \to Y$ in $\mathcal{C}$ as $f = ji$ where j is a Q-fibration and i is a Q-cofibration and Q-equivalence. First factor Qf as $Qf = vu$ where u is a weak equivalence and v is a fibration. Then let $f = v'u'$ be the factorization of f induced by $\eta: X \to QX$ and $\eta: Y \to QY$; and factor u' as $u' = ki$ where i is a cofibration and k is a trivial fibration. Then the factorization $f = (v'k)i$ has the desired properties, since $v'k$ satisfies the hypotheses of A.8(iii) and i is a Q-equivalence by (A.4)-(A.6). The "<==" part of A.7 is A.9, and the "==>" part follows by using the above procedure to factor f as $f = (v'k)i$, and then noting that f is a retract of $v'k$.

Appendix B. Bisimplicial sets

For convenience we have gathered here various definitions and results on bisimplicial sets which are used elsewhere in this paper. Much of this material is well-known, and the main innovation is the fibre square theorem (B.4) for diagonals of bisimplicial sets. As a consequence of that theorem we deduce a generalization of Quillen's spectral sequence ([Quillen 3]).

Let Δ be the category whose objects are the finite ordered sets $[m] = \{0,1,\ldots,m\}$ for $m \geq 0$, and whose morphisms are the non-decreasing maps. A bisimplicial set is a functor $\Delta^{o} \times \Delta^{o} \to (\text{sets})$, and these form a category (bis. sets). One can think of a bisimplicial set X as a collection of sets $X_{m,n}$ for $m,n \geq 0$ together with horizontal and vertical face and degeneracy operators $d_i^h\colon X_{m,n} \to X_{m-1,n}$, $s_i^h\colon X_{m,n} \to X_{m+1,n}$, $d_j^v\colon X_{m,n} \to X_{m,n-1}$, $s_j^v\colon X_{m,n} \to X_{m,n+1}$ for $0 \leq i \leq m$ and $0 \leq j \leq n$, where the horizontal and vertical operators commute, and the usual simplicial identities hold horizontally and vertically.

In practice, many constructions in algebraic topology can be achieved by first forming an appropriate bisimplicial set and then applying the diagonal functor

$$\text{diag}\colon (\text{bis. sets}) \to (\text{s.sets})$$

where diag X is given by the sets $X_{m,m}$ for $m \geq 0$ with operators $d_i = d_i^h d_i^v$ and $s_i = s_i^h s_i^v$. For example, if K and L are simplicial sets, there is an obvious bisimplicial set $K \underset{\sim}{\times} L$ with $(K \underset{\sim}{\times} L)_{m,n} = K_m \times L_n$, and $\text{diag}(K \underset{\sim}{\times} L) = K \times L$. Many other examples are given, at least implicitly, in [Artin-Mazur], [Bousfield-Kan, XII], [Dress], [May 2], [Segal 2], and elsewhere. Most of these examples lead to interesting homotopy or (co)homology spectral sequences.

The main results for bisimplicial sets involve the relation between the vertical simplicial terms and the diagonal, i.e. between the $X_{m,*}$ and diag X. (Of course, there are immediate corollaries with "vertical" replaced by "horizontal.") To understand these results one should first note that the construction of diag X is deceptively simple, and diag X may actually be viewed as the "total complex" or "realization" of X. Specifically, let Tot X be the simplicial set obtained from the disjoint union $\coprod_{m \geq 0} \Delta[m] \times X_{m,*}$ by identifying the simplex $(a, \theta^* x) \in \Delta[m] \times X_{m,*}$ with $(\theta_* a, x) \in \Delta[n] \times X_{n,*}$ for each $\theta\colon [m] \to [n]$ in Δ. Now the classical Eilenberg-Zilber-Cartier theorem ([Dold-Puppe, p. 213]) for bisimplicial abelian groups has the following well-known analogue for bisimplicial sets.

Proposition B.1. For a bisimplicial set X, there is a natural simplicial isomorphism ψ: Tot X $\approx$ diag X.

Proof. The desired map ψ: Tot X $\to$ diag X is induced by the maps $\Delta[m] \times X_{m,*} \to$ diag X sending $(\theta^*\iota^m, x) \in \Delta[m]_n \times X_{m,n}$ to $\theta^* x \in X_{n,n}$ for θ: $[n] \to [m]$ in Δ. One checks explicitly that ψ is iso whenever $X = \Delta[m] \underset{\sim}{\times} \Delta[n]$, i.e. X is freely generated by an (m,n)-simplex. The proposition then follows by a direct limit argument.

In view of B.1, the following fundamental theorem is not surprising.

Theorem B.2. Let f: X $\to$ Y be a map of bisimplicial sets such that $f_{m,*}$: $X_{m,*} \to Y_{m,*}$ is a weak equivalence for each $m \geq 0$. Then diag(f): diag X $\to$ diag Y is a weak equivalence.

This was proved in [Bousfield-Kan, p. 335], but a more direct proof using a patching argument is in [Tornehave] and [Reedy].

The diagonal functor not only preserves termwise weak equivalences of bisimplicial sets, but also clearly preserves termwise cofibre squares. To state a similar, but more complicated, result for termwise fibre squares, we will need

B.3. The π_*-Kan condition. This is a condition on a bisimplicial set X which holds automatically when each $X_{m,*}$ is connected, and in many other cases. Roughly speaking, it requires that the vertical homotopy groups of X satisfy Kan's extension condition horizontally. More precisely, for $m,t \geq 1$ and $a \in X_{m,0}$ consider the homomorphisms

$$(d_i^h)_*: \pi_t(X_{m,*}, a) \to \pi_t(X_{m-1,*}, d_i^h a) \qquad 0 \leq i \leq m$$

where the homotopy groups of a simplicial set are defined to be those of its geometric realization. We say X satisfies the π_t-Kan

<u>condition at</u> $a \in X_{m,0}$ if for every collection of elements

$$\{x_i \in \pi_t(X_{m-1,*}, d_i^h a)\}_{i=0,1,\ldots,k-1,k+1,\ldots,m}$$

which satisfy the compatibility condition $(d_i^h)_* x_j = (d_{j-1}^h)_* x_i$ for $i < j$, $i \neq k$, $j \neq k$, there exists an element $x \in \pi_t(X_{m,*}, a)$ such that $(d_i^h)_* x = x_i$ for all $i \neq k$. We say X satisfies the <u>π_*-Kan condition</u> if for each $m,t \geq 1$ it satisfies the π_t-Kan condition at each $a \in X_{m,0}$.

To see that X satisfies the π_*-Kan condition when each $X_{m,*}$ connected, one makes the following two observations. First, if $a,b \in X_{m,0}$ are in the same component of $X_{m,*}$, then the π_t-Kan condition for X at a is clearly equivalent to that at b. Second, if $a \in X_{m,0}$ can be expressed as $a = s_0^h \cdots s_0^h e$ for some $e \in X_{0,0}$, then X satisfies the π_t-Kan condition at a for all $t \geq 1$, because any simplicial group satisfies the ordinary Kan condition. Note also that if $X,Y \in$ (bis. sets) are related by a termwise weak equivalence $X \to Y$, then X satisfies the π_*-Kan condition if and only if Y does.

It is easy to show that a bisimplicial set X satisfies the π_*-Kan condition if it has a bisimplicial group structure. To give a more general criterion we use the following notation. For a simplicial set K and $t \geq 1$, let $\pi_t(K)_{free}$ denote the set of unpointed homotopy classes of maps from a t-sphere to $|K|$, and let $\beta: \pi_t(K)_{free} \to \pi_0 K$ be the obvious surjection. We call K <u>simple</u> if each component of $|K|$ is a simple space. It is now an easy exercise to prove

(B.3.1). Let X be a bisimplicial set with $X_{m,*}$ simple for $m \geq 0$. Then X satisfies the π_*-Kan condition if and only if the simplicial map $\beta: \pi_t^v(X)_{free} \to \pi_0^v X$ is a fibration for each $t \geq 1$.

To state our fibre square theorem, we recall that a commutative square

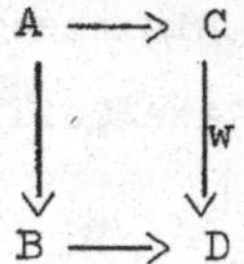

of simplicial sets is a homotopy fibre square (see A.2) if for some factorization $C \xrightarrow{i} W \xrightarrow{p} D$ of w with i a weak equivalence and p a (Kan) fibration, the map $A \to B \times_D W$ is a weak equivalence. Also, for a bisimplicial set X, we let $\pi_i^v X$ be the simplicial set with $(\pi_i^v X)_m = \pi_i X_{m,*}$.

Theorem B.4. Let

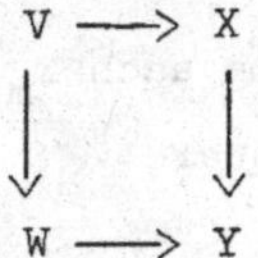

be a commutative square of bisimplicial sets such that the terms $V_{m,*}$, $W_{m,*}$, $X_{m,*}$, and $Y_{m,*}$ form a homotopy fibre square for each $m \geq 0$. If X and Y satisfy the π_*-Kan condition and if $\pi_0^v X \to \pi_0^v Y$ is a fibration, then

$$\begin{array}{ccc} \text{diag } V & \to & \text{diag } X \\ \downarrow & & \downarrow \\ \text{diag } W & \to & \text{diag } Y \end{array}$$

is a homotopy fibre square.

Note that the hypotheses on X and Y hold automatically when the terms $X_{m,*}$ and $Y_{m,*}$ are all connected. Some other interesting, but more specialized, versions of this theorem have been proved in [May 2, §12] and [Segal 2]; and some extensions and applications have been obtained by T. Gunnarson in his thesis work. Before starting to prove B.4, we apply it to generalize Quillen's spectral sequence for bisimplicial groups [Quillen 3].

Theorem B.5. Let X be a bisimplicial set satisfying the π_*-Kan condition, and let $* \in X_{0,0}$ be a base vertex (whose degeneracies are taken as the basepoints of the sets $X_{m,n}$.) Then there is a first quadrant spectral sequence $\{E^r_{s,t}\}_{r\geq 2}$ converging to $\pi_{s+t}(\text{diag } X)$ with $E^2_{s,t} = \pi^h_s \pi^v_t X$. The term $E^r_{s,t}$ is a set for $t + s = 0$, a group for $t + s = 1$, and an abelian group for $t + s \geq 2$. Convergence has the obvious meaning, e.g. there is an isomorphism of sets $E^\infty_{0,0} \cong \pi_0 \text{diag } X$ and a short exact sequence $1 \to E^\infty_{0,1} \to \pi_1 \text{diag } X \to E^\infty_{1,0} \to 1$ of groups.

Proof. By B.2 we can assume each $X_{m,*}$ is a Kan complex, and by B.4 there is a homotopy fibre square

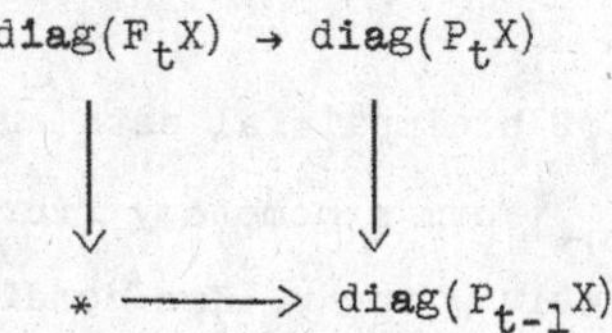

for $t \geq 0$ where $P_t X$ is the bisimplicial set given by the tth-Postnikov sections of the terms $X_{m,*}$ (taking $P_{-1}X = *$), and where $F_t X$ is the fibre of $P_t X \to P_{t-1}X$. Let $\{E^r_{s,t}\}$ be the associated spectral sequence with $E^2_{s,t} = \pi_{s+t} \text{diag}(F_t X)$. The convergence result follows since $\text{diag} X \to \text{diag} P_t X$ is iso in dimensions $\leq t$ and onto elsewhere, and it remains to show $\pi_{s+t} \text{diag}(F_t X) \cong \pi^h_s \pi^v_t X$. Since $(F_t X)_{m,n} = *$ for $n < t$, there is a natural bisimplicial map $F_t X \to K(\pi^v_t X, t)$ where $K(\pi^v_t X, t)$ is given by the minimal Eilenberg-MacLane complexes $K(\pi_t X_{m,*}, t)$. By B.2, we now have

$$\pi_{s+t}\mathrm{diag}F_tX \approx \pi_{s+t}\mathrm{diag}K(\pi_t^vX,t)$$

and the required isomorphism

$$\pi_{s+t}\mathrm{diag}K(\pi_t^vX,t) \approx \pi_s^h\pi_t^vX$$

follows for $t \geq 2$ from [Dold-Puppe, p. 213], and for $t = 0$ trivially. The remaining case $t = 1$ will follow by showing $\pi_*BG \approx \pi_{*-1}G$ for a simplicial group G, where

$$BG = \mathrm{diag}K(G_*,1)_*.$$

The natural principal fibrations

$$K(G_n,0) \to L(G_n,1) \to K(G_n,1)$$

with $|L(G_n,1)| \approx *$ induce a principal fibration

$$G = \mathrm{diag}K(G_*,0)_* \to \mathrm{diag}L(G_*,1)_* \to \mathrm{diag}K(G_*,1) = BG$$

and $|\mathrm{diag}L(G_*,1)_*| \approx *$ by an argument using B.2. Thus $\pi_*BG \approx \pi_{*-1}G$.

To prove B.4 we need a model category structure on (bis. sets). For $X,Y \in$ (bis. sets), let $HOM(X,Y)$ be the simplicial set whose n-simplices are the bisimplicial maps $X \otimes \Delta[n] \to Y$ where $(X \otimes \Delta[n])_{m,*} = X_{m,*} \times \Delta[n]$.

<u>Theorem B.6</u>. The category (bis. sets) is a proper closed simplicial model category when provided with the following additional structure: a map $f\colon X \to Y$ in (bis. sets) is called a <u>weak equivalence</u> if $f_{m,*}\colon X_{m,*} \to Y_{m,*}$ is a weak equivalence in (s.sets) for each $m \geq 0$; f is called a <u>cofibration</u> if it is injective; and f is called a

fibration if $f_{0,*}\colon X_{0,*} \to Y_{0,*}$ is a fibration and for each $m \geq 1$ the simplicial square

$$\begin{array}{ccc} X_{m,*} & \xrightarrow{d} & M_m X \\ \Big\downarrow f_{m,*} & & \Big\downarrow M_m f \\ Y_{m,*} & \xrightarrow{d} & M_m Y \end{array}$$

induces a fibration $X_{m,*} \to Y_{m,*} \times_{M_m Y} M_m X$ where $(M_m X)_n$ is the set of $(m+1)$-tuples $(x_0,\ldots,x_m)$ in $X_{m-1,n}$ such that $d_i^h x_j = d_{j-1}^h x_i$ for $i < j$, and where $d\colon X_{m,*} \to M_m X$ is given by $d(x) = (d_0^h x,\ldots,d_m^h x)$.

This theorem follows from [Reedy]; the proof is similar to that of 3.5. We remark that if $f\colon X \to Y$ is a fibration in (bis. sets), then each $f_{m,*}\colon X_{m,*} \to Y_{m,*}$ is a fibration in (s.sets), but not conversely.

<u>Proof of B.4</u>. By <u>CM5</u> (cf. §1) and B.2, we can suppose that the given square

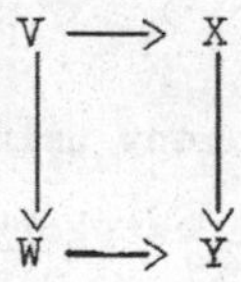

is a pull-back with $X \to Y$ a fibration and with X,Y fibrant. Since the diagonal functor preserves pull-backs, B.4 follows from

<u>Proposition B.7</u>. Let $X,Y \in$ (bis. sets) be fibrant objects satisfying the π_*-Kan condition, and let $f\colon X \to Y$ be a fibration. If $f_*\colon \pi_0^v X \to \pi_0^v Y$ is a fibration in (s.sets), then so is diag f: diag$X \to$ diag$Y$.

To prove B.7, we begin by noting that the diagonal functor has a left adjoint

$$L: (\text{s.sets}) \to (\text{bis.sets})$$

given by

$$L(K)_{m,n} = \underset{([m],[n])\to([i],[i])\in\Delta\times\Delta}{\operatorname{colim}} K_i .$$

To construct L(K) more explicitly, we use the bisimplicial map $c\colon L(K) \to K \underset{\sim}{\times} K$ adjoint to the diagonal $K \to K \times K$. Although c is not always injective, we have

Lemma B.8. If K is the simplicial set associated with an ordered simplicial complex (cf. [May 1, 1.4]), then $c\colon L(K) \to K \underset{\sim}{\times} K$ is an injection onto the bisimplicial subset generated by all $(x,x) \in K \underset{\sim}{\times} K$.

Proof. Suppose $(\theta_1^* x_1, \varphi_1^* x_1) = (\theta_2^* x_2, \varphi_2^* x_2)$ in $K \underset{\sim}{\times} K$ where x_1, x_2 are non-degenerate simplices of K and $\theta_1, \varphi_1, \theta_2, \varphi_2$ are maps in Δ. The injectivity of c follows because there exist factorizations $\theta_1 = \gamma_1\sigma$, $\varphi_1 = \gamma_1\tau$, $\theta_2 = \gamma_2\sigma$, $\varphi_2 = \gamma_2\tau$ in Δ such that $\gamma_1^* x_1 = \gamma_2^* x_2$. (Take $\gamma_1^* x_1$ to be the "largest common face" of x_1 and x_2.) The result on the image of c is obvious.

We next use B.8 to show

Lemma B.9. Let $f\colon X \to Y$ be a bisimplicial fibration such that $f_{*,n}\colon X_{*,n} \to Y_{*,n}$ is a fibration for each $n \geq 0$. Then diag f is a fibration.

Proof. It suffices to show that diag f has the right lifting property (RLP) for the maps $\Delta^k[n] \xrightarrow{\subset} \Delta[n]$ with $n \geq 1$ and $0 \leq k \leq n$, where $\Delta^k[n]$ is the simplicial subset of $\Delta[n]$ generated by the faces

d_i for $i \neq k$. By adjointness, it now suffices to show that f has the RLP for the bisimplicial maps $L\Delta^k[n] \to L\Delta[n]$ with $n \geq 1$ and $0 \leq k \leq n$. Using B.8 we factor these maps as

$$L\Delta^k[n] \xrightarrow{\subset} \Delta^k[n] \underset{\sim}{\times} \Delta[n] \xrightarrow{\subset} \Delta[n] \underset{\sim}{\times} \Delta[n] \cong L\Delta[n]$$

and we observe that the left map is a trivial cofibration in (bis.sets). The result now follows since f has the RLP for each of the factor maps.

Continuing with the proof of B.7, we must reformulate B.9 using "matching" objects. For $m \geq 1$, $0 \leq s_1 < \cdots < s_r \leq m$, and a bisimplicial set X, let $M_m^{(s_1,\ldots,s_r)}X$ denote the "matching" simplicial set whose n-simplices are the r-tuples $(x_{s_1},\ldots,x_{s_r})$ in $X_{m-1,n}$ such that $d_i^h x_j = d_{j-1}^h x_i$ for each $i < j$ in $\{s_1,\ldots,s_r\}$. Also let $d\colon X_{m,*} \to M_m^{(s_1,\ldots,s_r)}X$ be the simplicial map with $d(x) = (d_{s_1}^h x,\ldots,d_{s_r}^h x)$. It will be convenient to write $M_m^k X$ for $M_m^{(0,\ldots,\hat{k},\ldots,m)}X$.

<u>Lemma B.10</u>. Let $f\colon X \to Y$ be a bisimplicial fibration such that the square

$$\begin{array}{ccc} X_{m,*} & \xrightarrow{d} & M_m^k X \\ \big\downarrow{\scriptstyle f_{m,*}} & & \big\downarrow{\scriptstyle M_m^k f} \\ Y_{m,*} & \xrightarrow{d} & M_m^k Y \end{array}$$

induces a surjection

$$\pi_0 X_{m,*} \to \pi_0(Y_{m,*} \times_{M_m^k Y} M_m^k X)$$

for $m \geq 1$ and $0 \leq k \leq m$. Then diag f is a fibration.

Proof. Since f is a fibration and $\Delta^k[m] \underset{\sim}{\times} \Delta[0] \xrightarrow{\subset} \Delta[m] \underset{\sim}{\times} \Delta[0]$ is a cofibration, the map $X_{m,*} \to Y_{m,*} \times_{M_m^k Y} M_m^k X$, is a fibration by SM7 in [Quillen 1, II.2], and it is onto by our π_0-hypothesis. Hence, $X_{*,n} \to Y_{*,n}$ is a fibration for $n \geq 0$, and the result follows from B.9.

To verify the hypotheses of B.10 in our situation, we need

Lemma B.11. Let X be a fibrant bisimplicial set satisfying the π_*-Kan condition, and let $a = (a_{s_1}, \ldots, a_{s_r})$ be a vertex of $M_m^{(s_1,\ldots,s_r)} X$ where $1 \leq r \leq m$, $0 \leq s_1 < \cdots < s_r \leq m$. Then for $t \geq 0$ the obvious map

$$\pi_t(M_m^{(s_1,\ldots,s_r)} X, a) \to \pi_t(X_{m-1,*}, a_{s_1}) \times \cdots \times \pi_t(X_{m-1,*}, a_{s_r})$$

is an injection whose image consists of the elements $(u_{s_1}, \ldots, u_{s_r})$ such that $(d_i^h)_* u_j = (d_{j-1}^h)_* u_i$ for each $i < j$ in $\{s_1, \ldots, s_r\}$. Moreover, $d\colon X_{m,*} \to M_m^{(s_1,\ldots,s_r)} X$ is a fibration.

Proof. Using SM7 as in B.10, one shows that d is a fibration. Then the lemma follows by induction on r using the fibre squares

$$\begin{array}{ccc} M_m^{(s_1,\ldots,s_r)} X & \longrightarrow & X_{m-1,*} \\ \downarrow & & \downarrow \\ M_m^{(s_1,\ldots,s_{r-1})} X & \longrightarrow & M_{m-1}^{(s_1,\ldots,s_{r-1})} X \end{array}$$

for $r \geq 2$.

Finally we can give

Proof of B.7. Consider the square of simplicial sets

$$\begin{array}{ccc} X_{m,*} & \xrightarrow{d} & M_m^k X \\ \Big\downarrow f_{m,*} & & \Big\downarrow M_m^k f \\ Y_{m,*} & \xrightarrow{d} & M_m^k Y \end{array}$$

for $m \geq 1$ and $0 \leq k \leq m$. For each vertex $a \in Y_{m,0}$ we show that

$$d_*: \pi_1(Y_{m,*}, a) \to \pi_1(M_m^k Y, da)$$

is onto by using B.11 to compute $\pi_1(M_m^k Y, da)$ and using π_1-Kan condition for Y at a. Thus there is an isomorphism

$$\pi_0(Y_{m,*} \times_{M_m^k Y} M_m^k X) \cong \pi_0 Y_{m,*} \times_{\pi_0 M_m^k Y} \pi_0 M_m^k X$$

and we conclude that

$$\pi_0 X_{m,*} \to \pi_0(Y_{m,*} \times_{M_m^k Y} M_m^k X)$$

is onto by using B.11 in the case $t = 0$ and the hypothesis that $\pi_0^v X \to \pi_0^v Y$ is a fibration. Now B.7 follows from B.10.

References

D. W. Anderson: Chain functors and homology theories, Lecture Notes in Mathematics, Vol. 249, Springer-Verlag, New York, 1971.

M. Artin and B. Mazur: On the Van Kampen theorem, Topology 5 (1966), 179-189.

A. K. Bousfield and D. M. Kan: Homotopy limits, completions and localizations, Lecture Notes in Mathematics, Vol. 304, Springer-Verlag, New York, 1972.

K. S. Brown: Abstract homotopy theory and generalized sheaf cohomology, Trans. Amer. Math. Soc. 186 (1974), 419-458.

A. Dold and D. Puppe: Homologie nicht-additiver Funktoren, Anwendungen, Ann. Inst. Fourier 11 (1961), 201-312.

A. Dress: Zur Spectralsequenz von Faserungen, Invent. Math. 3 (1967), 172-178.

E. M. Friedlander: Stable Adams conjecture via representability theorems for Γ-spaces. To appear.

D. M. Kan: Semisimplicial spectra, Ill. J. of Math. 7 (1963), 479-491.

J. P. May 1: Simplicial Objects in Algebraic Topology, Van Nostrand, Princeton, 1967.

________ 2: The Geometry of Iterated Loop Spaces, Lecture Notes in Mathematics, Vol. 271, Springer-Verlag, New York, 1972.

D. G. Quillen 1: Homotopical Algebra, Lecture Notes in Mathematics, Vol. 43, Springer-Verlag, New York, 1972.

__________ 2: Rational homotopy theory, Ann. Math. 90 (1969), 205-295.

__________ 3: Spectral sequences of a double semi-simplicial group, Topology 5 (1966), 155-157.

C. L. Reedy: Homotopy theory of model categories. To appear.

G. Segal 1: Categories and cohomology theories, Topology 13 (1974), 293-312.

________ 2: Classifying spaces and spectral sequences, Pub. Math. I.H.E.S. no. 34 (1968), 105-112.

E. Spanier: Infinite symmetric products, function spaces, and duality, Ann. Math. 69 (1959), 142-198.

J. Tornehave: On BSG and the symmetric groups. To appear.

Algebraic and Geometric Connecting Homomorphisms in the Adams Spectral Sequence

R. Bruner

Let E be a commutative ring spectrum such that E_*E is flat over π_*E and such that, for any spectra X and Y, $[X, Y \wedge E] \cong \mathrm{Hom}_{E_*E}(E_*X, E_*Y \otimes_{\pi_*E} E_*E)$ (see, e.g., [1, §13 and §16]).

If $A \to B \to C$ is a cofiber sequence such that (1) is short exact

(1) $$0 \to E_*A \to E_*B \to E_*C \to 0$$

then there is an algebraically defined connecting homomorphism

$$\partial : \mathrm{Ext}^{s,t}_{E_*E}(M, E_*C) \to \mathrm{Ext}^{s+1,t}_{E_*E}(M, E_*A)$$

for any E_*E comodule M. When $M = E_*X$, these Ext groups are E_2 terms of Adams spectral sequences and we may ask:

(a) Does ∂ commute with differentials in the Adams spectral sequence?

(b) Does ∂ converge to the homomorphism $\delta_* : [X, C] \to [X, \Sigma A]$ induced by the geometric connecting map $\delta : C \to \Sigma A$?

It is possible to answer (b) without answering (a) (see [2, Theorem 1.7]). We show here that δ induces ∂ in the most natural possible way, answering (a) and (b) affirmatively.

The canonical Adams resolution of a spectrum Y with respect to E is defined by requiring that $Y_{i+1} \to Y_i \to Y_i \wedge E$ be a cofibration for each $i \geq 0$.

Lemma: The connecting map $\delta : C \to \Sigma A$ induces a map D of Adams resolutions with a shift of filtration:

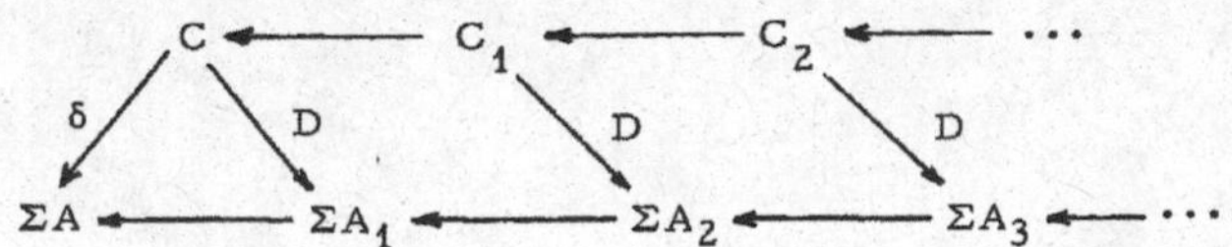

Proof. Since $E_*(\delta) = 0$, our assumptions on E imply that $C \to \Sigma A \to \Sigma A \wedge E$ is nullhomotopic. The existence of D now follows just as in the proof that a map of spectra induces a map of Adams resolutions.

Let $E_r^{**}(X,Y)$ be the E_r term of the Adams spectral sequence for $[X,Y]^E$ and let $F^s[X,Y] = \mathrm{Im}([X,Y_s] \to [X,Y])$ (so that $E_\infty^{s*} = F^s/F^{s+1}$).

By composing with D we obtain a map of exact couples and hence a map of Adams spectral sequences $\{D_r\}: \{E_r^{s,t}(X,C)\} \to \{E_r^{s+1,t}(X,A)\}$. By the lemma, $\delta_* F^s[X,C] \subset F^{s+1}[X,A]$ and therefore the ordinary associated graded homomorphism $E^0(\delta_*): E_\infty^{s*}(X,C) \to E_\infty^{s*}(X,A)$ is zero. Because of the filtration shift, δ_* induces a homomorphism $E_\infty^{s,t}(X,C) \to E_\infty^{s+1,t}(X,A)$ and this is clearly D_∞, the homomorphism induced by composition with D. It follows that in order to answer (a) and (b) affirmatively we need only show that D_2 is the connecting homomorphism for Ext.

Proposition. The connecting homomorphism

$$\mathrm{Ext}_{E_*E}^{s,t}(E_*X, E_*C) \to \mathrm{Ext}_{E_*E}^{s+1,t}(E_*X, E_*A)$$

induced by the short exact sequence (1) preserves all differentials and converges to δ_*.

Proof. Interpreting Ext as equivalence classes of exact sequences, the connecting homomorphism is Yoneda composite with (1). On the other hand, the homomorphism induced by D is the homomorphism induced by $D_*: E_*C \to E_*\Sigma A_1$ followed by Yoneda composite with $E_*A \to E_*(A \wedge E) \to E_*\Sigma A_1$. This is obvious from the following diagram if one keeps in mind both definitions of Ext: (i) cocycles modulo coboundaries, (ii) equivalence classes of exact sequences.

$$\begin{array}{ccccccccccc}
 & & 0 & \to & E_*C & \to & E_*(C\wedge E) & \to & E_*(\Sigma C_1\wedge E) & \to & \cdots \\
 & & & & \downarrow D_* & & \downarrow D_* & & \downarrow D_* & & \\
0 \to E_*A \to E_*(A\wedge E) & \to & E_*\Sigma A_1 & \to & E_*(\Sigma A_1\wedge E) & \to & E_*(\Sigma^2 A_2\wedge E) & \to & \cdots & &
\end{array}$$

(with $0 \to E_*\Sigma A_1 \to 0$ indicated at $E_*\Sigma A_1$)

Thus we need only show that there exists a commutative diagram

$$\begin{array}{ccccccccc}
0 & \to & E_*A & \to & E_*B & \to & E_*C & \to & 0 \\
 & & \| & & \downarrow & & \downarrow D_* & & \\
0 & \to & E_*A & \to & E_*(A\wedge E) & \to & E_*\Sigma A_1 & \to & 0
\end{array}$$

The existence of such a diagram follows immediately from the map of cofiber sequences induced by D

$$\begin{array}{ccccccc}
A & \to & B & \to & C & \xrightarrow{\delta} & \Sigma A \\
\| & & \downarrow & & \downarrow & & \| \\
A & \to & A\wedge E & \to & \Sigma A_1 & \to & \Sigma A
\end{array}$$

[1] J.F. Adams. Stable Homotopy and Generalized Homology. Univ. Chicago Lect. Notes in Math. 1974

[2] Johnson, Miller, Wilson, Zahler. Boundary Homomorphisms in the Generalized Adams Spectral Sequence and the Nontriviality of Infinitely Many γ_t in Stable Homotopy. Proc. of the Conf. on Homotopy Theory, Northwestern Univ., 1974, Notas de Matematica y Simposia, Sociedad Matematica Mexicana.

OBSTRUCTION THEORY AND K-THEORY

Donald M. Davis and Mark Mahowald

1. INTRODUCTION

In [8] we sketched a method of employing the spectrum bo ([4]) in obstruction theory to determine some monimmersion theorems for real projective spaces. Many of the nonimmersion results which we announced there were incorrect due to an incorrect analysis of the indeterminancy (see 3.13). However, the method is valid and does yield some known results on the generalized vector field problem, i.e. the determination of the geometric dimension (gd) of stable vector bundles over real projective spaces, including the nonimmersions first proved by James ([14]). In this paper we present the method in some detail, illustrating it on a particular case. All results which can be obtained by this method were already established in [4]. Thus the importance of this paper lies in its methods, and not its results. Consequently, we shall not describe all results which we have obtained by this method.

The main result is that in the stable range symplectic vector bundles are bo-orientable.

1.1 Definition. A fibration $F \to E \xrightarrow{p} B$ is <u>principal through dimension M</u> (or M-principal) if there is a map from the M-skeleton $B^{(M)} \xrightarrow{c} Y$ and an M-equivalence $p^{-1}(B^{(M)}) \to F_c$, where F_c is the homotopy theoretic fibre of c, i.e. the pullback over $B^{(M)}$ of the path space PY. If b is a connected φ-spectrum, let $F \wedge b \to E \wedge_B b \xrightarrow{p'} B$ denote the fibrewise smash product as defined in [7]. We say p is <u>b-orientable</u> through dimension M if p' is principal through dimension M.

If M is less than twice the connectivity of F, then $Y = \Sigma F$ (if p principal) or $Y = \Sigma F \wedge b$ (if p b-orientable).

As in [7] we let BSp_N denote the classifying space for stable symplectic vector bundles of real geometric dimension N, i.e. it is the pullback over BSp of BO_N.

1.2 Theorem. If $N \not\equiv 0(4)$ $\widetilde{BSp}_N \to BSp$ is bo-orientable through dimension 2N.

The main theorem of [19] implies that if p is b-orientable through dimension M, then in this range p can be written as a composite of principal fibrations such that all fibres are of the form $-\wedge b$. This enables one to do obstruction theory. The difficulty is to determine the higher-order obstructions.

We can determine some bo-secondary obstructions by using the main result of [7] and the spectrum bJ. Let $\overline{bo}$ denote the cofibre of the inclusion $S^0 \xrightarrow{\iota} bo$. By [20], there is a splitting $bo \wedge \overline{bo} \simeq \Sigma^4 bsp \vee X$, where X is a 7-connected space explicitly described in [20] and bsp is the connected Ω-spectrum whose $8k^{th}$ space is the (8k-1)-connected covering BSp[8k], localized at 2. bsp was denoted bo^4 in [8]. $\Sigma^4 bsp$ may be interpreted as bo[4], the spectrum formed from bo by killing π_i for $i < 4$. Milgram's splitting can be chosen so that if q collapses X, then

$$\theta: bo = S^0 \wedge bo \xrightarrow{\iota \wedge k} bo \wedge \overline{bo} \xrightarrow{q} \Sigma^4 bsp$$

is a lifting of the Adams operation $\Psi^3 - 1$: bo → bo. Let bJ denote the fibre of θ. ([5], [17]).

Let $B_N^0 = \widetilde{BSp}_N \wedge_{BSp} bo$ denote the space which was called E_N^0 in [7]. Recall that $P_N = RP^\infty/RP^{N-1}$ has the same 2N-type as $V_N = \bigcup_k V_{N+k,k}$, which is the fibre of $\widetilde{BSp}_N \to BSp$. Theorem 1.2 is used to prove

1.3 Theorem. There is a fibration $B_N^J \to BSp$ which can be written as the composite of two fibrations which are principal through dimension 2N. Indeed, through dimension 2N there are maps

$$\begin{array}{ccccccc}
P_N \wedge bJ & \longrightarrow & & & B_N^J & & \\
\downarrow & & & & \downarrow & & \\
P_N \wedge bo & \xrightarrow{i} & (P_N \wedge bo) \times B_N^O & \xrightarrow{\mu} & B_N^O & \xrightarrow{c_1} & P_N \wedge \Sigma^4 bsp. \\
 & & & & \downarrow & & \\
 & & & & BSp & \xrightarrow{c_O} & \Sigma P_N \wedge bo
\end{array}$$

such that $B_N^O = F_{c_O}$ and $B_N^J = F_{c_1}$. Moreover, $c_1\mu i \simeq 1 \wedge \theta$. Also there is a map of fibrations

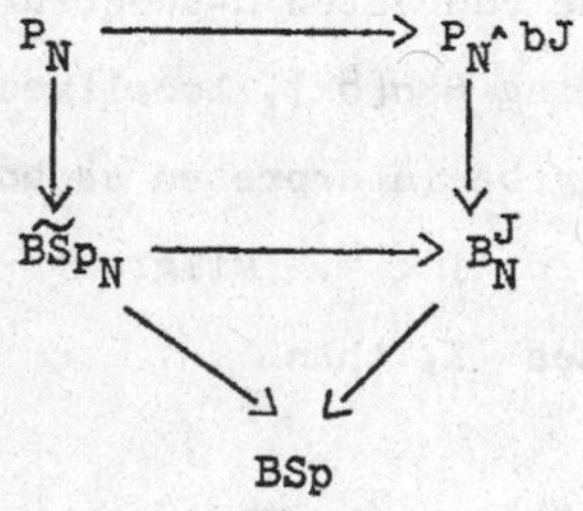

1.4 Remark. In the above diagram the maps c_O and c_1 are actually defined only on the 2N-skeleton. In order to keep our diagrams readable, we will often take the liberty of calling such a situation a diagram through dimension 2N.

We let g: $RP^n \xrightarrow{\text{"}4pg\text{"}} BSp$ be the map classifying the 4p-fold Whitney sum of the Hopf bundle. By the methods of [4; Ch 3] (or an easy indeterminacy argument in the Adams resolution) any map

$P^n \to \Sigma P_N \wedge bo$ which is trivial in $\mathbb{Z}_2$-cohomology is null homotopic. Thus, unless a nonlifting is detected by Stiefel-Whitney classes, g lifts to ℓ: $RP^n \to B_N^O$. For certain values of p,n, and N we can compute the class $[c,\ell] \in [P^n, P_N \wedge \Sigma^4 bsp]$ to be nonzero and not in the indeterminacy of the lifting ℓ. For such values, g does not lift to B_N^J and hence not to $\widetilde{BSp}_N$, proving $gd(4p\xi_n) > N$.

There is some chance that by applying similar methods to other parts of the decomposition of $bo \wedge \overline{bo}$, some new results might be obtained, but this now seems rather unlikely. Indeed, it seems that the results of [4] are the only results which the Adams operations can tell us about the generalized vector field problem. This observation has been independently substantiated in [10]. Somewhat more promising is the possibility of applying these techniques to do BP-obstruction theory, where BP is the Brown-Peterson spectrum [6].

In Section 2 we prove the orientability results, Theorems 1.2 and 1.3. In Section 3 we present the proof of a geometric dimension result which is illustrative of the general situation. Section 4 provides some technical details of this proof.

2. bo-ORIENTABILITY OF SYMPLECTIC VECTOR BUNDLES

In this section we prove Theorems 1.2 and 1.3.

Throughout the paper it will be convenient to abbreviate $\mathrm{Ext}_A(H^*(X;\mathbb{Z}_2),\mathbb{Z}_2)$ to $\mathrm{Ext}_A(X)$, where X is any topological space and A is any subalgebra of the mod 2 Steenrod algebra $\mathcal{A}$. Recall from [4] that if X is any space, the Adams spectral sequence which converges to $\pi_*(X \wedge bo)$ has $E_2^{s,t} \approx \mathrm{Ext}_{\mathcal{A}}^{s,t}(X \wedge bo) \approx \mathrm{Ext}^{s,t}_{\mathcal{A}_1}(X)$, where $\mathcal{A}_1$ denotes the subalgebra of $\mathcal{A}$ generated by Sq^1 and Sq^2. The proof which follows will make frequent use of the computations of $\pi_*(P_n^{n+k} \wedge bo)$ of [4; P. 3]. In particular, some groups are (if N is odd)

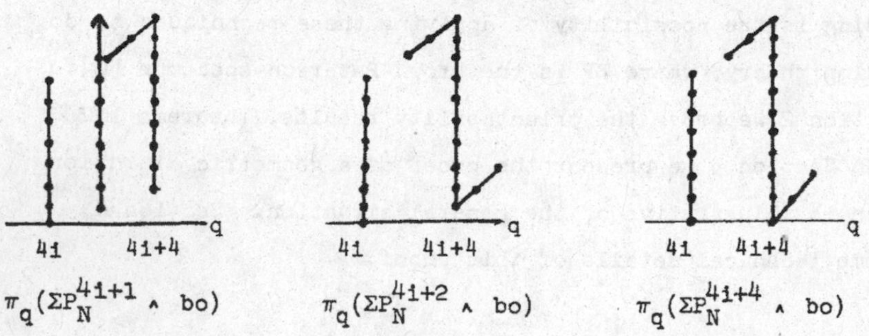

$\pi_q(\Sigma P_N^{4i+1} \wedge bo)$ $\pi_q(\Sigma P_N^{4i+2} \wedge bo)$ $\pi_q(\Sigma P_N^{4i+4} \wedge bo)$

The relevant part of these charts is their bottom. The height of these towers depends upon the value of N. The charts pictured above correspond to $N = 4i - 15$. Adams spectral sequence charts of this type will be employed frequently throughout the paper. Dots indicate nonzero classes, vertical lines indicate multiplication by h_0 in Ext which corresponds to (up to elements of higher filtration) multiplication by 2 in homotopy groups. Diagonal lines (/) indicate multiplication by h_1 in Ext which corresponds to the nonzero element η in $\pi_{n+1}(S^n)$.

2.1 Theorem. If $N \not\equiv 0(4)$ there is a 2N-equivalence

$$((BSp/\widetilde{BSp}_N) \wedge bo)^{(2N)} \to \bigvee_I \Sigma^{4|I|+1} P_N \wedge bo$$

where I ranges over $\{0\}$ and all sets of positive integers, and $|I|$ is the sum of all elements of I.

Proof. The proof is very similar to that of [7; Theorem 2.1]. The $\mathbb{Z}_2$-cohomology Serre spectral sequence of the fibre pair $(CV_N, V_N) \to (BSp, \widetilde{BSp}_N) \to BSp$ has no nonzero differentials in the stable range, because it is mapped onto by the Serre spectral sequence of $(CV_N, V_N) \to (BSO, BSO_N) \to BSO$ which clearly has no nonzero differentials in that range. Thus $H^*(BSp, \widetilde{BSp}_N; \mathbb{Z}_2)$ as a vector space is isomorphic to $\tilde{H}^*(\Sigma P_N) \otimes H^*(BSp)$ in the stable range. They are also isomorphic as $\mathcal{A}_1$-modules, because by [15, Section 7] $H^*(BSp, \widetilde{BSp}_N)$ as an $\mathcal{A}_1$-module is an extension of modules $\tilde{H}^*(\Sigma^{4|I|+1} P_N)$ and by the techniques of [4; Theorem 3.9] any such extension splits as $\mathcal{A}_1$-modules. Alternatively, if $k\colon BSp/\widetilde{BSp}_N \to BO/BO_N$, then $\{k^*(w_{4I} w_{N+j})\}$ form a basis with desired $\mathcal{A}_1$-action.

Thus in the Adams spectral sequence for $\pi_*(BSp/\widetilde{BSp}_N \wedge bo)$, $E_2 \approx \mathrm{Ext}_{\mathcal{A}_1}(\bigvee_I \Sigma^{4|I|+1} P_N)$ through dimension 2N. Since $N \not\equiv 0(4)$, the only possible nonzero differentials are zero by naturality with respect to h_0 and h_1.

We now form maps $\Sigma^{4|I|+1} P_N^{2N-4|I|} \wedge bo \to BSp/\widetilde{BSp}_N \wedge bo$ which send $\pi_{4i}(\Sigma^{4|I|+1} P_N^{2N-4|I|} \wedge bo)$ isomorphically to the corresponding summand in $\pi_{4i}(BSp/\widetilde{BSp}_N \wedge bo)$ for all $i \le N/2$.

The map is easily begun. For example if $N \equiv 1(4)$ and $|I| = 2$, the relevant part of $\pi_*(BSp/\widetilde{BSp}_N \wedge bo)$ is

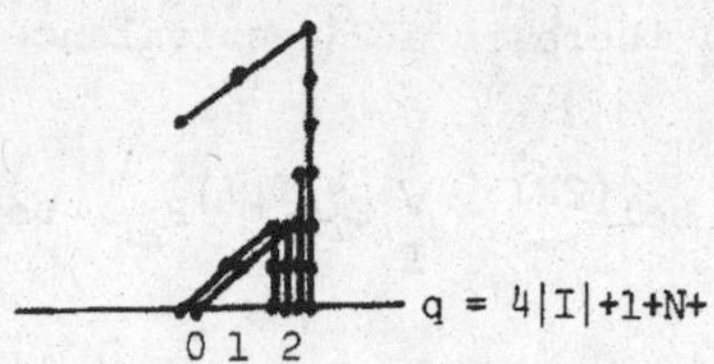

Thus $\pi_{N+9}(BSp/\widetilde{BSp}_N \wedge bo) \approx \mathbb{Z}_2 \oplus \mathbb{Z}_2 \oplus \mathbb{Z}_2$, since there cannot be a nontrivial extension because η times the filtration 4 class is nonzero but $2\eta = 0$. Therefore there are maps $\Sigma^9 P_N^{N+1} = S^{N+9} \cup_2 e^{N+10} \to BSp/\widetilde{BSp}_N \wedge bo$ inducing the desired morphisms in π_{N+9} and π_{N+10}.

Suppose the map has been defined on $\Sigma^{4|I|+1} P_N^{4i} \wedge bo$. We shall delete the $\Sigma^{4|I|}$ in the subsequent discussion. Since $\pi_{4i+1}(BSp/\widetilde{BSp}_N \wedge bo) = 0$, the composite

$$c\colon S^{4i+1} \xrightarrow{att} \Sigma P_N^{4i} \xrightarrow{1\wedge\iota} \Sigma P_N^{4i} \wedge bo \xrightarrow{f_N^{4i}} BSp/\widetilde{BSp}_N \wedge bo$$

is trivial, where att indicates the attaching map for the top cell of ΣP_N^{4i+1}. Thus there is a map $\bar{f}_N^{4i+1}\colon \Sigma P_N^{4i+1} \to BSp/\widetilde{BSp}_N \wedge bo$ which extends $f_N^{4i} \circ (1\wedge\iota)$. Let f_N^{4i+1} denote the composite

$$\Sigma P_N^{4i+1} \wedge bo \xrightarrow{\bar{f}_N^{4i+1}\wedge bo} BSp/\widetilde{BSp}_N \wedge bo \wedge bo \xrightarrow{1\wedge\mu} BSp/\widetilde{BSp}_N \wedge bo.$$

Then $f_N^{4i+1} | \Sigma P_N^{4i} \wedge bo$ is homotopic to $(1\wedge\mu)\circ(f_N^{4i}\wedge 1)\circ(1\wedge\iota\wedge 1)$, which sends π_{4j} $j \le i$ in the desired fashion. (This is easily seen by considering the induced homomorphism of Adams spectral sequences.)

The extension over $\Sigma P_N^{4i+2} \wedge bo$ follows similarly once we have shown that the composite

$$S^{4i+2} \xrightarrow{att} \Sigma P_N^{4i+1} \xrightarrow{1\wedge\iota} \Sigma P_N^{4i+1} \wedge bo \xrightarrow{f_N^{4i+1}} BSp/\widetilde{BSp}_N \wedge bo$$

is trivial. If $N \equiv 2(4)$, $\pi_{4i+2}(BSp/\widetilde{BSp}_N \wedge bo) = 0$, so we now consider N odd. Since
$\eta_*: \pi_{4i+2}(BSp/\widetilde{BSp}_N \wedge bo) \to \pi_{4i+3}(BSp/\widetilde{BSp}_N \wedge bo)$ is injective, it suffices to show that

$$S^{4i+3} \xrightarrow{\eta} S^{4i+2} \xrightarrow{(1\wedge\iota)att} \Sigma P_N^{4i+1} \wedge bo$$

is trivial. This is true since following it by the inclusion into $\Sigma P_N^{4i+2} \wedge bo$ is certainly trivial, but
$i_*: \pi_{4i+3}(\Sigma P_N^{4i+1} \wedge bo) \to \pi_{4i+3}(\Sigma P_N^{4i+2} \wedge bo)$ is an isomorphism.

Let g denote a generator of $\pi_{4i+4}(\Sigma P_N^{4i+2} \wedge bo)$ and k denote the collapsing map of the $\mathbb{Z}_2$-Moore space. The composite
$M_{4i+3} \xrightarrow{k} S^{4i+4} \xrightarrow{g} \Sigma P_N^{4i+2} \wedge bo \xrightarrow{f_N^{4i+2}} BSp/\widetilde{BSp}_N \wedge bo$ is trivial since $f_N^{4i+2}{}_*(g)$ is divisible by 2. Thus f_N^{4i+2} extends over the mapping cone $MC(h)$ of

$$h: M_{4i+3} \wedge bo \xrightarrow{gk\wedge 1} \Sigma P_N^{4i+2} \wedge bo \wedge bo \xrightarrow{1\wedge\mu} \Sigma P_N^{4i+2} \wedge bo$$

because $f_N^{4i+2}(1\wedge\mu)(gk\wedge 1) = (1\wedge\mu)(\bar{f}_N^{4i+2}\wedge 1)(1\wedge\mu)(gk\wedge 1)$

$$= (1\wedge\mu)(1\wedge 1\wedge\mu)(\bar{f}_N^{4i+2}\wedge 1\wedge 1)(gk\wedge 1)$$

$$= (1\wedge\mu)(1\wedge\mu\wedge 1)(\bar{f}_N^{4i+2}\wedge 1\wedge 1)(gk\wedge 1) = (1\wedge\mu)(f_N^{4i+2}gk\wedge 1) = 0.$$

Next we show that $MC(h) \simeq \Sigma P_N^{4i+4} \wedge bo$. If i denotes the inclusion $\Sigma P_N^{4i+2} \to \Sigma P_N^{4i+4}$, then $\pi_{4i+4}(i\wedge bo)$ maps onto elements divisible by 2 so that $(i\wedge bo)gk = 0$ and hence $(i\wedge bo)h = ((1\wedge bo)gk\wedge bo) = 0$. Thus there is a map $\ell: MC(h) \to \Sigma P_N^{4i+4} \wedge bo$. That

$\pi_*(MC(h)) \approx \pi_*(\Sigma P_N^{4i+4} \wedge bo)$ follows from the commutative diagram of exact sequences

$$\begin{array}{ccccc}
\to \pi_*(M_{4i+3} \wedge bo) & \xrightarrow{(k\wedge 1)_*} & \pi_*(S^{4i+4} \wedge bo) & \to & \pi_*(S^{4i+4} \wedge bo) \to \\
 & & & & \| \\
\downarrow 1 & & \downarrow & & \pi_*(MC(k\wedge 1)) \\
 & & & & \downarrow \\
\to \pi_*(M_{4i+3} \wedge bo) & \xrightarrow{h_*} & \pi_*(\Sigma P_N^{4i+2} \wedge bo) & \to & \pi_*(MC(h)) \quad \to
\end{array}$$

h_* is nontrivial in $\pi_{4i+8j+5}$ and $\pi_{4i+8j+6}$ because $(k\wedge 1)_*$ is, and the nontrivial extensions in $\pi_{4i+4j}(MC(h))$ follow from those in $\pi_{4i+4j}(MC(k\wedge 1))$. That ℓ induces this isomorphism of homotopy groups follows from the diagram

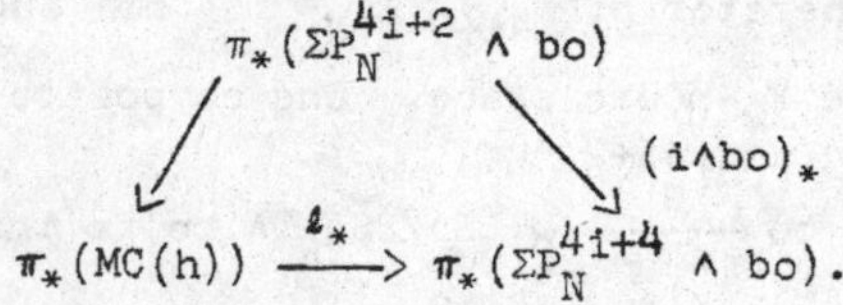

Finally, the diagram

$$\begin{array}{ccccc}
 & & \pi_{4i+4}(\Sigma P_N^{4i+2} \wedge bo) & & \\
 & \swarrow & \downarrow & \searrow f_{N*}^{4i+2} & \\
\pi_{4i+2}(\Sigma P_N^{4i+4} \wedge bo) & \xleftarrow[\approx]{\ell_*} & \pi_{4i+4}(MC(h)) & \xrightarrow{e_*} & \pi_{4i+4}(BSp/\widetilde{BSp}_N \wedge bo)
\end{array}$$

implies that $e_*\ell_*^{-1}$ sends generator to desired generator.

Theorem 1.2 follows from Theorem 2.1 by the proof of [7; Theorem 2.2(i)] applied to the composite

$$BSp^{(2N)} \to BSp/\widetilde{BSp}_N^{(2N)} \to (BSp/\widetilde{BSp}_N \wedge bo)^{(2N)} \to$$

$$\to \bigvee_I \Sigma^{4|I|+1} P_N \wedge bo \to \Sigma P_N \wedge bo.$$

2.2 <u>Theorem</u>. Suppose $F \to E \xrightarrow{p} B$ is a fibration and b is a ring spectrum with unit $\iota: \mathcal{S} \to b$. Let $\bar{b}$ denote the cofibre of ι and $\gamma: b \to \bar{b}$ the collapsing map. Suppose $t \leq$ twice the connectivity of F. If there is a map $B/E^{(t)} \xrightarrow{c} \Sigma F \wedge b$ such that the composite $CF/F^{(t)} \to B/E^{(t)} \to \Sigma F \wedge b$ is homotopic to $i \wedge \iota$, then there is a map $E \wedge_B b^{(t)} \to F \wedge \bar{b} \wedge b$ such that the restriction to $F \wedge b^{(t)}$ is homotopic to $1_F \wedge \gamma \wedge \iota$.

<u>Proof</u>. This is essentially a restatement of [19; Theorems 7.2,7.3]. Let $X^{[t]}$ denote the homotopy t-skeleton as in [19, Ch. 3]. If p does not have a section, $E \wedge_B b$ is formed as in [7; Ch. 2]. If $X \xrightarrow{f} E$ is a map and E is a B-sectioned space, we denote by $\text{fibre}_B(f)$ the pullback of

$$\begin{array}{c} P_B E \\ \downarrow \\ X \to E \end{array}$$

. Corresponding to c is a map $S_B E \to S_B E/B \xrightarrow{\approx} B/E \to (\Sigma F \wedge b)^{[t]}$. Our desired map is obtained from the composite

$$E \wedge_B b \to E \wedge_B b/_B E \to \text{fibre}_B(S_B E \to S_B E \wedge_B b)$$

$$\to \text{fibre}((\Sigma F \wedge b)^{[t]} \to (\Sigma F \wedge b \wedge \bar{b})^{[t]}),$$

noting that the latter space has the same t-type as $F \wedge b \wedge \bar{b}^{[t]}$, and using the bijection $[X,Y^{[t]}] \approx \text{im}([X^{(t+1)},Y] \to [X^{(t)},Y])$. That the restriction to the fibre is as claimed is clear from the construction. ∎

Proof of Theorem 1.3. By Theorem 1.2, Theorem 2.2 applies to $V_N \to \widetilde{BSp}_N \to BSp$, $b = bo$, to give $c_1 \colon B_N^{O(t)} \to V_N \wedge \overline{bo} \wedge bo$. (The one hypothesis of Theorem 2.2 requiring verification follows since in the proof of Theorem 2.1 the first part of the map $\vee \Sigma^{4|I|+1} P_N \wedge bo \to BSp/\widetilde{BSp}_N \wedge bo$ can be chosen to be $j \wedge 1$, where j is the standard map $CP_N/P_N \to BSp/\widetilde{BSp}_N$.) By [20; Theorem C] there is a map $\overline{bo} \wedge bo \xrightarrow{q} \Sigma^4 bsp$ such that $q(\gamma \wedge \iota) = 1 \wedge \theta$. Thus $(1_{V_N} \wedge q) \circ c_1 \circ \mu \circ i = (1 \wedge q) \circ (1 \wedge \gamma \wedge \iota) = 1 \wedge \theta$. We let $B_N^J = \text{fibre}(c_1)$. Then $\text{fibre}(B_N^J \to BSp) = \text{fibre}(1 \wedge \theta) = P_N \wedge bJ$. The map $\widetilde{BSp}_N \to B_N^J$ exists because $\widetilde{BSp}_N \to B_N^O \to P_N \wedge \Sigma^4 bsp$ is trivial by construction. ∎

3. AN EXAMPLE OF bo-OBSTRUCTION THEORY

In this section we shall present a proof of the following result. Let $\nu(2^a(2b+1)) = a$.

3.1 Theorem. If p,k, and i are integers > 1 such that

i) p is even and k is odd

ii) $\nu(p) \geq \nu\binom{p}{k} = 4i - 1$

iii) $\nu\binom{p}{k-1} = 4i - 2$

iv) $\nu\binom{p}{j} \geq 4i - 2(k-j) - \begin{cases} 0 & k-j \text{ odd} \\ 1 & k-j \text{ even} \end{cases}$ for all $j \leqslant k - 2$,

then $gd(4p\xi_{4k+2}) > 4k - 8i + 1$.

One easily checks by the methods of [7; P. 4] that this implies the result of [14] that $RP^{2^{4i+1}-1}$ cannot be immersed in $R^{2^{4i+2}-8i-4}$. As noted in the introduction, Theorem 3.1 and all other geometric dimension results which can be proved by this method are contained in [4]. In particular, the condition in [4; 1.1] that certain binomial coefficients be odd is not so restrictive as one might have thought.

Let $N = 4k - 8i + 1$ throughout this section and let $g: RP^{4k+2} \to BSp$ classify $4p\xi$. Theorem 3.1 will be proved by proving that in the diagram

$$\begin{array}{ccccc} & & B_N^J & & \\ & & \downarrow & & \\ (P_N \wedge bo) \times B_N^0 & \xrightarrow{\mu} & B_N^0 & \xrightarrow{c_1} & P_N \wedge \Sigma^4 bsp \\ & & \downarrow & & \\ RP^{4k+2} & \xrightarrow{g} & BSp & \xrightarrow{c_0} & \Sigma P_N \wedge bo \end{array}$$

(3.2) There is a lifting ℓ: $RP^{4k+2} \to B_N^0$ of g such that $[c_1\ell]$ is a nonzero element in

$$[RP^{4k+2}, P_N \wedge \Sigma^4 bsp] \approx \mathbb{Z}_{2^{4i-1}} \oplus (\mathbb{Z}_2)^{2i}.$$

(3.3) If f: $RP^{4k+2} \to P_N \wedge bo$ is any map, then

$$[c_1\mu(f\times\ell)] \neq 0 \in [RP^{4k+2}, P_N \wedge \Sigma^4 bsp].$$

We begin with some computations.

3.4 <u>Proposition</u> i) $\pi_q(P_N \wedge bo)$ is given by the chart

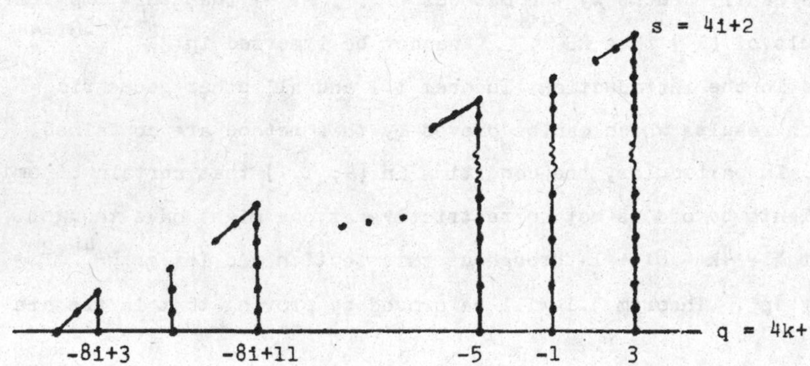

ii) $[P^{4k+2}, P_N \wedge bo] \approx \mathbb{Z}_{2^{4i+2}}$ generated by the class of the map f_0: $P^{4k+2} \to P_N^{4k+2} \to P_N \xrightarrow{1\wedge\iota} P_N \wedge bo$

iii) $[P^{4k+2}, \Sigma P_N \wedge bo] \approx (\mathbb{Z}_2)^{2i+1}$. Any map $P^{4k+2} \to \Sigma P_N \wedge bo$ which is trivial in $\mathbb{Z}_2$-cohomology is null-homotopic.

iv) $\pi_q(P_N \wedge \Sigma^4 bsp)$ is given by the chart

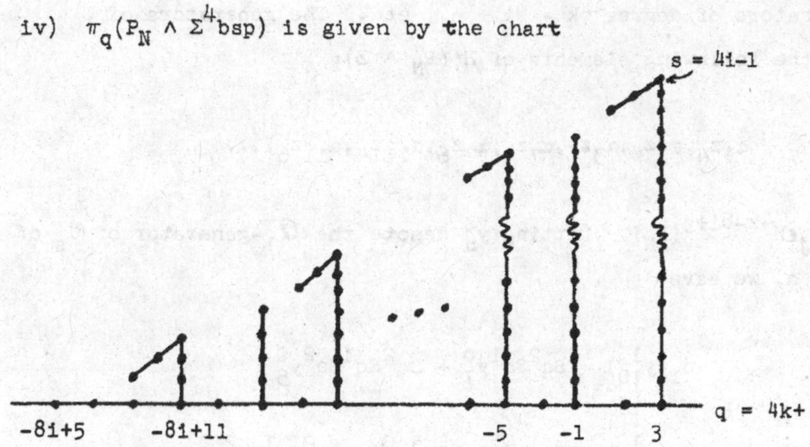

v) $[P^{4k+2}, P_N \wedge \Sigma^4 bsp] \approx \mathbb{Z}_{2^{4i-1}} \oplus (\mathbb{Z}_2)^{2i}$

Proof i) is proved in [13] or [4; 3.4]. By [18; 1.5] $\pi_q(P_N \wedge \Sigma^4 bsp) \approx \pi_q(P_N \wedge Z \wedge bo)$, where $Z = S^4 \cup_\eta e^6 \cup_2 e^7$, and E_2 of of the Adams spectral sequence for this is computed in [4; 3.8 and 3.10]. (ii), (iii), and (v) are computed by dualizing to obtain $[P^{4k+2}, Y \wedge bo] \approx \pi_{2^i-1}(P^{2^i-2}_{2^i-4k-3} \wedge Y \wedge bo) \Leftarrow \mathrm{Ext}_{\mathcal{Q}_1}(P^{2^i-2}_{2^i-4k-3} \wedge Y)$, which is computed for the relevant Y by [4; 3.7, 3.9, and 3.11].

An alternative approach which avoids the use of duality is to compute $\mathrm{Ext}_{\mathcal{Q}_1}(Y, P^{4k+2})$ using the minimal $\mathcal{Q}_1$-resolution for $H^*(Y)$. For example, a minimal $\mathcal{Q}_1$-resolution of $H^*(P_N \wedge Z)$ begins

$$C_2 \xrightarrow{d_2} C_1 \xrightarrow{d_1} C_0 \to H^*(P_N \wedge Z) \to 0$$

with $C_0 = \mathcal{Q}_1(5,7,9,11,13,15,\ldots)$, $C_1 = \mathcal{Q}_1(10,12,16,\ldots)$, and $C_2 = \mathcal{Q}_1(12,13,17,\ldots)$. Here $\mathcal{Q}_1(n_1,n_2,\ldots)$ denotes a free $\mathcal{Q}_1$-module on generators of degree $4k - 8i + n_1$, etc. The generators of C_0 map to the following elements of $H^*(P_N \wedge Z)$:

$$\alpha_1 z_4, \alpha_3 z_4, \alpha_3 z_6, \alpha_7 z_4, \alpha_7 z_6, \alpha_{11} z_4, \alpha_{11} z_6, \ldots$$

where $\alpha_j \in H^{4k-8i+j}(P_N)$. Letting y_n^s denote the $\mathcal{A}_1$-generator of C_s of degree n, we have

$$d_1(y_{10}^1) = Sq^2 Sq^1 y_7^0 + Sq^2 Sq^1 Sq^2 y_5^0$$

$$d_1(y_{12}^1) = Sq^1 y_{11}^0 + Sq^3 y_9^0 + Sq^2 Sq^1 Sq^2 y_7^0$$

$$d_1(y_{16}^1) = Sq^1 y_{15}^0 + Sq^3 y_{13}^0 + Sq^2 Sq^1 Sq^2 y_{11}^0$$

$$d_2(y_{12}^2) = Sq^2 y_{10}^1$$

$$d_2(y_{13}^2) = Sq^1 y_{12}^1 + Sq^2 Sq^1 y_{10}^1$$

$$d_2(y_{17}^2) = Sq^1 y_{16}^1 + Sq^2 Sq^1 Sq^2 y_{12}^1.$$

Then $\mathrm{Ext}_{\mathcal{A}_1}^{0,0}(P_N \wedge Z, P^{4k+2}) \cong \ker(\mathrm{Hom}_{\mathcal{A}_1}(C_0, H^*P^{4k+2}) \xrightarrow{d_1^*} \mathrm{Hom}_{\mathcal{A}_1}(C_1, H^*P^{4k+2}))$ with $d_1^*(\hat{y}_7^0) = \hat{y}_{12}^1$

$$d_1^*(\hat{y}_{11}^0) = \hat{y}_{12}^1 + \hat{y}_{16}^1$$

$$d_1^*(\hat{y}_{15}^0) = \hat{y}_{16}^1$$

$$d_1^*(\hat{y}_{4i+1}^0) = 0$$

so that $\mathrm{Ext}_{\mathcal{A}_1}^{0,0}(P_N \wedge Z, P^{4k+2})$ has a generator for each $\hat{y}_{4i+1}^0$ and one for the sum of all the $\hat{y}^0_{4i+3}$.

$$\mathrm{Ext}^{1,1}_{\mathcal{A}_1}(P_N \wedge Z, P^{4k+2}) \simeq \frac{\ker(\mathrm{Hom}(\Sigma^{-1}C_1, H^*P) \xrightarrow{(\Sigma^{-1}d_2)^*} \mathrm{Hom}(\Sigma^{-1}C_2, H^*P))}{\mathrm{im}(\mathrm{Hom}(\Sigma^{-1}C_0, H^*P) \xrightarrow{(\Sigma^{-1}d_1)^*} \mathrm{Hom}(\Sigma^{-1}C_1, H^*P))}$$

The kernel is generated by the sum of all $\hat{y}^1_j$, while the image is 0. The h_0-extension from $\mathrm{Ext}^{0,0}$ to $\mathrm{Ext}^{1,1}$ can also be shown by this method. ∎

The homotopy groups of $P_N \wedge bJ$ can be computed from the homotopy exact sequence of the fibration $P_N \wedge bJ \to P_N \wedge bo \to P_N \wedge \Sigma^4 bsp$. It is convenient to indicate this on a chart whose entry $e_N(i,s)$ in position (i,s) is $\mathrm{Ext}^{s,s+i}_{\mathcal{A}}(P_N \wedge bo) \oplus \mathrm{Ext}^{s-1,s+i}_{\mathcal{A}}(P_N \wedge \Sigma^4 bsp)$ with differentials inserted to indicate the homomorphism $\pi_i(P_N \wedge bo) \xrightarrow{\theta_*} \pi_i(P_N \wedge \Sigma^4 bsp)$. ([5],[17]) For our value of N this gives

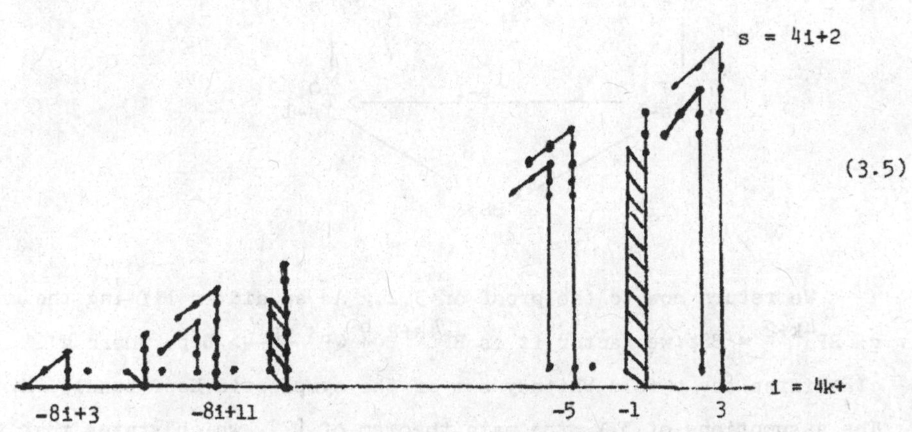

(3.5)

There are also d_{i-1} differentials emanating from the towers in dimension $2^i(2b+1)-1$. This can be seen as in [16; Ch. 7] or [4; Ch. 4].

We prove in Section 4 that even though this is not an Ext-chart, we can form a modified Postnikov tower (MPT) (see [11]) for the fibration $B_N^J \to BSp$ corresponding to it.

3.6 Theorem If $t \leq 2N$, there is a t-MPT for the fibration $V_N \wedge bJ \to B_N^J \to BSp$, i.e. a sequence of t-principal fibrations

$$E_r^J \xrightarrow{p_{r-1}} --- \xrightarrow{p_1} E_1^J \xrightarrow{p_0} BSp$$

and a t-equivalence $B_N^J \to E_r^J$, such that $\text{fibre}(p_s) = \bigtimes_{\substack{e_N(i,s)\neq 0 \\ i<t}} K(Z_2,i)$.

There is a map from this MPT into the usual (determined by Ext) MPT for $V_N \wedge bo \to B_N^O \to BSp$

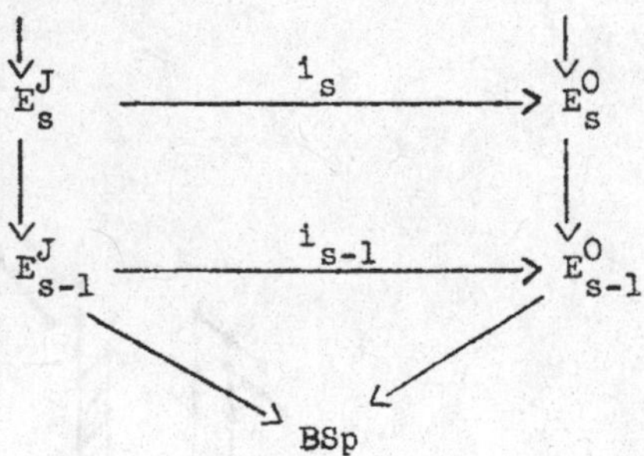

We return now to the proof of 3.2. As an aid in lifting the map $g: RP^{4k+2} \to BSp$ we factor it as $RP^{4k+2} \xrightarrow{h} QP^k \xrightarrow{g'} BSp$, where g' classifies the p-fold Whitney sum of the symplectic Hopf bundle H. The assumptions of 3.1, the main theorem of [7], which states that the bo-primary obstructions for pH are given by the binomial coefficients $\binom{p}{j}$, and the naturality methods of [9; P. 2] imply

(3.7) g' lifts to E^J_{4i-2}.

We would like to show $B^J_N \to E^J_{4i-3}$ is (4k+2)-principal, but unfortunately this may not be true in the presence of w_4.

3.8 <u>Definition</u>. Let $BSp\langle w_4\rangle$ denote the space obtained from BSp by killing the $\mathbb{Z}_2$-Stiefel-Whitney class w_4, and then killing the resulting $\mathbb{Z}_2$-cohomology in degrees 5,6, and 7. Let $\overline{B^J_N}$, $\overline{E^O_i}$, etc., denote the pullbacks over $BSp\langle w_4\rangle$ of B^J_N, E^O_i, etc.

We prove in Section 4

3.9 <u>Proposition</u> $\overline{B^J_N} \to \overline{E^J_{4i-3}}$ and $\overline{B^O_N} \to \overline{E^O_{4i-3}}$ are (4k+2)-principal.

g' lifts to $BSp\langle w_4\rangle$ since p is even, and hence it lifts to ℓ': $QP^k \to \overline{E^J_{4i-3}}$. The fibre F^J of $\overline{B^J_N} \to \overline{E^J_{4i-3}}$ is the space obtained from $P_N \wedge bJ$ by killing the Ext classes at height less than 4i-3 in our chart. We can compute the class in $[P^{4k+2}, \Sigma F^J]$ of the composite

$$P^{4k+2} \overset{h}{\to} QP^k \xrightarrow{\ell'} \overline{E^J_{4i-3}}^{(4k+2)} \overset{c}{\to} \Sigma F^J$$

by following it by $\Sigma F^J \overset{j}{\to} \Sigma P_N \wedge bJ$. We prove at the end of this section

3.10 <u>Theorem</u>. The class in $[P^{4k+2}, \Sigma P_N \wedge bJ]$ of $jc\ell'h$ equals the image of 2^{4i-2} times a generator of $[P^{4k+2}, P_N \wedge \Sigma^4 bsp]$.

Note that $[P^{4k+2}, P_N \wedge \Sigma^4 bsp] \to [P^{4k+2}, \Sigma P_N \wedge bJ]$ is an isomorphism. This follows from the exact sequence obtained by applying $[P^{4k+2},\ \ ]$ to

$$P_N \wedge bo \xrightarrow{\theta} P_N \wedge \Sigma^4 bsp \to \Sigma P_N \wedge bJ \xrightarrow{i} \Sigma P_N \wedge bo \qquad (3.11)$$

using Proposition 3.4 and the fact that $\theta_* = 0$ by the commutative diagram

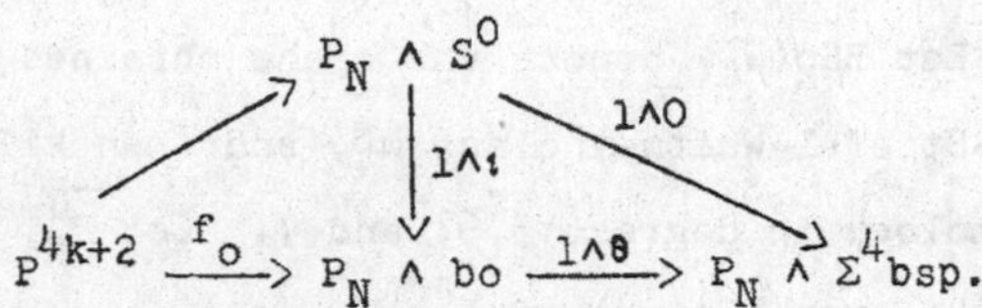

Thus 3.10 implies that $[c\ell'h] \neq 0$ and hence $\ell'h$ does not lift to $\overline{B_N^J}$.

Similarly it is easily shown that the composite

$$P^{4k+2} \xrightarrow{\ell'h} \overline{E^J_{4i-3}} \xrightarrow{i} \overline{E^0_{4i-3}} \to \Sigma F^0$$

is null homotopic. Thus there is a map ℓ: $P^{4k+2} \to \overline{B_N^0}$ which lifts $i\ell'h$. In Section 4 we use the fact that $\ell'h$ does not lift to $\overline{B_N^J}$ to prove

3.12 Theorem. ℓ does not lift to $\overline{B_N^J}$.

This is equivalent to (3.2).

(3.3) is proved by the techniques of [9; P. 4]. By 3.4(ii) $[f] = m[f_0]$ for some integer m. $\Sigma(c_1\mu(f\times\ell))$ is the homotopy sum of three maps (which we write without the Σ, since Σ: $[P^{4k+2}, P_N \wedge \Sigma^4 bsp] \to [\Sigma P^{4k+2}, \Sigma P_N \wedge \Sigma^4 bsp]$ is an isomorphism):

a) $P^{4k+2} \xrightarrow{f} P_N \wedge bo \xrightarrow{c_1 i = 1\wedge\theta} P_N \wedge \Sigma^4 bsp$

b) $P^{4k+2} \xrightarrow{\ell} B_N^0 \xrightarrow{c_1} P_N \wedge \Sigma^4 bsp$

c) $P^{4k+2} \xrightarrow{f\wedge\ell} (P_N \wedge bo) \wedge B_N^0 \xrightarrow{\Sigma^{-1}H(\mu)} B_N^0 \xrightarrow{c_1} P_N \wedge \Sigma^4 bsp$

(a) is trivial by the argument following 3.10, (b) is non-trivial by (3.2), while (c) is trivial because $N = 4k - 8i + 1$ and so $[P^{4k+2}, P_N \wedge bo \wedge B_N^0] \simeq [P^{4k+2}, P_N \wedge bo \wedge BSp^{[8i+1]}]$. But 3.1(ii) implies $4p\,\xi_{8i+1}$ is trivial (since $KO(P^{8i+1}) \simeq Z/2^{4i+1} Z$) and hence so is our map $P^{4k+2} \to BSp^{[8i+1]}$.

Remark 3.13 The mistake in [8] was to ignore (c), which can cancel (b) if 3.1(ii) is not present.

Proof of Theorem 3.10. It is well-known ([21]) that $KU(QP^m) \simeq Z[\gamma]/\gamma^{m+1}$ and it is not hard to see that $\psi^3(\gamma) = \gamma(3+\gamma)^2$. (This follows from the fact that when γ is pulled back to CP^{2m+1} one obtains $\eta + \eta^{-1} - 2$ and $\psi^3(\eta) = \eta^3$.) As in [1] or [4; 4.1] $ku^*(QP^m) \simeq Z[e,\pi]/e^{m+1}$ and $ku^*(RP^{2a}_{2b-1}) \simeq (A^b\mathbb{Z}[A,\pi])/(A^{a+1}, \pi A^2 + 2A)$, where $\deg A = 2$, $\deg e = 4$, and $\deg \pi = -2$. Under the isomorphism $ku^0(QP^m) \simeq \widetilde{KU}(QP^m)$ $\pi^2 e$ corresponds to γ, so $9\pi^2\psi^3 e = \psi^3\pi^2 e = \pi^2 e(3+\pi^2 e)^2$ and hence $\psi^3 e = e(1 + \frac{1}{3}\pi^2 e)^2$. By the Atiyah-Hirzebruch spectral sequence [21] and the Kunneth theorem [3]

$$ku^*(QP^m_{m'} \wedge RP^{2a}_{2b-1}) \simeq (A^b e^{m'} \mathbf{Z}[A,e,\pi])/(A^{a+1}, e^{m+1}, \pi A^2 + 2A).$$

In particular $ku^{2^L}(QP^k_{k-1} \wedge RP^{2^L-4k+8i-2}_{2^L-4k-1}) \simeq \mathbf{Z}_{2^{4i-2}} \oplus \mathbf{Z}_{2^{4i}}$ with generators $e^{k-1}A^{2^{L-1}-2(k-1)}$ and $e^k A^{2^{L-1}-2k}$ satisfying $(\psi^3-1)(e^{k-1}A^{2^{L-1}-2(k-1)}) = (3^{-(2^{L-1}-2(k-1))}-1)e^{k-1}A^{2^{L-1}-2(k-1)} + 2(k-1)\pi^2 e^k A^{2^{L-1}-2k} = 8(k-1)(e^{k-1}A^{2^{L-1}-2(k-1)} + e^k A^{2^{L-1}-2k})$ and $(\psi^3-1)(e^k A^{2^{L-1}-2k}) = 8k e^k A^{2^{L-1}-2k}$ (up to units in $\mathbf{Z}_{(2)}$).

Complexification $ko^{2^L}(QP^k_{k-1} \wedge RP^{2^L-4k+8i-2}_{2^L-4k-1}) \to ku^{2^L}(QP^k_{k-1} \wedge RP^{2^L-4k+8i-2}_{2^L-4k-1})$ is surjective and the first group is $\mathbb{Z}_{2^{4i-1}} \oplus \mathbb{Z}_{2^{4i}}$ with generators $\overline{G}_1$ and $\overline{G}_2$ satisfying $(\psi^3-1)\overline{G}_1 = 8(k-1)(\overline{G}_1 + \overline{G}_2)$ and $(\psi^3-1)\overline{G}_2 = 8k\overline{G}_2$. By [22] and [2] there are isomorphisms $[QP^k_{k-1}, \Sigma P_{4k-8i+1} \wedge bo] \approx [QP^k_{k-1}, \Sigma P^{4k}_{4k-8i+1} \wedge bo] \approx ko^{2^L}(QP^k_{k-1} \wedge P^{2^L-4k+8i-2}_{2^L-4k-1})$; we shall denote by G_i the elements in the first group corresponding to $\overline{G}_i$.

$\ker([QP^k_{k-1}, \Sigma P_{4k-8i+1} \wedge bo] \xrightarrow{\theta_*} [QP^k_{k-1}, \Sigma P_{4k-8i+1} \wedge \Sigma^4 bsp]) =$
$= \ker(\psi^3-1) \approx \mathbb{Z}_{2^{\nu+3}} \oplus \mathbb{Z}_8$ generated by $2^{4i-4-\nu}G_1 - 2^{4i-4}G_2$ and $2^{4i-3}G_2$, where $\nu = \nu(k-1)$. (The theory is easily modified if $\nu > 4i - 4$.)
By (3.11) $[QP^k_{k-1}, \Sigma P_{4k-8i+1} \wedge bJ] \approx \ker \theta_*$; let g_1 and g_2 be i_*^{-1} of the above generators.

In the commutative diagram of short exact sequences

$$\begin{array}{ccccccccc}
0 \to & [S^{4k}, \Sigma P_N \wedge bJ] & \to & [QP^k_{k-1}, \Sigma P_N \wedge bJ] & \to & [S^{4k-4}, \Sigma P_N \wedge bJ] & \to 0, \\
& \downarrow h^* & & \downarrow h^* & & \downarrow h^* & \\
0 \to & [P^{4k+2}_{4k-1}, \Sigma P_N \wedge bJ] & \to & [P^{4k+2}_{4k-5}, \Sigma P_N \wedge bJ] & \to & [P^{4k-2}_{4k-5}, \Sigma P_N \wedge bJ] & \to 0,
\end{array}$$

the groups are (excluding some $\mathbb{Z}_2$'s in the lower one)

$$\begin{array}{ccccc}
\mathbb{Z}_8 & \to & \mathbb{Z}_{2^{\nu+3}} \oplus \mathbb{Z}_8 & \to & \mathbb{Z}_{2^{\nu+3}} \\
\downarrow \phi_1 & & \downarrow h^* & & \downarrow \phi_2 \\
\mathbb{Z}_8 & \longrightarrow & \mathbb{Z}_{16} & \longrightarrow & \mathbb{Z}_2
\end{array}$$

with ϕ_1 and ϕ_2 surjective (by the Atiyah-Hirzebruch spectral sequence).

This implies $h^*(g_1) = g$, $h^*(g_2) = 2g$.

3.14 Lemma. In the diagram below $[j'i'\ell''] = 2^{4i-2}G_1 + 2^{4i-1}G_2$

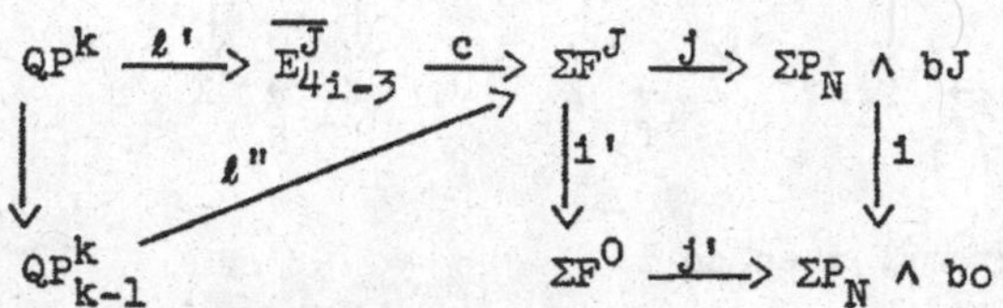

Proof. The proof is similar to that of (3.7). Since [7; 1.8,3.1] implies pH_{k-1} lifts to B_N^u but not to B_N^o, the coefficient of G_1 is 2^{4i-2}, while since pH_k lifts to B_{N+2}^u but not to B_N^u, the coefficient of G_2 is 2^{4i-1}. ∎

The proof of Theorem 3.10 follows from 3.14 and the preceding paragraphs by $[jc\ell'h] = h^*i_*^{-1}[j'i'\ell''] = h^*i_*^{-1}(2^{4i-2}G_1 + 2^{4i-1}G_2) = h^*(2^{\nu+2}g_1 + (1 - \delta_{\nu,1})4g_2) = 8g$, where $\delta_{\nu,1}$ is the usual Kronecker delta.

4. PROOF OF 3.6, 3.9, and 3.12

<u>Lemma 4.1</u>. If
$$\begin{array}{ccc} X_1 & \rightarrow & PK_1 \\ \downarrow p_1 & & \downarrow \\ X & \underset{f_1}{\rightarrow} & K_1 \end{array} \qquad \begin{array}{ccc} X_2 & \rightarrow & PK_2 \\ \downarrow p_2 & & \downarrow \\ X & \underset{f_2}{\rightarrow} & K_2 \end{array} \quad , \text{ and } \quad \begin{array}{ccc} X_3 & \overset{q_2}{\rightarrow} & X_2 \\ \downarrow q_1 & & \downarrow p_2 \\ X_1 & \underset{p_1}{\rightarrow} & X \end{array}$$

are pullback diagrams, then so is

$$\begin{array}{ccc} X_3 & \rightarrow & P(K_1 \times K_2) \\ \downarrow & & \downarrow \\ X & \rightarrow & K_1 \times K_2 \end{array} \quad .$$

<u>Lemma 4.2</u>. If $F \rightarrow E \rightarrow B$ is a fibration such that $H^q(F; \mathbb{Z}_2)$ is transgressive for $q \leq t \leq$ twice the connectivity of F, and $B \rightarrow K$ is a map into a generalized Eilenberg-MacLane space corresponding to the transgressions of a set of $\mathcal{a}$-generators of $H^*(F;\mathbb{Z}_2)$ through dimension t, and $E_1 = \text{fibre}(B \rightarrow K)$, then $E \rightarrow E_1$ is surjective in cohomology through dimension t and $F_1 \rightarrow E \rightarrow E_1$ is t-totally-transgressive (in the sense of [11]) so that we can form a t-MPT for it.

<u>Proof</u>. Surjectivity is proved by considering the map of Serre spectral sequences induced by

$$\begin{array}{ccc} F & \rightarrow & \Omega K \\ \downarrow & & \downarrow \\ E & \rightarrow & E_1 \\ \downarrow & & \downarrow \\ B & \rightarrow & B \end{array}$$

$F_1 \rightarrow E \rightarrow E_1$ is shown to be transgressive as in [12]. ∎

Proof of Theorem 3.6. Consider the diagram

$$
\begin{array}{cccccccccccc}
 & & & \tilde{K}_{i+1} & & \tilde{K}_i & & \tilde{K}_2 & & \tilde{K}_1 & & \tilde{K}_o \\
 & & a_{i+1} & \nearrow & a_i & \nearrow & a_2 & \nearrow & a_1 & \nearrow & a_o & \nearrow \\
B_N^J \to \tilde{E}_{r-\Delta} & \to \cdots \to & & \tilde{E}_{i+1} & \to & \tilde{E}_i & \to \cdots \to & \tilde{E}_2 & \to & \tilde{E}_1 & \to & B_N^o \\
\downarrow & & & \downarrow & & \downarrow & & \downarrow & & \downarrow & & \downarrow \\
E_{r,r-\Delta} & \to \cdots \to & & E_{r,i+1} & \to & E_{r,i} & \to \cdots \to & E_{r,2} & \to & E_{r,1} & \to & E_r^o \\
\downarrow & & & \downarrow & & \downarrow & & \downarrow & & \downarrow & & \downarrow \\
E_{r-1,r-\Delta} & \to \cdots \to & & \vdots & & \vdots & & \vdots & & \vdots & & \vdots \to K_{r-1} \\
\downarrow & & & \downarrow & & \downarrow & & \downarrow & & \downarrow & & \downarrow \\
 & & & E_{i+1,i+1} & \to & E_{i+1,i} & \to \cdots \to & E_{i+1,2} & \to & E_{i+1,1} & \to & E_{i+1}^o \to K_{i+1} \\
 & & & \tilde{K}_i \swarrow & & \downarrow & & \downarrow & & \downarrow & & \downarrow \\
 & & & & & E_{i,i} & \to \cdots \to & E_{i,2} & \to & E_{i,1} & \to & E_i^o \to K_i \\
 & & & & & \tilde{K}_{i-1} \swarrow & & \downarrow & & \downarrow & & \downarrow \\
 & & & & & & & \vdots & & \vdots & & \vdots \\
 & & & & & & & \downarrow & & \downarrow & & \downarrow \\
 & & & & & & \to & E_{3,2} & \to & E_{3,1} & \to & E_3^o \to K_3 \\
 & & & & & & \tilde{K}_2 \swarrow & \downarrow & & \downarrow & & \downarrow \\
 & & & & & & & E_{2,2} & \to & E_{2,1} & \to & E_2^o \to K_2 \\
 & & & & & & & & \tilde{K}_1 \swarrow & \downarrow & & \downarrow \\
 & & & & & & & & & E_{1,1} & \to & E_1^o \to K_1 \\
 & & & & & & & & & & \tilde{K}_o \swarrow & \downarrow \\
 & & & & & & & & & & & BSp \to K_o
\end{array}
$$

The top and right sides are ordinary (based on Ext) t-MPT's for $B_N^J \to B_N^O$ and $B_N^O \to BSp$. These exist by Lemma 4.2; the cohomology is transgressive since the fibrations are induced.

$E_{i,i} = \text{fibre}(E_{i,i-1} \to K_{i-1})$. All squares

$$\begin{array}{ccc} E_{i,j} & \to & E_{i,j-1} \\ \downarrow & & \downarrow \\ E_{i-1,j} & \to & E_{i-1,j-1} \end{array}$$

are pullbacks. The maps $\tilde{E}_i \to E_{i,i}$ exist since $\tilde{E}_i \to \tilde{E}_{i-1} \to E_{i,i-1} \to \tilde{K}_{i-1}$ is trivial. The maps $\tilde{E}_i \xrightarrow{p_{i+k,i}} E_{i+k,i}$ exist by the universal property of pullbacks. We show the existence of the map $b_i\colon E_{i+1,i} \to \tilde{K}_i$ such that $b_i p_{i+1,i} = a_i$ as follows: The vertical maps

$$\begin{array}{ccccccccc} \tilde{E}_i & \to & \tilde{E}_{i-1} & \to & \tilde{E}_{i-2} & \to \cdots \to & \tilde{E}_1 & \to & B_N^O \\ \downarrow{\scriptstyle p_{i,i}} & & \downarrow{\scriptstyle p_{i,i-1}} & & \downarrow & & \downarrow{\scriptstyle p_{i,1}} & & \downarrow \\ E_{i,i} & \to & E_{i,i-1} & \to & E_{i,i-2} & \to \cdots \to & E_{i,1} & \to & E_i^O \end{array}$$

have homotopy equivalent fibres because $\tilde{E}_j \to \tilde{E}_{j-1}$ and $E_{i,j} \to E_{i,j-1}$ are both induced by maps into $\tilde{K}_{j-1}$. Thus the fibre of $p_{i,i}$ is transgressive and $E_{i,i} \to K_i$ kills the image of the transgressions of $H^*(\text{fibre}(p_{i,i}))$. Hence by Lemma 4.2 $p^*_{i+1,i}$ is surjective, and thus we can form b_i.

By Lemma 4.1 $E_{i+1,i} = \text{fibre}(E_{i,i-1} \to K_i \times \tilde{K}_{i-1})$. Then $E_1^J = E_1^O$, $E_i^J = E_{i,i-1}$ $(2 \leq i \leq r-\Delta+1)$, $B_i^J = E_{i,r-\Delta}$ $(r-\Delta+1 \leq i \leq r)$ provides the desired resolution.

Proof of Theorem 3.9. Let $F = F^J = \text{fibre}(\overline{B_N^J} \to \overline{E_{4i-3}^J})$. $\text{Ext}_{\mathcal{A}}(F)$ is given by the chart

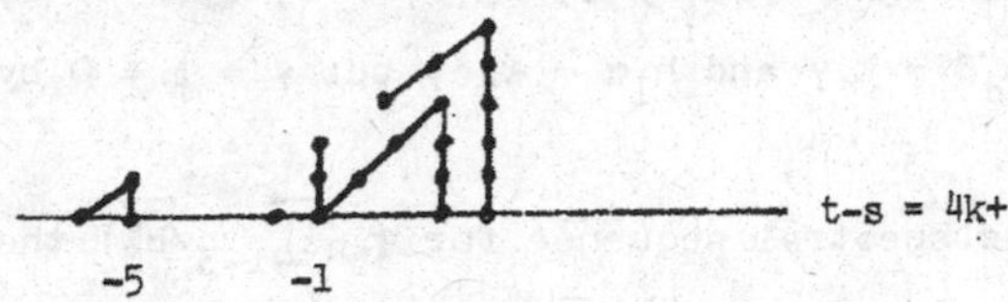

The relative Serre spectral sequence $(CF,F) \to (\overline{E_{4i-3}^J},\overline{B_N^J}) \to \overline{E_{4i-3}^J}$ is trivial in the stable range. By [15; Ch. 7] the $\mathcal{A}$-module $H^*(\overline{E_{4i-3}^J},\overline{B_N^J})$ is an extension of $\tilde{H}^*(\Sigma F)$, $\tilde{H}^*(\Sigma^5 F)$, and $\tilde{H}^*(\Sigma^9 F)$ through degree $4k+6$. The boundary homomorphisms in the $\text{Ext}_{\mathcal{A}}$-sequences must be zero, and so $\text{Ext}_{\mathcal{A}}(\overline{E_{4i-3}^J}/\overline{B_N^J})$ is (up to possible exotic h_0- or h_1-extensions)

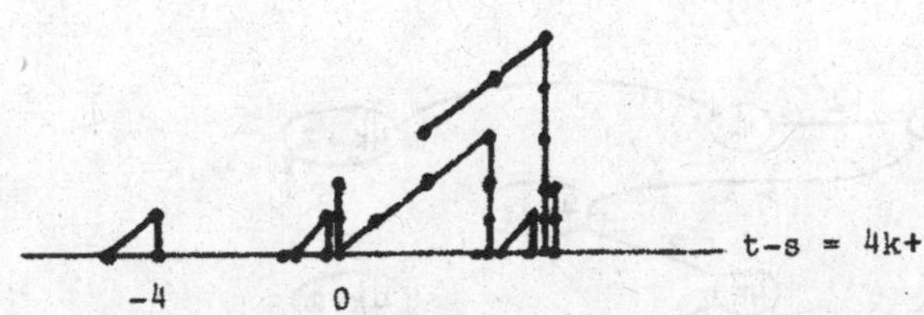

We now show the above chart is accurate by showing there exist $\alpha \in \text{Ext}^{0,4k}$, $\beta \in \text{Ext}^{0,4k-1}$ such that $h_0\alpha = h_1\beta$ and $h_1\alpha = 0$. This follows

from the commutative diagram of exact sequences (for $t - s \leq 4k + 2$)

$$\begin{array}{ccccc}
0 \to \mathrm{Ext}(\Sigma F) & \xrightarrow{i_*} & \mathrm{Ext}(\overline{E^J_{4i-3}}/\overline{B^J_N}) & \xrightarrow{k_*} & \mathrm{Ext}(\overline{E^J_{4i-3}}/\overline{B^J_N}/\Sigma F) \\
\downarrow 1 & & \downarrow q_* & & \downarrow 0 \\
0 \to \mathrm{Ext}(\Sigma F) & \xrightarrow{j_*} & \mathrm{Ext}(E^J_{4i-3}/B^J_N) & \longrightarrow & \mathrm{Ext}(E^J_{4i-3}/B^J_N/\Sigma F)
\end{array}$$

by choosing α, β so that $k_*(\alpha) \neq 0$, $k_*(\beta) \neq 0$, $q_*(\alpha) = 0$, $q_*(\beta) = 0$. for then $h_0\alpha + h_1\beta = k_*\gamma$ and $h_1\alpha = k_*\delta$, but $\gamma = \delta = 0$ by diagram chasing.

In the Adams spectral sequence for $\pi_*(\overline{E^J_{4i-3}}/\overline{B^J_N})$ the first possible nonzero Adams differential might hit the element in $s = 2$, $t - s = 4k + 2$.

We will construct a map $\Sigma^5 F^{(4k+2)} \xrightarrow{h} \overline{E^J_{4i-3}}/\overline{B^J_N}$ which induces an injection of homotopy groups through degree $4k + 1$. The composite $CF/F \to \overline{E^J_{4i-3}}/\overline{B^J_N} \to \mathrm{cofibre}(h)$ induces an isomorphism in π_i for $i \leq 4k + 1$ and is epic in π_{4k+2}. Thus there is a map $\mathrm{cofibre}(h)^{(4k+2)} \to CF/F$ such that it followed by the above composite is homotopic to the inclusion. Then $\overline{E^J_{4i-3}}^{(4k+2)} \to (\overline{E^J_{4i-3}}/\overline{B^J_N})^{(4k+2)}$ $\mathrm{cofibre}(h)^{(4k+2)} \to \Sigma F$ induces $\overline{B^J_N}$ through dimension $4k + 2$ by the type of argument used in proving [7; 2.2(i)].

The cell structure of $(\Sigma^5 F)^{(4k+2)}$ is depicted below

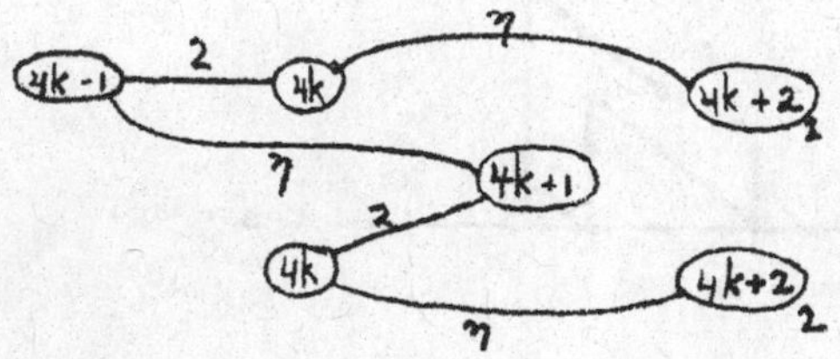

Let $f = f_1 \vee f_2\colon S^{4k-1} \vee S^{4k} \to \overline{E^J_{4i-3}}/\overline{B^J_N}$ be a map nontrivial in H^{4k-1} and H^{4k} such that $2[f_2] = [f_1 \circ \eta]$. f can be extended over $(\Sigma^5 F)^{(4k+1)}$. Since h_1 times the class in Ext^0 corresponding to f_2^* is zero, and since $\pi_{4k+1}(\overline{E^J_{4i-3}}/\overline{B^J_N})$ contains no elements of filtration greater than 1, $f_2 \circ \eta = 0$, so the map can be extended over $(4k+2)_2$. The extension over $(4k+2)_1$ exists similarly because its attaching map has filtration 1 and is mapped trivially in

$$\mathrm{Ext}_{\mathcal{Q}}(\Sigma^5 F^{(4k+1)}) \to \mathrm{Ext}_{\mathcal{Q}}(\overline{E^J_{4i-3}}/\overline{B^J_N}).$$

The proof for $\overline{B^o_N}/\overline{E^o_{4i-3}}$ is similar but easier. ∎

Proof of 3.12. As in [9; 2.2] let $X\langle 0,i\rangle$ denote a space whose 2-primary stable homotopy groups are those obtained by considering the elements of the Adams spectral sequence (or when $X = P_N \wedge bJ$ the chart 3.5) of X of filtration $\leq i$. We form the diagram

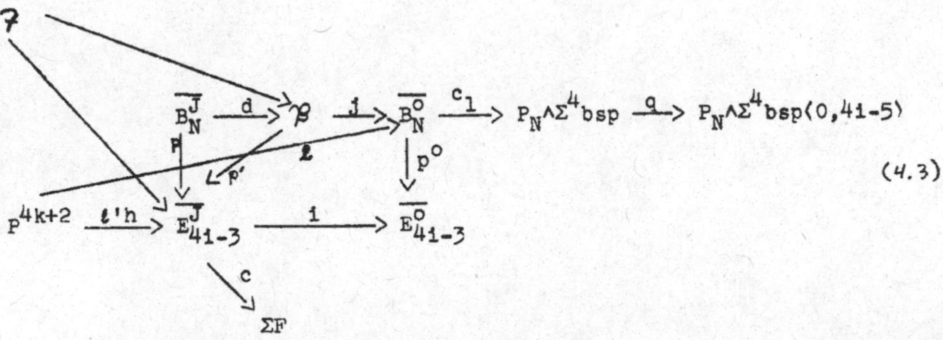

(4.3)

where $\mathcal{B}$ is the pullback of i and p^o, and
$\mathcal{F}$ = fibre(j) = fibre(i) = fibre($P_N \wedge bJ\langle 0,4i-4\rangle \to P_N \wedge bo\langle 0,4i-4\rangle$) = $P_N \wedge \Sigma^3 bsp\langle 0,4i-5\rangle$. There is a unique $b\colon P^{4k+2} \to \mathcal{B}$ such that $jb = \ell$ and $p'b = \ell' h$. Suppose there exists $\ell''\colon P^{4k+2} \to \overline{B^J_N}$ such that $jd\ell'' = \ell$. Then since j is induced by qc_1, there is $D\colon P^{4k+2} \to \mathcal{F}$ such that the composite $P^{4k+2} \xrightarrow{b \times} \mathcal{B} \times \mathcal{F} \xrightarrow{\mu} \mathcal{B}$ equals $d\ell''$.

However, $cp'd\ell'' = cp\ell''$ is trivial, so the existence of ℓ'' will be contradicted once we show $cp'\mu(b \times D)$ is nontrivial.

$[P^{4k+2}, \mathcal{F}]$ is a sum of filtration zero $\mathbb{Z}_2$'s. Thus to show that D cannot cancel our class it is convenient to try to consider $[c\ell'h]$ to be of order > 2. This is accomplished by noting that if N is replaced by N - 2 one obtains a diagram analogous to 4.3 and a map of diagrams. We denote by primes (') the analogous spaces and maps. $[P^{4k+2}, \mathcal{F}']$ is a sum of $\mathbb{Z}_2$'s mapping onto $[P^{4k+2}, \mathcal{F}]$. But $c\ell'h$ factors through $c'\ell''h$, a class of order 4. If the reader has made it this far, he can undoubtedly fill in the remaining details to deduce that D cannot cancel our class. ∎

REFERENCES

1. J. F. Adams, "Vector fields on spheres," Ann. of Math. 75(1962) 603-632.

2. M. F. Atiyah, "Thom complexes," Proc. London Math. Soc. 11(1961) 291-310.

3. M. F. Atiyah, "Vector bundles and the Kunneth formula," Topology 1 (1962) 245-248.

4. D. M. Davis, "Generalized homology and the generalized vector field problem," Quar. Jour. Math Oxford 25(1974) 169-193.

5. D. M. Davis, "The cohomology of the spectrum bJ," Bol. Soc. Mat. Mex. 1976.

6. D. M. Davis, "The BP-coaction for projective spaces," to appear.

7. D. M. Davis and M. Mahowald, "The Geometric dimension of some vector bundles over projective spaces," Trans. Amer. Math. Soc. 205(1975) 295-315.

8. D. M. Davis and M. Mahowald, "A strong nonimmersion theorem for $RP^{8\ell+7}$," Bull. Amer. Math. Soc. 81(1975) 155-156.

9. D. M. Davis and M. Mahowald, "The immersion conjecture is false," to appear.

10. S. Gitler, K. Y. Lam, and M. Mahowald, to appear.

11. S. Gitler and M. Mahowald, "The geometric dimension of real stable vector bundles," Biol. Soc. Mat. Mex. 11(1966) 85-107.

12. ______________________, Addendum, 12(1967) 32-34.

13. S. Gitler, M. Mahowald, and R. J. Milgram, "The nonimmersion problem for RP^n and higher-order cohomology operations," Proc. Nat. Acad. Sci. U. S. A. 60(1968), 432-437.

14. I. M. James, "On the immersion problem for real projective spaces," Bull. Amer. Math. Soc. 69(1963), 231-238.

15. L. Kristensen, "On the cohomology of 2-stage Postnikov systems," Acta Math. 107(1962), 73-123.

16. M. Mahowald, "The metastable homotopy of S^n," Memoirs Amer. Math. Soc. 72(1967).

17. M. Mahowald, "The order of the image of the J-homomorphism," Bull. Amer. Math. Soc. 76(1970), 1310-1313.

18. M. Mahowald and R. J. Milgram, "Operations which detect Sq^4 in connective K-theory and their applications, to appear.

19. M. Mahowald and R. Rigdon, "Obstruction theory with coefficients in a spectrum," Trans. Amer. Math. Soc. 204 (1975) 365-384.

20. R. J. Milgram, "The Steenrod algebra and its dual for connective K-theory," Notas de Matematicas y Simposia, 1(1975) Soc. Mat. Mex. 127-158.

21. B. Sanderson, "Immersions and embeddings of projective spaces," Proc. London Math. Soc. 53(1964), 137-153.

22. G. W. Whitehead, "Generalized homology theories," Trans. Amer. Math. Soc. 102(1962), 227-283.

The tame homotopy groups of a suspension

by

W. G. Dwyer*

The purpose of this note is to show that the machinery of [1] leads directly to a formula for some of the homotopy groups of a suspension.

For each $k \geq 0$, let S_k be the smallest subring of $\mathbb{Q}$ containing $1/p$ for each prime p such that $2p - 3 \leq k$. Let $r \geq 3$ be a fixed positive integer. If X is an $(r-1)$-connected CW-complex, let $C_*'(X)$ denote the natural chain complex which agrees with the integral cellular chain complex $C_*(X)$ in dimensions greater than r, is zero in dimensions less than r, and in dimension r itself contains kernel $(\partial: C_r(X) \to C_{r-1}(X))$.

<u>Theorem</u>: <u>Suppose that</u> X <u>is an</u> $(r-1)$-<u>connected</u> CW-<u>complex which has the homotopy type of a suspension</u>. <u>Then for all</u> $k \geq 0$ <u>there are isomorphisms</u>

$$(\pi_{r+k}X) \otimes S_k \cong H_{r+k-1}(L^g s^{-1} C_*'(X)) \otimes S_k$$

Here s^{-1} is the functor which shifts all of the groups in a chain complex down by one in dimension. L^g is the free differential graded Lie algebra functor, that is, the left adjoint to the forgetful functor from the category of differential graded Lie algebras over $\mathbb{Z}$ [1] to the category of chain complexes.

*Partially supported by NSF grant #MCS76-08795.

Remark: Since two free chain complexes over $\mathbb{Z}$ with the same homology groups are chain homotopy equivalent, the argument in Lemma 5.2 of [1] shows that the complex $C_*'(X)$ above can be replaced by any free chain complex over $\mathbb{Z}$ which vanishes below dimension r and has the same homology groups as X.

Remark. The isomorphisms in the theorem are not natural with respect to arbitrary cellular maps.

Proof of Theorem: Let Y be a pointed CW-complex such that the suspension of Y is homotopy equivalent to X and let $K = E_r(\mathrm{Sing}(Y))$ be the r'th Eilenberg subcomplex of the singular complex of Y [4, p. 237]. Let J be the simplicial suspension of K [2, p. 311]. Then K has no non-basepoint simplices below dimension $r - 1$, the geometric realization of K is homotopy equivalent to Y, and the geometric realization of J is homotopy equivalent to X. By the argument of [1, Lemma 5.2] (see the first remark above) it is enough to show that there are isomorphisms

$$(\pi_{r+k}J) \otimes S_k \approx H_{r+k-1}(L^g\tilde{C}_*(K)) \otimes S_k$$

where $\tilde{C}_*(K)$ is the reduced normalized integral simplicial chain complex of K.

At this point we need some auxiliary functors. In general, we will use the same notation for a functor defined on some category C and its prolongation to the category of simplicial objects over C.

G will denote Kan's loop group functor [2]. F is the free group functor from the category of pointed sets to the category of groups, and $\mathbb{Z}\otimes$- the analogous functor with its range in the category of abelian groups. Log is the left adjoint to the exponential functor from the category of Lazard algebras to the category of groups and U is the left adjoint to the forgetful functor from the category of

Lazard algebras to the category of Lie algebras. (A Lazard algebra [1, §7] is a Lie algebra with enough divisibility and completeness so that the Campbell-Hausdorff formula makes sense.) L denotes the free Lie algebra functor from the category of abelian groups to that of Lie algebras. N is the normalization functor from the category of simplicial abelian groups to the category of chain complexes and N^{-1} is its inverse. N* is the left adjoint to the composite of N with the forgetful functor from the category of simplicial Lie algebras to the category of simplicial abelian groups (See [4], [1] and [3].)

The isomorphisms

$$\mathrm{Log} \cdot F \approx U \cdot L \cdot (\mathbb{Z} \otimes -)$$

and

$$L \cdot N^{-1} \approx N^* \cdot L^g$$

follow from the easily verified adjoint identities. Note also that there are isomorphisms

$$\tilde{C}_*(K) \approx N(\mathbb{Z} \otimes K)$$

and

$$G(J) \approx F(K).$$

The first is by definition; the second expresses the fact that the G-construction is a generalization of Milnor's construction "FK" [2, p. 311].

To prove the theorem, note that it follows from [1, 6.1] that there are isomorphisms

$$(\pi_{r+k}J) \otimes S_k \approx \pi_{r+k-1}(\mathrm{Log}\ GJ) \otimes S_k.$$

The identities above give isomorphisms

$$\begin{aligned} \mathrm{Log}\ GJ &\approx \mathrm{Log}\ FK \\ &\approx UL(\mathbb{Z} \otimes K) \\ &\approx ULN^{-1}N(\mathbb{Z} \otimes K) \\ &\approx UN^*L^gN(\mathbb{Z} \otimes K). \end{aligned}$$

Finally, [1, 6.2] provides additional isomorphisms

$$\pi_{r+k-1}(UN^*L^gN(\mathbb{Z} \otimes K)) \otimes S_k \approx H_{r+k-1}(L^gN(\mathbb{Z} \otimes K)) \otimes S_k.$$

Remark: A closer examination of the above proof shows a little bit more, namely, that the differential graded Lie algebra $L^g s^{-1} C'_*(X)$ can be taken as a model in the sense of [1] for the tame homotopy type of X. Consequently, by the first remark above, the abstract homology groups of the space X determine its tame homotopy type. It follows that from the point of view of tame homotopy theory any 2-connected suspension is equivalent to a wedge of Moore spaces.

References

[1] W. G. Dwyer, Tame homotopy theory, to appear.

[2] D. M. Kan, A combinatorial definition of homotopy groups, Ann. of Math. 67 (1958), pp. 282-312.

[3] J. P. May, Simplicial Objects in Algebraic Topology, Van Nostrand, Princeton, 1967.

[4] D. G. Quillen, Rational homotopy theory, Ann. of Math. 90 (1969), pp. 205-295.

Yale University
New Haven, Connecticut 06520

SOME TABLES FOR FORMAL GROUPS AND BP

V. Giambalvo

University of Connecticut
Storrs, Connecticut 06268

The following is a collection of formulas describing some of the power series associated with the formal groups for BP, and the structure maps for $BP_*(BP)$ as a comodule over BP_*, for the primes 2 and 3. The calculations were carried out on an IBM 360 using PL/1 and FORMAC. Details of the programs used and of possible extensions may be obtained from the author.

The task of constructing these tables was suggested by Steve Wilson. The actual computations were carried out in, and with support from, the University of Connecticut Computer Center.

To each formal group $F(X,Y) = X + Y + \sum_{i,j>1} a_{ij}X^iY^j$ there is an associated power series $\log(X) = \sum_{i>0} \ell_i X^i$ and its inverse $\exp(X)$ such that $\exp(\log(X)) = X$ and $F(X,Y) = \exp(\log(X) + \log(Y))$. We will be concerned with the formal group for BP, given by $\log(X) = \sum_{i>0} m_i X^{p^i}$, $m_0 = 1$, for $p = 2$ and 3.

Part 1 deals with this formal group. The first table computes $\exp(X)$ in terms of the coefficients of the log series. The second table gives the series for the homomorphism $[p](X)=\exp(p \log(X))$ both in terms of the m_i and in terms of the Hazewinkel generators v_i. The v_i and m_i are related by the formula $pm_n = v_n + \Sigma m_i v_{n-i}^{p^i}$. Finally some of the coefficients a_{ij} of the series $F(X,Y)$ are given, both in terms of m_i and v_i. Note that $a_{ij} = a_{ji}$.

Part 2 of the tables computes the structure maps for $BP_*(BP)$. Recall $BP_* = Z_{(p)}[v_1,v_2,\ldots]$ and $BP_*(BP) = BP_*[t_1,t_2,\ldots]$. The structure maps are the co-unit $\eta_R\colon BP_* \to BP_*(BP)$ given by

$$\eta_R(m_n) = \sum_{0\leq i\leq n} m_i t_{n-i}^{p^i},$$

the conjugation $c\colon BP_*(BP) \to BP_*(BP)$ given by

$$\sum_{h+i+j=k} m_k t_i^{p^h} c(t_j)^{p^{h+i}} = m_k$$

and the coproduct $\psi\colon BP_*(BP) \to BP_*(BP)$ given by

$$\sum_{i+j=k} m_i \psi(t_j)^{p^i} = \sum_{h+i+j=k} m_h t_i^{p^h} \otimes t_j^{p^{h+i}} .$$

Actually $\eta_R(v_i)$, $c(t_i)$ and $\psi(t_i)$ are computed up through the first case which is not easy $(i \leq 3)$. It would not be difficult to compute more, but the number of terms involved make it seem unlikely to be useful. However modification to the programs to select specific terms would not be difficult.

PART 1: $\exp(X) \bmod X^{17}$ for $p = 2$

$$\begin{aligned}
\exp(X) = {} & X - m_1X^2 + 2m_1^2X^3 - (m_2 + 5m_1^3)X^4 + (6m_1m_2 + 14m_1^4)X^5 - (28m_1^2m_2 - 42m_1^5)X^6 \\
& + (120m_1^3m_2 + 132m_1^6 + 4m_2^2)X^7 - (m_3 + 45m_1m_2^2 + 495m_1^4m_2 + 429m_1^7)X^8 \\
& + (10m_1m_3 + 330m_1^2m_2^2 + 2002m_1^5m_2 + 1430m_1^8)X^9 \\
& - (66m_1^2m_3 + 2002m_1^3m_2^2 + 8008m_1^6m_2 + 4862m_1^9 + 22m_2^3)X^{10} \\
& + (364m_1m_2^3 + 12m_2m_3 + 364m_1^3m_3 + 10920m_1^4m_2^2 + 31824m_1^7m_2 + 16796m_1^{10})X^{11} \\
& - (182m_1m_2m_3 + 3640m_1^2m_2^3 + 1820m_1^4m_3 + 55692m_1^5m_2^2 + 125970m_1^8m_2 + 58786m_1^{11})X^{12} \\
& + (1680m_1^2m_2m_3 + 28560m_1^3m_2^3 + 8568m_1^5m_3 + 271320m_1^6m_2^2 + 497420m_1^9m_2 \\
& \qquad + 208012m_1^{12} + 140m_2^4)X^{13} \\
& - (3060m_1m_2^4 + 12240m_1^3m_2m_3 + 193800m_1^4m_2^3 + 38760m_1^6m_3 + 1279080m_1^7m_2^2 \\
& \qquad + 1961256m_1^{10}m_2 + 120m_2^2m_3 + 742900m_1^{13})X^{14} \\
& + (2448m_1m_2^2m_3 + 38760m_1^2m_2^4 + 77520m_1^4m_2m_3 + 1193808m_1^5m_2^3 + 170544m_1^7m_3 \\
& \qquad + 5883768m_1^8m_2^2 + 7726160m_1^{11}m_2 + 2674440m_1^{14} + 8m_3^2)X^{15} \\
& - (m_4 + 153m_1m_3^2 + 29070m_1^2m_2^2m_3 + 373065m_1^3m_2^4 + 447678m_1^5m_2m_3 + 6864396m_1^6m_2^3 \\
& \qquad + 735471m_1^8m_3 + 26558675m_1^9m_2^2 + 30421775m_1^{12}m_2 + 9694845m_1^{15} \\
& \qquad + 969m_2^5)X^{16}
\end{aligned}$$

$\exp(X) \bmod X^{28}$ for $p = 3$

$$\begin{aligned}
\exp(X) = {} & X - m_1X^3 + 3m_1^2X^5 - 12m_1^3X^7 - (m_2 - 55m_1^4)X^9 + (12m_1m_2 - 273m_1^5)X^{11} \\
& - (105m_1^2m_2 - 1428m_1^6)X^{13} + (816m_1^3m_2 - 7752m_1^7)X^{15} \\
& - (5985m_1^4m_2 - 43263m_1^8 - 9m_2^2)X^{17} - (210m_1m_2^2 - 42504m_1^5m_2 + 246675m_1^9)X^{19} \\
& + (3036m_1^2m_2^2 - 296010m_1^6m_2 + 1430715m_1^{10})X^{21}
\end{aligned}$$

$$- (35100m_1^3m_2^2 - 2035800m_1^7m_2 + 8414640m_1^{11})X^{23}$$
$$+ (356265m_1^4m_2^2 - 13884156m_1^8m_2 + 50067108m_1^{12} - 117m_2^3)X^{25}$$
$$- (m_3 - 4060m_1m_2^3 + 3322704m_1^5m_2^2 - 94143280m_1^9m_2 + 300830572m_1^{13})X^{27}$$

$[2](X)$ for $p = 2 \bmod X^{17}$ in terms of the m_i

$$[2](X) = 2X - 2m_1X^2 + 8m_1^2X^3 - (14m_2 + 36m_1^3)X^4 + (120m_1m_2 + 176m_1^4)X^5$$
$$- (888m_1^2m_2 + 912m_1^5)X^6 + (6240m_1^3m_2 + 4928m_1^6 + 448m_2^2)X^7$$
$$- (254m_3 + 7172m_1m_2^2 + 42848m_1^4m_2 + 27472m_1^7)X^8$$
$$+ (3064m_1m_3 + 80496m_1^2m_2^2 + 290816m_1^5m_2 + 156864m_1^8)X^9$$
$$- (28632m_1^2m_3 + 775024m_1^3m_2^2 + 1961472m_1^6m_2 + 912832m_1^9 + 19040m_2^3)X^{10}$$
$$+ (451264m_1m_2^3 + 22464m_2m_3 + 239392m_1^3m_3 + 6850240m_1^4m_2^2 + 13183744m_1^7m_2 + 5394176m_1^{10})X^{11}$$
$$- (442632m_1m_2m_3 + 6814928m_1^2m_2^3 + 1882320m_1^4m_3 + 57356064m_1^5m_2^2 + 88443968m_1^8m_2 + 32282240m_1^{11})X^{12}$$
$$+ (5811360m_1^2m_2m_3 + 83369280m_1^3m_2^3 + 14245952m_1^5m_3 + 462765184m_1^6m_2^2 + 592746752m_1^9m_2 + 195264000m_1^{12} + 932288m_2^4)X^{13}$$
$$- (29319872m_1m_2^4 + 63378208m_1^3m_2m_3 + 900627520m_1^4m_2^3 + 105098688m_1^6m_3 + 3635159424m_1^7m_2^2 + 3970848000m_1^{10}m_2 + 1699520m_2^2m_3 + 1191825920m_1^{13})X^{14}$$
$$+ (46522752m_1m_2^2m_3 + 556855040m_1^2m_2^4 + 620457600m_1^4m_2m_3 + 8960342272m_1^5m_2^3 + 7331457024m_1^{14} + 761561344m_1^7m_3 + 27987144192m_1^8m_2^2 + 26598675456m_1^{11}m_2 + 260096m_3^2)X^{15}$$
$$- (65534m_4 + 6063108m_1m_3^2 + 789099408m_1^2m_2^2m_3 + 8262562000m_1^3m_2^4 + 5659676160m_1^5m_2m_3 + 84080594304m_1^6m_2^3 + 5446592576m_1^8m_3 + 212145331840m_1^9m_2^2 + 178193961216m_1^{12}m_2 + 45406194944m_1^{15} + 49590800m_2^5)X^{16}$$

$[2](X)$ for $p = 2 \bmod X^{17}$ in terms of the v_i

$$[2](X) = 2X - v_1X^2 + 2v_1^2X^3 - (7v_2 + 8v_1^3)X^4 + (30v_1v_2 + 26v_1^4)X^5$$
$$- (111v_1^2v_2 + 84v_1^5)X^6 + (502v_1^3v_2 + 300v_1^6 + 112v_2^2)X^7$$

$$- (127v_3 + 960v_1v_2^2 + 2299v_1^4v_2 + 1140v_1^7)X^8$$
$$+ (766v_1v_3 + 5414v_1^2v_2^2 + 9958v_1^5v_2 + 4334v_1^8)X^9$$
$$- (3579v_1^2v_3 + 29579v_1^3v_2^2 + 43118v_1^6v_2 + 16692v_1^9 + 2380v_2^3)X^{10}$$
$$+ (31012v_1v_2^3 + 5616v_2v_3 + 17770v_1^3v_3 + 161034v_1^4v_2^2 + 189976v_1^7v_2$$
$$+ 65744v_1^{10})X^{11}$$
$$- (55329v_1v_2v_3 + 240631v_1^2v_2^3 + 86487v_1^4v_3 + 838452v_1^5v_2^2 + 837637v_1^8v_2$$
$$+ 262400v_1^{11})X^{12}$$
$$+ (363210v_1^2v_2v_3 + 1600786v_1^3v_2^3 + 404198v_1^5v_3 + 4232750v_1^6v_2^2 + 368550v_1^9v_2$$
$$+ 1056540v_1^{12} + 58268v_2^4)X^{13}$$
$$- (1022466v_1v_2^4 + 2193009v_1^3v_2v_3 + 10071369v_1^4v_2^3 + 1864478v_1^6v_3$$
$$+ 21110372v_1^7v_2^2 + 1625450v_1^{10}v_2 + 212440v_2^2v_3 + 4292816v_1^{13})X^{14}$$
$$+ (2972696v_1v_2^2v_3 + 10170952v_1^2v_2^4 + 12667346v_1^4v_2v_3 + 60190566v_1^5v_2^3$$
$$+ 8581604v_1^7v_3 + 104219628v_1^8v_2^2 + 71867828v_1^{11}v_2 + 17587492v_1^{14}$$
$$+ 65024v_3^2)X^{15}$$
$$- (32767v_4 + 774272v_1v_3^2 + 25417245v_1^2v_2^2v_3 + 80952889v_1^3v_2^4 + 69633465v_1^5v_2v_3$$
$$+ 344343134v_1^6v_2^3 + 39306153v_1^8v_3 + 509125669v_1^9v_2^2$$
$$+ 318135602v_1^{12}v_2 + 72547972v_1^{15} + 1566096v_2^5)X^{16}$$

$[3](X) \bmod X^{28}$ for $p = 3$ in terms of the m_i

$$[3](X) = 3X - 24m_1X^3 + 648m_1^2X^5 - 22680m_1^3X^7 - (19680m_2 - 906120m_1^4)X^9$$
$$+ (1948536m_1m_2 - 39161880m_1^5)X^{11} - (144725616m_1^2m_2 - 1782778248m_1^6)X^{13}$$
$$+ (9647551656m_1^3m_2 - 84205559448m_1^7)X^{15}$$
$$- (609973825536m_1^4m_2 - 4088238304392m_1^8 - 11620843m_2^2)X^{17}$$
$$- (2242944969 84m_1m_2^2 - 37443174594264m_1^5m_2 + 202766127578136m_1^9)X^{19}$$
$$+ (27276329260728m_1^2m_2^2 - 2257636316956560m_1^6m_2 + 10229293584254088m_1^{10})X^{21}$$
$$- (2678311213085008m_1^3m_2^2 - 134569318031340552m_1^7m_2$$
$$+ 523275105375281304m_1^{11})X^{23}$$
$$+ (232282027220756400m_1^4m_2^2 - 7959882821409557280m_1^8m_2$$
$$+ 27079096740643969416m_1^{12} - 99113009568480 0m_2^3)X^{25}$$
$$- (762559748498 4m_3 - 28477285430249016m_1m_2^3$$

$$+ 1858509832616005l112 m_1^5 m_2^2 - 4683552282259668414000 m_1^9 m_2$$

$$+ 1415075117768856781848 m_1^{13}) X^{27}$$

$[3](X)$ for $p = 3$ mod X^{28} in terms of the v_i

$$[3](X) = 3X - 8v_1X^3 + 72v_1^2X^5 - 840v_1^3X^7 - (6560v_2 - 9000v_1^4)X^9$$
$$+ (216504v_1v_2 - 88992v_1^5)X^{11} - (5360208v_1^2v_2 - 658776v_1^6)X^{13}$$
$$+ (119105576v_1^3v_2 + 1199088v_1^7)X^{15} - (2424100032v_1^4v_2 + 199267992v_1^8$$
$$- 129120480v_2^2)X^{17}$$
$$- (8307202392v_1v_2^2 - 45824243688v_1^5v_2 - 5896183992v_1^9)X^{19}$$
$$+ (336744805688v_1^2v_2^2 - 807801733088v_1^6v_2 - 13344934881 6v_1^{10})X^{21}$$
$$- (11021856839856v_1^3v_2^2 - 13162584394728v_1^7v_2 - 2658275605728v_1^{11})X^{23}$$
$$+ (314960186505360v_1^4v_2^2 - 193206850384 0v_1^8v_2 - 4857972537146 4v_1^{12}$$
$$- 3670852206240v_2^3)X^{25}$$
$$- (2541865828328v_3 - 350724136455360v_1v_2^3 + 8146415921144640v_1^5v_2^2$$
$$- 2382655204483352v_1^9v_2 - 824825727922536v_1^{13})X^{27}$$

The coefficients a_{ij} of $F(X,Y)$ for $p = 2$

$$a_{11} = - 2m_1 = - v_1$$
$$a_{12} = 4m_1^2 = v_1^2$$
$$a_{13} = - 4m_2 - 8m_1^3 = - 2v_2 - 2v_1^3$$
$$a_{14} = 16m_2m_1 + 16m_1^4 = 4v_2v_1 + 3v_1^4$$
$$a_{15} = - 48m_1^2m_2 - 32m_1^5 = - 6v_1^2v_2 - 4v_1^5$$
$$a_{16} = 128m_1^3m_2 + 64m_1^6 + 16m_2^2 = 4v_2^2 + 12v_1^3v_2 + 6v_1^2$$
$$a_{17} = - 8m_3 - 96m_1m_2^2 - 320m_2m_1^4 - 128m_1^7 = - 4v_3 - 14v_2^2v_1 - 24v_2v_1^4 - 10v_1^7$$
$$a_{18} = 32m_3m_1 + 384m_2^2m_1^2 + 768m_2m_1^5 + 256m_1^8 = 8v_3v_1 + 28v_2^2v_1^2 + 40v_1^5v_2 + 15v_1^8$$
$$a_{22} = - 6m_2 - 20m_1^3 = - 3v_2 - 4v_1^3$$
$$a_{23} = 44m_2m_1 + 72m_1^4 = 11v_2v_1 + 10v_1^4$$
$$a_{24} = - 224m_2m_1^2 - 224m_1^5 = - 28v_2v_1^2 - 21v_1^5$$
$$a_{25} = 72m_2^2 + 912m_2m_1^3 + 640m_1^6 = 18v_2^2 + 75v_1^3v_2 + 43v_1^6$$

$a_{26} = -28m_3 - 656m_2^2m_1 - 3232m_2m_1^4 - 1728m_1^7 = -14v_3 - 89v_2^2v_1 - 190v_2v_1^4 - 88v_1^7$

$a_{27} = 184m_3m_1 + 3744m_2^2m_1^2 + 10432m_2m_1^5 + 4480m_1^8 = 46v_3v_1 + 257v_2^2v_1^2 + 420v_2v_1^5 + 169v_1^8$

$a_{33} = -344m_2m_1^2 - 400m_1^5 = -43v_2v_1^2 - 34v_1^5$

$a_{34} = 136m_2^2 + 2080m_2m_1^3 + 1760m_1^6 = 34v_2^2 + 164v_2v_1^3 + 101v_1^6$

$a_{35} = -56m_3 - 1696m_2^2m_1 - 10400m_2m_1^4 - 6720m_1^7 = -28v_3 - 226v_2^2v_1 - 551v_2v_1^4 - 275v_1^7$

$$a_{36} = 504m_3m_1 + 13056m_2^2m_1^2 + 45248m_2m_1^5 + 23296m_1^8$$
$$= 126v_3v_1 + 879v_2^2v_1^2 + 1586v_2v_1^5 + 680v_1^8$$

$a_{44} = -70m_3 - 2276m_2^2m_1 - 14944m_1^4m_2 - 10320m_1^7 = -35v_3 - 302v_2^2v_1 - 769v_2v_1^4 - 394v_1^7$

$$a_{45} = 812m_3m_1 + 23064m_2^2m_1^2 + 88960m_1^5m_2 + 50400m_1^8$$
$$= 203v_3v_1 + 1543v_2^2v_1^2 + 2933v_2v_1^5 + 1303v_1^8$$

The coefficients a_{ij} of $F(X,Y)$ for $p = 3$

$a_{12} = -3m_1 = -v_1$

$a_{14} = 9m_1^2 = v_1^2$

$a_{16} = -27m_1^3 = -v_1^3$

$a_{18} = -9m_2 + 81m_1^4 = -3v_2$

$a_{23} = 27m_1^2 = 3v_1^2$

$a_{25} = -162m_1^3 = -6v_1^3$

$a_{27} = -36m_2 + 810m_1^4 = -12v_2 + 6v_1^4$

$a_{34} = -351m_1^3 = -13v_1^3$

$a_{36} = -84m_2 + 2943m_1^4 = -28v_2 + 27v_1^4$

$a_{45} = -126m_2 + 5346m_1^4 = -42v_2 + 52v_1^4$

PART 2: The structure maps for $BP_*(BP)$

For $p = 2$

$\eta_R(v_1) = v_1 + 2t_1$

$\eta_R(v_2) = v_2 + 2t_2 - 5v_1t_1^2 - 4t_1^3 - 3v_1^2t_1$

$$\eta_R(v_3) = v_3 + 2t_3 - (v_2^2 + v_2v_1^3 + 2v_1^6)t_1 - (v_1^2v_2 + 11v_1^5)t_1^2 - (2v_2v_1 + 36v_1^4)t_1^3$$
$$- (2v_2v_1 + v_1^4)t_2 - (70v_1^3 - v_2)t_1^4 - (4v_2 + 2v_1^3)t_2t_1 - 85v_1^2t_1^5$$

$$- 2v_1^2t_2t_1^2 - 56v_1t_1^6 - 4v_1t_2t_1^3 - v_1t_2^2 - 16t_1^7 - 4t_1t_2^2$$

$$c(t_1) = - t_1$$

$$c(t_2) = - t_2 - v_1t_1^2 - t_1^3$$

$$c(t_3) = - t_3 - t_2^2t_1 - 3t_2t_1^4 - t_1^7 - v_1t_2^2 - 3v_1t_2t_1^3 - 3v_1t_1^6 - v_1^2t_2t_1^2 - 2v_1^2t_1^5$$
$$- v_1^3t_1^4 - v_2t_1^4$$

$$\psi(t_1) = t_1 \otimes 1 + 1 \otimes t_1$$

$$\psi(t_2) = t_2 \otimes 1 + 1 \otimes t_2 + t_1 \otimes t_1^2 - v_1t_1 \otimes t_1$$

$$\psi(t_3) = t_3 \otimes 1 + 1 \otimes t_3 + t_1 \otimes t_2^2 + t_2 \otimes t_1^4 - v_1t_2 \otimes t_2 - v_1t_1 \otimes t_2t_1^2$$
$$- v_1t_2t_1 \otimes t_1^2 + v_1^2t_1 \otimes t_2t_1 + v_1^2t_2t_1 \otimes t_1 + v_1^2t_1^2 \otimes t_1^3$$
$$- (2v_2 + v_1^3)t_1^3 \otimes t_1 - (2v_2 + v_1^3)t_1 \otimes t_1^3 - (3v_2 + 2v_1^3)t_1^2 \otimes t_1^2$$

<u>For p = 3</u>

$$\eta_R(v_1) = v_1 + 3t_1$$

$$\eta_R(v_2) = v_2 - 4v_1^3t_1 - 18v_1^2t_1^2 - 35v_1t_1^3 - 27t_1^4 + 3t_3$$

$$\eta_R(v_3) = v_3 + (-v_2^3 + 4v_2^2v_1^4 - 9v_2v_1^8 - 3v_1^{12})t_1 + (30v_2^2v_1^3 - 124v_2v_1^7 - 36v_1^{11})t_1^2$$
$$+ (89v_2^2v_1^2 - 948v_1^6v_2 - 231v_1^{10})t_1^3 - (3v_2^2v_1 + v_1^9)t_2$$
$$+ (132v_2^2v_1 - 4438v_1^5v_2 - 791v_1^9)t_1^4 - (9v_2^2 - 24v_2v_1^4 + 27v_1^8)t_2t_1$$
$$+ (81v_2^2 - 13494v_2v_1^4 - 790v_1^8)t_1^5 + (180v_1^3v_2 - 372v_1^7)t_2t_1^2$$
$$- (27037v_2v_1^3 - 5424v_1^7)t_1^6 + (534v_2v_1^2 - 2844v_1^6)t_2t_1^3$$
$$- (34725v_2v_1^2 - 30226v_1^6)t_1^7 - 9v_2v_1t_2^2 + (792v_2v_1 - 13314v_1^5)t_2t_1^4$$
$$- (26082v_2v_1 - 81975v_1^5)t_1^8 - (27v_2 - 36v_1^4)t_2^2t_1$$
$$+ (486v_2 - 40482v_1^4)t_2t_1^5 - (8747v_2 - 143703v_1^4)t_1^9 + 270v_1^3t_2^2t_1^2$$
$$- 81111v_1^3t_2t_1^6 + 173636v_1^3t_1^{10} + 801v_1^2t_2^2t_1^3 - 104175v_1^2t_2t_1^7$$
$$+ 144423v_1^2t_1^{11} - 8v_1t_2^3 + 1118v_1t_2^2t_1^4 - 78246v_1t_2t_1^8 + 76545v_1t_1^{12}$$
$$+ 3t_3 - 27t_2^3t_1 + 729t_2^2t_1^5 - 26244t_1^2t_1^9 + 19683t_1^{13}$$

$$c(t_1) = - t_1$$

$$c(t_2) = - t_2 + t_1^4$$

$$c(t_3) = - t_3 + t_2^3t_1 - v_1t_2^2t_1^4 - 3t_2^2t_1^5 + v_1t_2t_1^8 + 4t_2t_1^9 - t_1^{13}$$

$$\psi(t_1) = t_1 \otimes t_1$$

$$\psi(t_2) = t_2 \otimes 1 + 1 \otimes t_2 - v_1 t_1^2 \otimes t_1 - v_1 t_1 \otimes t_1^2 + t_1 \otimes t_1^3$$

$$\begin{aligned}
\psi(t_3) = {} & t_3 \otimes 1 + 1 \otimes t_3 + 2v_1^2 t_2 t_1^2 \otimes t_2 t_1 + 2v_1^2 t_2 t_1 \otimes t_2 t_1^2 - v_1^3 t_1^4 \otimes t_2 t_1^2 \\
& - 2v_1 t_2 t_1 \otimes t_2 t_1^3 - 2v_1^3 t_1^3 \otimes t_2 t_1^3 - v_1^3 t_1^2 \otimes t_2 t_1^4 + 2v_1^2 t_1^3 \otimes t_2 t_1^4 \\
& + 2v_1^2 t_1^2 \otimes t_2 t_1^5 - v_1 t_1^2 \otimes t_2 t_1^6 - v_1 t_2^2 \otimes t_2 + v_1^2 t_2^2 t_1^2 \otimes t_1 \\
& - 3v_2 t_1^8 \otimes t_1 - v_1^4 t_1^8 \otimes t_1 + v_1^2 t_1^2 \otimes t_2^2 t_1 - v_1 t_2 \otimes t_2^2 \\
& + v_1^2 t_1 \otimes t_2^2 t_1^2 - v_1 t_1 \otimes t_2^2 t_1^3 + t_1 \otimes t_2^3 + v_1^2 t_2^2 t_1 \otimes t_1^2 \\
& - v_1^3 t_2 t_1^4 \otimes t_1^2 - 12v_2 t_1^7 \otimes t_1^2 - 4v_1^4 t_1^7 \otimes t_1^2 - v_1 t_2^2 t_1 \otimes t_1^3 \\
& - 2v_1^3 t_2 t_1^3 \otimes t_1^3 - 28v_2 t_1^6 \otimes t_1^3 - 9v_1^4 t_1^6 \otimes t_1^3 - v_1^3 t_2 t_1^2 \otimes t_1^4 \\
& + 2v_1^2 t_2 t_1^3 \otimes t_1^4 - 42v_2 t_1^5 \otimes t_1^4 - 13v_1^4 t_1^5 \otimes t_1^4 + 2v_1^2 t_2 t_1^2 \otimes t_1^5 \\
& - 42v_2 t_1^4 \otimes t_1^5 - 13v_1^4 t_1^4 \otimes t_1^5 - v_1^3 t_1^5 \otimes t_1^5 - v_1 t_2 t_1^2 \otimes t_1^6 \\
& - 28v_2 t_1^3 \otimes t_1^6 - 9v_1^4 t_1^3 \otimes t_1^6 - 2v_1^3 t_1^4 \otimes t_1^6 - 12v_2 t_1^2 \otimes t_1^7 \\
& - 4v_1^4 t_1^2 \otimes t_1^7 - v_1^3 t_1^3 \otimes t_1^7 + v_1^2 t_1^4 \otimes t_1^7 - 3v_2 t_1 \otimes t_1^8 - v_1^4 t_1 \otimes t_1^8 \\
& + v_1^2 t_1^3 \otimes t_1^8 + t_2 \otimes t_1^9
\end{aligned}$$

On a modified Eilenberg-Moore Theorem

by

V. K. A. M. Gugenheim

1. Introduction and main result.

We shall consider the following pull-back diagrams in the category of topological spaces with base-point:

$$\begin{array}{ccc} E' & \xrightarrow{p_0'} & Y_0 \\ \downarrow{\scriptstyle p_1'} & & \downarrow{\scriptstyle f_0} \\ Y_1 & \xrightarrow{f_1} & X \end{array} \qquad \begin{array}{ccc} E & \xrightarrow{\bar{f}} & PX \\ \downarrow{\scriptstyle p} & & \downarrow{\scriptstyle p_X} \\ Y_0 \times Y_1 & \xrightarrow{f_0 \times f_1} & X \times X \end{array} \tag{1.1}$$

where PX is the free space of paths in X and p_X the "end-point" map $p_X(\xi) = (\xi(0)), \xi(1))$. $E' \subset Y_0 \times Y_1$ consists of the points (y_0, y_1) with $f_0 y_0 = f_1 y_1$ and $E \subset Y_1 \times Y_2 \times PX$ of the points (y_0, y_1, ξ) with $f_0 y_0 = \xi(0)$, $f_1 y_1 = \xi(1)$. A map $h\colon E' \to E$ is defined by $h(y_0, y_1) = (y_0, y_1, \widetilde{f_0 y_0} = \widetilde{f_1 y_1})$ where $\tilde{x}$ denotes the constant path at $x \in X$.

By standard arguments one easily proves

1.2 THEOREM

If f_0 (or f_1) is a Serre fibration then h is a weak equivalence (i.e. induces isomorphisms in all homotopy and homology groups).

The main result of this paper is the following

1.3 THEOREM

There is a natural commutative diagram of chain-maps

Supported in part by NSF Grant No. 346-32-54-306

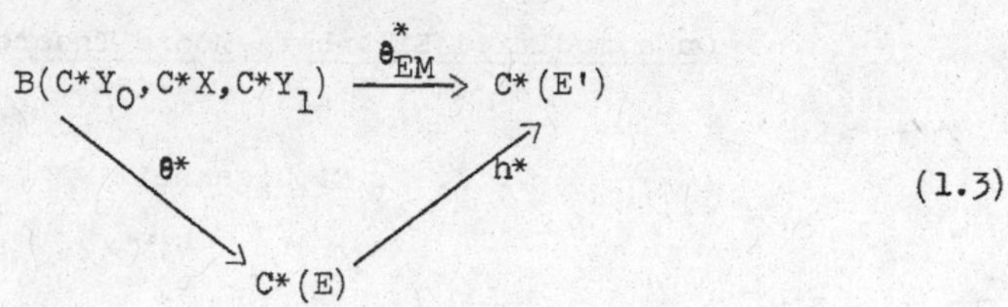

(1.3)

where C^* denotes the normalised singular cochain functor over a given commutative Ring and B denotes the "two sided bar construction", $h^* = C^*(h)$, and θ^*_{EM} denotes the "classical" map introduced by Eilenberg and Moore, cf [6], [9], [14].

1.4 THEOREM

If X, Y_0, Y_1 have integral homology groups of finite type, Y_0, Y_1 are connected, X is simply connected and R is Noetherian, then $H(\theta^*)$ is an isomorphism.

In view of 1.2, 1.3 this implies the Eilenberg-Moore theorem; conversely that theorem and 1.3 imply 1.4, cf. Chapter 4 below. The significance of 1.4 lies in the existence of the map θ^*; thus, for example, suppose $Y_0 = Y_1$ = a point. Then θ^* is a chain-map $B(C^*X) \to C^*(\Omega X)$ where ΩX is the loop-space; the classical approach merely provides chain-maps

$$B(C^*X) \leftarrow B(R, C^*X), C^*P(X,*)) \to C^*(\Omega X)$$

where $P(X,*)$ is the space of paths with $\xi(0) = *$. Thus, the new construction answers the question: Given a cycle in $B(C^*X)$, what is the corresponding element in $H^*(\Omega X, R)$?

As will be seen, θ^* is derived from a functor $I: B(C^*X) \to C^*PX$ which is an analogue of the "iterated integrals" introduced by Kuo Tsai Chen, cf [3], [4], [7]. The present paper grew out of my study of Chen's work. There is, however, no analogue to the interesting theorem of [5] or [4, 4.3.1]. This theorem applies to the more general case when $f_0 \times f_1: Y_0 \times Y_1 \to X \times X$ is replaced by any map $f: Y \to X \times X$; it seems to depend essentially on the use of commutative cochains.

I would like to acknowledge several helpful conversations with A. K. Bousfield.

2. The morphism I.

If A is a differential positively graded algebra with differential of grading +1, where for the moment A need have neither unit nor augmentation, we denote by $\mathbb{B}(A)$ the "bar construction" $\mathbb{B}(A) = \coprod_{p\geq 0} \mathbb{B}_p(A)$ where $\mathbb{B}_p(A)$ is the p-fold tensor product $\otimes^p A$ and the grading of $a_1 \otimes \cdots \otimes a_p$, written $[a_1,\ldots,a_p]$ is $\Sigma|a_i| - p$, and the differential is given by the usual formula, cf., e.g., [10].

$\mathbb{B}(A)$ also has the usual diagonal

$$\psi[a_1,\ldots,a_p] = \Sigma_{0\leq i\leq p}[a_1,\ldots,a_i] \otimes [a_{i+1},\ldots,a_p].$$

If A has a unit $\eta: R \to A$ and augmentation $\epsilon: A \to R$ we write $B(A) = \mathbb{B}(\ker \epsilon)$ for the "usual" bar-construction. If C is a differential algebra and $f,g: \mathbb{B}(A) \to C$ (or $B(A) \to C$), $f \cup g$ will denote $\emptyset(f \otimes g)\psi$, where $\emptyset$ is the product of C. By $\tau: \mathbb{B}(A) \to A$ we denote the "twisting cochain" $\tau[a] = a$, $\tau|\mathbb{B}_p(A) = 0$ if $p \neq 1$. Then $D\tau = \tau \cup \tau$ where, as usual, $Df = d\cdot f - (-1)^k f\cdot d$ if $|f| = k$; cf. [10].

Let X be a topological space, without base-point for the moment. By $p_0,p_1: PX \to X$ we denote the end-point maps and by $i: X \to PX$ the map $ix = \tilde{x}$, the constant path. Note that $p_0 i = p_1 i = 1$. We shall also use the following notations:

$$\pi_t = i\, p_t: PX \to PX$$

$$\tau_t = p_t^* \tau: \mathbb{B}(C^*X) \to C^*PX$$

$$\sigma_t = \pi_t^* \tau : \mathbb{B}(C^*PX) \to C^*PX$$

when t = 0,1.

2.1 THEOREM

There is a functorial morphism

$$I: \mathbb{B}(C^*X) \to C^*PX$$

of grading 0 such that I[ ] = 1 and

$$DI = \tau_0 \cup I - I \cup \tau_1$$

$$i^*I = \eta\epsilon$$

where $\eta: R \to C^*X$ is the unit.

Remark. Compare this with 1.2 of [7] or 4.1.2 of [3]. I is the analogue of Chen's "iterated integral".

We shall derive 2.1 from

2.2 THEOREM

There is a functorial morphism $J: \mathbb{B}(C^*PX) \to C^*PX$ of grading 0 such that J[ ] = 1 and

$$DJ = \sigma_0 \cup J - J \cup \tau$$

$$i^*J = \eta\epsilon .$$

Now, since $\sigma_0\mathbb{B}(p_1^*) = \pi_0^*\tau\mathbb{B}(p_1^*) = \pi_0^*p_1^*\tau = (p_1\pi_0)^*\tau = p_0^*\tau = \tau_0$ and $\tau\mathbb{B}(p_1^*) = p_1^*\tau = \tau_1$, it is clear that $I = J\cdot\mathbb{B}(p_1^*)$ will satisfy the conditions of 2.1. It remains to prove 2.2:

We begin by proving that π_0 and 1_P, the identity map on PX, are homotopic: Define the map

$$j:\ PX \times [0,1] \to PX$$

by $j(\xi,t) = \xi_t$ where $\xi_t(s) = \xi(ts)$, $t,s\in[0,1]$. Then $j(\xi,0) = \pi_0\xi$, $j(\xi,1) = \xi$.

Hence there is a chain-homotopy $j^{\#}$: $C^*PX \to C^*PX$ such that $\pi_0^* = 1 + Dj^{\#}$. Also, we easily see that

$$j^{\#}p_0^* = 0,\ i^*j^{\#} = 0.$$

This follows because the "prisms" p_0j and $j|iX \times [0,1]$ are degenerate and C* is the <u>normalized</u> cochain functor. Thus, considering the diagram

$$C^*PX \underset{p_0^*}{\overset{i^*}{\rightleftarrows}} C^*X$$

we are in the situation of Theorem 4.1_* of [10] and the existence of the "homotopy in DASH" J follows. We remark that according to this theorem we should also have $j^{\#}j^{\#} = 0$, which is not the case: This condition, however, is actually <u>not</u> needed; nor, indeed, is the condition $j^{\#}p_0^* = 0$. Since the proof in [10] is further obscured by typographical errors, it seems best to complete the proof here: J is defined inductively by the formula

$$J = -j^{\#}(\sigma_0 \cup J - J \cup \tau) + \eta\epsilon.$$

This works because $\sigma_0[\ ] = \tau[\ ] = 0$ and we get $J[\ ] = 1 = \eta\epsilon$, $J[w] = -j^{\#}(\pi_0^*w \pm w) = \pm j^{\#}w$, etc.

Calculating inductively and remembering $\pi_0^*\tau = \pi_0^*\sigma_0 = \sigma_0$,

$i^*j^{\#} = 0$ so that $i^*J = \eta E$, we easily obtain $DJ = \sigma_0 \cup J - J \cup \tau$, q.e.d.

Now suppose that a base-point $* \in X$ has been chosen so that there is an augmentation $C^*(X) \xrightarrow{\epsilon} R$. We can then restrict I to the "usual" bar-construction $B(C^*X) = \mathbb{B}(\ker \epsilon)$. We shall denote the restricted map by the same symbol I; it still satisfies the relations of 2.1.

By $C^*_{(s)}(X)$ we shall denote the cochain complex based on the singular subcomplex having only the simplex "at the base-point" in dimension $\leq s$. If X is s-connected the restriction $C^*(X) \to C^*_{(s)}(X)$ is a homology-isomorphism. IN PX we take $\tilde{*}$, the constant path at $*$, as our base-point. Since $j(\tilde{*},t) = \tilde{*}$ for all $t \in [0,1]$, we easily see that our entire construction can be carried out in $C^*_{(s)}$ and leads to a commutative diagram

$$\begin{array}{ccc} B(C^*X) & \xrightarrow{I} & C^*PX \\ \downarrow & & \downarrow \\ B(C^*_{(s)}X) & \xrightarrow{I_{(s)}} & C^*_{(s)}PX \end{array} \qquad (2.3)$$

Since $B(C^*_{(1)}X)$ has only the element [ ] in dimension 0, we easily obtain the following:

2.4 LEMMA

If X is simply connected, then $H^0(BC^*X) = H^0(BC^*_{(0)}X) = R$.

3. The chain-map θ^*

We refer to the diagram 1.1 and observe that the map p can be written $pe = (p_0e, p_1e)$ where $p_t\colon E \to Y_t$; it is hoped that the generic use of the symbols p_0, p_1 will cause no confusion. As an R-module, $B(C^*Y_0, C^*X, C^*Y_1)$ is $C^*Y_0 \otimes B(C^*X) \otimes C^*Y_1$ and we can therefore define $\theta^*\colon B(C^*Y_0, C^*X, C^*Y_1) \to C^*E$ as the composition

$$C^*Y_0 \otimes B(C^*X) \otimes C^*Y_1 \xrightarrow{p_0^*\otimes \bar{f}^* I \otimes p_1^*} C^*E \otimes C^*E \otimes C^*E \xrightarrow{\cup} C^*E$$

where $\cup$ denotes the iterated cup-product.

3.1 LEMMA

θ^* is a chain-map.

This follows by a straightforward calculation from 2.1 and the definition of the differential in the two-sided bar-construction, cf. [9] or [10].

Now, using the notations of 1.1, 2.1 and writing $q = f_0 p_0' = f_1 p_1'$ we easily see that $\bar{f}h = iq$ and hence $h^*\bar{f}^*I = q^*i^*I = \eta\varepsilon$ by 2.1; we are using the symbols η, ε generically. Hence:

$$\begin{aligned} h^*\theta^* &= h^* \cup (p_0^* \otimes \bar{f}^* I \otimes p_1^*) \\ &= \cup(h^*p_0^* \otimes h^*\bar{f}^* I \otimes h^*p_1^*) \\ &= \cup((p_0')^* \otimes \eta\varepsilon \otimes (p_1')^*) \end{aligned}$$

since $p_t h = p_t'$. The last composition is exactly (the dual of) the map introduced by Eilenberg and Moore. We have thus proved the commutativity of 1.3.

Using 2.3 above we obtain a commutative diagram

$$\begin{array}{ccc} B(C^*Y_0, C^*X, C^*Y_1) & \xrightarrow{\theta^*} & C^*E \\ \downarrow & & \downarrow \\ B(C^*_{(0)}Y_0, C^*_{(0)}X, C^*_{(0)}Y_1) & \xrightarrow{\theta^*_{(0)}} & C^*_{(0)}E \end{array}$$

If Y_0, Y_1 are connected and X is simply connected, then (it is easily seen) E is connected. Hence, then, the vertical maps in the above diagram are homology-isomorphisms. To prove 1.4 it therefore suffices to prove that $\theta^*_{(0)}$ is a homology isomorphism. To simplify the typography we shall write X^* for $C^*_{(0)}X$ etc. in what follows. We shall now filter $E^* = C^*_{(0)}E$ by the (decreasing!) Serre-Filtration and $B(Y^*_0, X^*, Y^*_1)$ by

$$F^u = \Sigma_{u_0+u_1\geq u} Y_0^{u_0} \otimes B(X^*) \otimes Y_1^{u_1}$$

Then it is easily seen that $\theta^*_{(0)}$ is filtration preserving. We write it as the top-line of the following diagram

$$\begin{array}{ccccc} Y^*_0 \otimes B(X^*) \otimes Y^*_1 & \xrightarrow{1\otimes I\otimes 1} & Y^*_0 \otimes P^* \otimes Y^*_1 & \xrightarrow{U(p^*_0\otimes \overline{F}^*\otimes p^*_1)} & E^* \\ & & \Big\downarrow{\scriptstyle 1\otimes\omega^*\otimes 1} & & \\ & & Y^*_0 \otimes \Omega^* \otimes Y^*_1 & & \end{array}$$

where $P^* = C^*_0(PX)$, $\Omega^* = C^*_0(\Omega X)$, ΩX is the loop-space and $\omega\colon \Omega X \to PX$ the injection. Since $\omega^* p^*_t|\overline{C}^*_{(0)}(X) = 0$ $(t = 0,1)$ it follows easily from 2.1 that $D(I_\Omega) = 0$ where $I_\Omega = \omega^*\circ I\colon B(X^*) \to \Omega^*$.

Now we observe that at the level E_0 of the induced spectral sequences any term of filtration > 0 in the image of I will be mapped to zero in $E_0(E^*)$. Hence $E_0(\theta^*_{(0)})$ can be factored through $1 \otimes \omega^* \otimes 1$ and is the following composition

$$Y^*_0 \otimes B(X^*) \otimes Y^*_1 \xrightarrow{1\otimes I_\Omega\otimes 1} Y^*_0 \otimes \Omega^* \otimes Y^*_1 \xrightarrow{\alpha_0} E_0(E^*)$$

where α_0 is essentially (apart from a shuffle and an Eilenberg-Zilber map) the map $(Y_1 \times Y_2)^* \otimes \Omega^* \to E_0(E^*)$ introduced by Serre. For a compatible account, see pp. 23, 24 of [9] where our present α is denoted by τ.

Since the non-zero entries $B(C^*_{(0)}X)$ all have grading > 0, it follows that the "twisting terms" in the differential of $B(Y^*_0, X^*, Y^*_1)$

are zero at level E_0. Hence $E_1(\theta^*_{(0)})$ is the composition

$$Y_0^* \otimes H\,B(X^*) \otimes Y_1^* \xrightarrow{1\otimes H(I_\Omega)\otimes 1} Y_0^* \otimes H\Omega^* \otimes Y_1^* \xrightarrow{\alpha_1} E_1(E^*) \quad (3.3)$$

By Serre's basic theorem, α_1 is a homology isomorphism. Now, let us consider the special case $Y_0 = *$, $Y_1 = X$, f_1 the identity on X. Then $E = P(X,*)$, the space of paths beginning at $*$; thus in this case E^* and $B(R,X^*,X^*)$ are both acyclic, andhence θ^* is a homology isomorphism. $E_1(\theta^*_{(0)})$ reduces to the composition

$$H(BX^*) \otimes X^* \xrightarrow{H(I_\Omega)\otimes 1} H^*(\Omega) \otimes X^* \xrightarrow{\alpha_1} E_1(E^*)$$

In passing to $E_2(\theta^*)$, $H(BX^*)$ and $H^*\Omega$ are just coefficient groups,* and so $E_2^{u,v}(\theta^*_{(0)})$ is the composition

$$H^u(X,H^v(BX^*)) \to H^u(X,H^v\Omega) \xrightarrow{\alpha_2} E_2^{u,v}(E^*)$$

which reduces in the cases u = 0, v = 0 to

$$H^v(BX^*) \xrightarrow{H(I_\Omega)} H^v(\Omega) \xrightarrow{\alpha_2} E_2^{0,v}(E^*)$$
$$H^u(X) \xrightarrow{1} H^u(X) \xrightarrow{\alpha_2} E_2^{u,0}(E^*)$$

by the universal coefficient theorem and 2.4.

Sonce α_2 is an isomorphism, it follows that $E_2^{u,0}(\theta^*_{(0)})$ and $H(\theta^*_{(0)})$ are isomorphisms. By the Moore-Zeeman comparison theorem, it follows that $E_2^{0,v}(\theta^*_{(0)})$ is an isomorphism, and hence we have

3.4. PROPOSITION. If X is simply connected,

$$H(I_\Omega): H^*(B(C^*X)) \to H^*(\Omega X)$$

*See NOTE at the end of the paper.

is an isomorphism.

3.4 is a form of the original theorem of Adams [1]; it is also, in view of the map I_Ω, an analogue of the theorem of Chen, [3], [7].

Since α_1 is a homology isomorphism, it follows immediately from 3.3 that $E_2(\theta^*_{(0)})$ is an isomorphism, and hence, from the completeness of the spectral sequences involved, that $H(\theta^*_{(0)})$ is an isomorphism, q.e.d.

A remark should, perhaps, be made on the applicability of the Moore-Zeeman comparison theorem. The chomological form is given in [15] and follows easily from Lemma 3.8 in [12]. The exactness condition

$$0 \to E_2^{u,0} \otimes E_2^{0,v} \to E_2^{u,v} \to \mathrm{Tor}_1(E_2^{u+1,0}, E_2^{0,v}) \to 0$$

follows from Serre's theorem because all homology groups in sight are finitely generated.

4. Products and some Comments.

It is known that, using Eilenberg-Zilber maps, $B(C^*Y_0, C^*X, C^*Y_1)$ has a natural product structure and that the map $H(\theta^*_{EM})$ is multiplicative, cf. [6], [14], [8]. From this and 1.2, 1.3 we deduce that $H(\theta^*)$ is multiplicative, at least if one of f_0, f_1 is a Serre fibration. This condition, however is unnecessary:

4.1 THEOREM

$H(\theta^*)$ is multiplicative.

To see this, we observe that the map f_0 (say) can be factorized as

$$Y_0 \xrightarrow{a} Z_0 \xrightarrow{b} X$$

where b is a Serre-fibration and a a weak equivalence. This leads to a diagram of induced fibrations

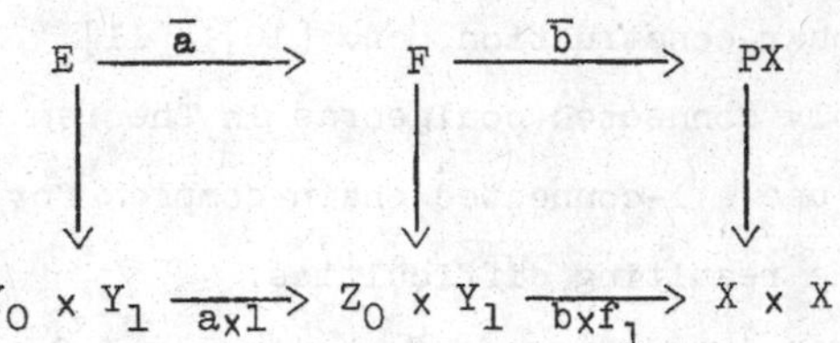

and, due to the naturality of θ^*, to the commutative diagram

$$\begin{array}{ccc} B(C^*Z_0, C^*X, C^*Y_1) & \xrightarrow{\theta_F^*} & C^*(F) \\ \Big\downarrow{\scriptstyle B(a^*,1,1)} & & \Big\downarrow{\scriptstyle \bar{a}^*} \\ B(C^*Y_0, C^*X, C^*Y_1) & \xrightarrow{\theta^*} & C^*(E) \end{array}$$

where $\bar{a}^*$, $H(\theta_F^*)$ and $B(a^*,1,1)$ are multiplicative and $HB(a^*,1,1)$ is an isomorphism. Hence $H(\theta^*)$ is multiplicative.

Comments

(1) A direct proof of 4.1 would be desirable. The result, at least in the case $Y_0 = Y_1 =$ a point, is the analogue of the fact that Chen's iterated integral is multiplicative, cf [13], [3], [4].

(2) Once 1.3 has been proved the trick of 4.1 can be used to deduce 1.4 from the "classical" Eilenberg Moore theorem. In particular, one can transfer known results in the case when X is not simply connected: It is then necessary to make sure that whatever conditions are needed will apply to b in the above. In spite of this, it seemed of interest to give the direct proof of 1.4 which returns to the pattern introduced by Adams and Hilton, [1], [2].

(3) A result dual to 1.3, 1.4 in homology would require the replacement of I by a map

$$C_*PX \to \Omega(C_*X)$$

where Ω is the cobar construction, cf. [10], [11]. Due to the restriction to simply connected coalgebras in Theorem 4.1* of [10], it seems necessary to use a 1-connected chain-complex for $C_* PX$: I have not investigated the resulting difficulties.

NOTE. There is a gap in the proof of chapter 3: I forgot to prove that the differential d^1 in $Y_0^* \otimes HB(X^*) \otimes Y_1^*$ in Formula 3.3 is the untwisted one: Since $X^* = C^*_{(0)}(X)$ contains 1-dimensional elements this could introduce "twisting terms". To deal with this, replace X by a "minimal subcomplex" $X_{(m)}$ of the singular complex. There are injection and retraction $X_{(m)} \xrightarrow{i} X \xrightarrow{r} X_{(m)}$ (cf. p. 220 of [6]) leading to chain equivalences

$$X^*_{(m)} \xrightarrow{r^*} X^* \xrightarrow{i^*} X^*_{(m)} \text{ with } i^*r^* = 1$$

where $X^*_{(m)} = C^*(X_{(m)})$. Since X is simply connected, $X^*_{(m)}$ has no non-zero elements of dimension 1. Using the chain maps

$$X^*_{(m)} \xrightarrow{r^*} X^* \xrightarrow{f_t^*} Y_t^* = C^*_{(0)}(Y_t) \quad (t = 0,1)$$

we can consider the morphism of filtered objects

$$B(Y_0^*, X^*_{(m)}, Y_1^*) \xrightarrow{B(1,r^*,1)} B(Y_0^*, X^*, Y_1^*)$$

At the level E^1 this induces the isomorphism

$$Y_0^* \otimes HB(X^*_{(m)} \otimes Y_1^* \xrightarrow{1\otimes HB(r^*)\otimes 1} Y_0^* \otimes HB(X^*) \otimes Y_1^*.$$

Since $X^*_{(m)}$ has no non-zero elements of dimension 1, the left hand side has the "untwisted" differential d^1; and, since the shown map commutes with d^1, so does the right hand side, q.e.d.

References

[1] J. F. Adams, "On the cobar construction", Colloque de topologie algébrique, Louvain, (1956), pp. 81-87.

[2] J. F. Adams and P. J. Hilton, "On the Chain algebra of a loop space", Comm. Math. Helv. Vol. 30, (1956), pp. 305-330.

[3] Kuo-Tsai Chen, "Iterated integrals of differential forms and loop-space homology", Ann. of Math., Vol. 97 (1973), pp. 217-246.

[4] Kuo-Tsai Chen, "Iterated path integrals", Bulletin of the Am. Math. Soc. (1977) (to appear).

[5] Kuo-Tsai Chen, "Pullback de Rham Cohomology of the Free Path Fibration" (to appear).

[6] S. Eilenberg and J. C. Moore, "Homology and fibrations I", Comm. Math. Helv. 40 (1966), pp. 398-413.

[7] V.K.A.M. Gugenheim, "On Chen's Iterated Integrals", Ill. J. of Math. (to appear).

[8] V.K.A.M. Gugenheim, "On the Multiplicative Structure of the de Rham Cohomology of Induced Fibrations", Ill. J. of Math. (to appear).

[9] V.K.A.M. Gugenheim and J. Peter May, "On the Theory and Applications of Differential Torsion Products", Memoirs of the Am. Math. Soc., 142 (1974).

[10] V.K.A.M. Gugenheim and H. J. Munkholm, "On the extended functoriality of Tor and Cotor", J. of Pure and Applied Algebra, (1974), pp. 9-29.

[11] D. Husemoller, J. C. Moore, J. Stasheff, "Differential Homological Algebra and Homogeneous Spaces", J. of Pure and Applied Algebra, (1974), pp. 113-185.

[12] D. G. Quillen, "An application of simplicial profinite groups", Comm. Math. Helv. 44, (1969), pp. 45-60.

[13] Rimhak Ree, "Lie elements and an Algebra associated with shuffles", Ann. of Math., Vol. 68, (1958), pp. 210-220.

[14] L. Smith, "Homological Algebra and the Eilenberg Moore spectral sequence", Trans. Am. Math. Soc., 129 (1967), pp. 58-93.

[15] E. C. Zeeman, "A proof of the comparison theorem for spectral sequences", Proc. Cambridge Phil. Soc. 53, part 1, (1957), pp. 57-62.

University of Illinois
at Chicago Circle

ON THE HOPF CONSTRUCTION ASSOCIATED WITH A COMPOSITION

C. H. Hanks
Millikin University
Decatur, Illinois 62522

Introduction

Let $\alpha: \Sigma A \to BG$ denote the classifying map of a fibre bundle with fibre X, a left G-space. By adjointing we obtain a map $\text{adj}\,\alpha: A \to \Omega BG \simeq G$ which provides an action of A on X given by $A \times X \xrightarrow{\text{adj}\,\alpha \times \underline{X}} G \times X \xrightarrow{\gamma} X$, where γ denotes the action of G on X.

<u>Definition</u>. The Hopf construction associated with α, denoted $J(\alpha): \Sigma A \wedge X \to \Sigma X$, is the Hopf construction on the map $\gamma\cdot(\text{adj}\,\alpha \times \underline{X})$.

The purpose of this paper is to describe the Hopf construction $J(\alpha\cdot\beta)$ associated with a composition $\Sigma B \xrightarrow{\beta} \Sigma A \xrightarrow{\alpha} BG$. The Hopf construction associated with the classifying map for a bundle over a suspension is of interest because it appears as the attaching map in a certain description of the <u>Thom space</u> (mapping cone) of the bundle. Specifically, given a space Y and an X - G-bundle over ΣY, $X \to E \xrightarrow{p} \Sigma Y$, classified by a map $\xi: \Sigma Y \to BG$, we have the well-known result that $\Sigma Y \cup_p C(E) \simeq \Sigma X \cup_{J(\xi)} C(\Sigma Y \wedge X)$. Throughout spaces are assumed to be countable CW-complexes with basepoints, (so in particular joins and suspensions are reduced), and all maps are basepoint-preserving. For notational convenience the identity map of a space Y will be denoted by $\underline{Y}$.

The paper is divided into three sections. In §1 the definition of the Hopf construction is reviewed and a combinatorial description of its adjoint is obtained (Lemma 1). Lemma 1 makes possible the recognition of certain maps appearing in §3. as "iterated" Hopf constructions. In §2. the map $J(\alpha\cdot\beta)$ is factored through the space $\Sigma\Omega\Sigma A$, thereby giving rise to the appearnace of certain Hopf invariants of β in the description of $J(\alpha\cdot\beta)$. In §3. the theorem describing $J(\alpha\cdot\beta)$ is obtained, following the analysis and recognition of certain maps defined by J. Milnor in [8] as iterated Hopf constructions, modulo certain "generalized signs" which are precisely described. Under the hypotheses that A is connected, B is finite-dimensional, and the fibre X is a suspension with the action of G basepoint-preserving, the theorem expresses $J(\alpha\cdot\beta)$ as a sum of maps involving certain Hopf invariants of β and iterated Hopf constructions on α.

The material in the paper represents portions of the author's thesis written under Professor Michael Barratt of Northwestern University whom the author would like to thank for his advice and help. The author would also like to thank the faculty and staff of the mathematics department at the Centro de Investigación y Estudios Avanzados in Mexico City, where the paper was written during a visit in the summer of 1977, for their generaous hospitality and the use of their facilities.

§1. A combinatorial description of the Hopf construction

The classical Hopf construction $\chi(f)$ on a map $f\colon A \times B \to C$ is the map from the join to the suspension $\chi(f)\colon A*B \to \Sigma C$ defined by $\chi(f)(a,t,b) = (f(a,b),t)$ where $t\in[0,1]$ represents the join parameter on the left and the suspension parameter on the right, with the usual identifications. (In particular, in the join: $(a,0,b) \sim (a,0,b')$.) It is easy to see that the homotopy class of $\chi(f)$ depends only on the homotopy class of f. The classical Hopf construction

$\chi(\mu)$: $A*B \to \Sigma A \wedge B$ on the identification map μ: $A \times B \to A \wedge B$, which pinches $A \vee B$ to a point, is in fact a homotopy equivalence; thus if we let $\chi(\mu)^{-1}$ denote a homotopy inverse of $\chi(\mu)$, the classical Hopf construction on a map f: $A \times B \to C$ defines a unique homotopy class $J(f) \in [\Sigma A \wedge B, \Sigma C]$, represented by $\chi(f) \cdot \chi(\mu)^{-1}$, which we will call the Hopf construction on f (to be distinguished from the classical Hopf construction).

We are interested in the adjoint of J(f) and to that end will take as a model for $\Omega\Sigma C$ the free group FC on the points of C, where the functor F is the topological analogue of Milnor's semi-simplicial free group functor F as defined in [8]. For a countable complex C, FC, with the basepoint of C as the identity in the group, may be topologized so that it is a topological group such that there is a natural homotopy equivalence of H-spaces $\Omega\Sigma C \simeq FC$.

If we compose f: $A \times B \to C$ with the inclusion of generators $C \xrightarrow{i} FC$, we obtain a map which we will also call f. In addition we have two maps f_A, f_B: $A \times B \to C \xrightarrow{i} FC$ defined by $f_A(a,b) = f(a,b_0)$ and $f_B(a,b) = f(a_0,b)$ where the subscript 0 indicates the appropriate basepoint. Utilizing the group operation in FC (written multiplicatively) we now form the map $f_A^{-1} f\, f_B^{-1}$: $A \times B \to FC$ defined by the correspondence $(a,b) \to \overline{f(a,b_0)}f(a,b)\overline{f(a_0,b)}$ where $\bar{c}$ denotes the inverse in FC of a generator $c \in C$. This map carries $A \vee B \subset A \times B$ to the identity in FC and therefore induces a map $j(f)$: $A \wedge B \to FC$ such that $f_A^{-1} f\, f_B^{-1} = j(f) \cdot \mu$. Letting $[j(f)]$ denote the class of the map $j(f)$ in the group $[A \wedge B, FC]$ of homotopy classes of pointed maps $A \wedge B \to FC$, we are now ready for

<u>Lemma 1</u>. Given a map f: $A \times B \to C$, the homotopy classes $[j(f)] \in [A \wedge B, FC]$ and $J(f) \in [\Sigma A \wedge B, \Sigma C]$ are adjoint classes. That is, the adjoint of the Hopf construction on f is homotopic to the map $A \wedge B \to FC$ defined by the correspondence $(a,b) \to \overline{f(a,b_0)}f(a,b)\overline{f(a_0,b)}$

where $\overline{c}$ denotes the inverse in FC of a generator $c \epsilon C$ and $(a,b) \epsilon A \wedge B$ denotes the image under μ: $A \times B \to A \wedge B$ of the corresponding element in $A \times B$.

An immediate consequence of the lemma is the

Corollary. Given maps $f: A \times B \to C$ and $g: W \to Z$, the adjoint of the map $g \wedge J(f): W \wedge \Sigma A \wedge B \to Z \wedge \Sigma C$ is homotopic to the map $W \wedge A \wedge B \to F(Z \wedge C)$ defined by the correspondence
$(w,a,b) \to (\overline{g(w),f(a,b_0)})(g(w),f(a,b))(\overline{g(w),f(a_0,b)})$.

Proof of Lemma 1. It suffices, by the nature of the adjoint isomorphism, to show that the adjoint of the map $j(f)$ is homotopic to $\chi(f)\cdot\chi(\mu)^{-1}$ and therefore that adj $j(f)\cdot\chi(\mu) \simeq \chi(f)$. By inspection it may be seen that the classical Hopf construction $\chi(f)$ on any map $f: A \times B \to C$ factors through $\Sigma(A \times B)$ as $\Sigma f\cdot\chi(\underline{A \times B})$. Hence adj $j(f)\cdot\chi(\mu)$ = adj $j(f)\cdot\Sigma\mu\cdot\chi(\underline{A \times B})$. We now obtain a description of adj $j(f)\cdot\Sigma\mu$ by adjointing $j(f)\cdot\mu$ twice. When we adjoint $j(f)\cdot\mu$ the first time we get

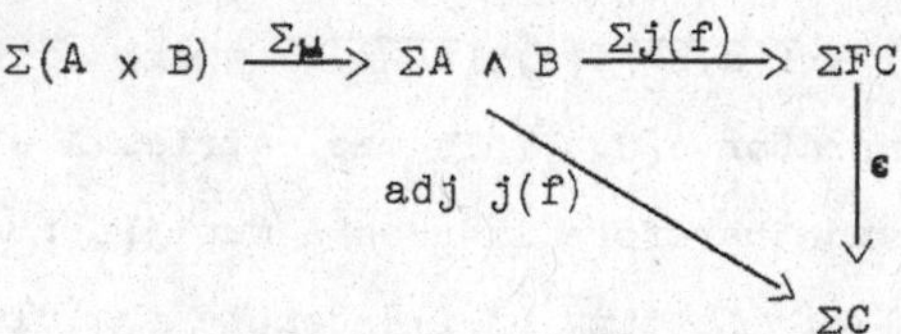

where ϵ denotes the evaluation map $\Sigma\Omega(\ ) \to (\ )$. Hence adj $j(f)\cdot\Sigma\mu$ is adjoint to $j(f)\cdot\mu = f_A^{-1} f f_B^{-1}$. Each of the maps f_A, f, f_B: $A \times B \to FC$ is in fact a map into the set of generators C. Whenever this is the case the adjoint map is just the suspension of the map into C:

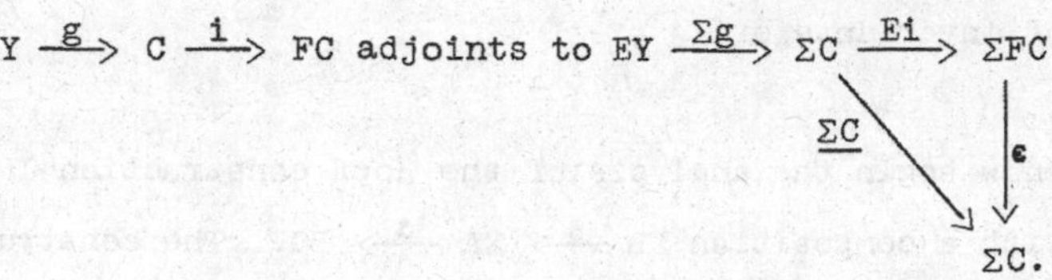

Applying the adjoint homomorphism a second time, therefore, we get $\operatorname{adj} j(f)\circ\Sigma\mu = -\Sigma f_A + \Sigma f - \Sigma f_B$, where the + operation on the right denotes track addition in the track group $[\Sigma(A \times B),\Sigma C]$. We now have $\operatorname{adj} j(f)\circ\chi(\mu) = (-\Sigma f_A + \Sigma f - \Sigma f_B)\circ\chi(\underline{A \times B})$, and we note that the map on the right factors as indicated in the diagram:

$$A*B \longrightarrow \Sigma(A \times B) \xrightarrow{\ \rho\ } \Sigma(A \times B) \vee \Sigma(A \times B) \vee \Sigma(A \times B)$$

$$\downarrow \Sigma\pi_A \qquad \downarrow \Sigma(\underline{A\times B}) \qquad \downarrow \Sigma\pi_B$$

$$\Sigma(A\times\{b_0\}) \vee \Sigma(A \times B) \vee \Sigma(\{a_0\}\times B)$$

$$-\Sigma(f|A\times\{b_0\}) \searrow \qquad \downarrow \Sigma f \qquad \swarrow -\Sigma(f|\{a_0\}\times B)$$

$$\Sigma C$$

where ρ denotes the pinching map for track addition and the maps $\Sigma\pi_A$, $\Sigma\pi_B$ denote the suspensions of the obvious projections. It is now easy to construct a homotopy between $(\Sigma\pi_A \vee \Sigma(A \times B) \vee \Sigma\pi_B)\circ\rho\circ\chi(\underline{A \times B})$ and the composition of $\chi(\underline{A \times B})$ with the inclusion of $\Sigma(A \times B)$ into the middle suspension in $\Sigma(A \times \{b_0\}) \vee \Sigma(A \times B) \vee \Sigma(\{a_0\} \times B)$. We get as a result that $\operatorname{adj} j(f)\circ\chi(\mu) \simeq \Sigma f\circ\chi(\underline{A \times B}) = \chi(f)$ and the lemma is proved.

§2. The Hopf invariants of β

We can now begin the analysis of the Hopf construction $J(\alpha\cdot\beta)$ associated with a composition $\Sigma B \xrightarrow{\beta} \Sigma A \xrightarrow{\alpha} BG$. The construction requires first that we adjoint, which yields the map $B \xrightarrow{\text{adj}\,\beta} \Omega\Sigma A \xrightarrow{\Omega\alpha} G$, and then use that map to obtain the action of B on X, which we immediately see factors as $B \times X \xrightarrow{\text{adj}\,\beta\times\underline{X}} (\Omega\Sigma A) \times X \xrightarrow{(\Omega\alpha)\times\underline{X}} G \times X \xrightarrow{\gamma} X$. We now perform the Hopf construction on this action to obtain $J(\alpha\cdot\beta)$ and notice that it factors also:

$$\Sigma B \wedge X \xrightarrow{\Sigma\text{adj}\,\beta\wedge\underline{X}} \Sigma(\Omega\Sigma A) \wedge X \xrightarrow{\Sigma(\Omega\alpha)\wedge\underline{X}} \Sigma G \wedge X \xrightarrow{J(\gamma)} \Sigma X. \quad (2.1)$$

At this point we can remark that if the map $\beta\colon \Sigma B \to \Sigma A$ is in fact a suspension, say $\beta = \Sigma g$ where $g\colon B \to A$, then $J(\alpha\cdot\beta)$ has a very simple description given by

<u>Lemma 2</u>. If $\beta = \Sigma g$, then $J(\alpha\cdot\beta) = J(\alpha)\cdot(\beta \times \underline{X})$.

<u>Proof</u>. We consider the diagram:

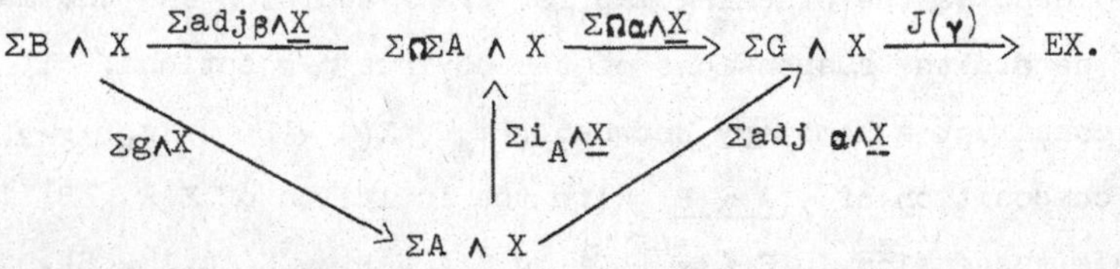

$$\begin{array}{ccccccc}
\Sigma B \wedge X & \xrightarrow{\Sigma\text{adj}\beta\wedge\underline{X}} & \Sigma\Omega\Sigma A \wedge X & \xrightarrow{\Sigma\Omega\alpha\wedge\underline{X}} & \Sigma G \wedge X & \xrightarrow{J(\gamma)} & \Sigma X. \\
& {\scriptstyle \Sigma g\wedge X}\searrow & \uparrow{\scriptstyle \Sigma i_A\wedge\underline{X}} & \nearrow{\scriptstyle \Sigma\text{adj}\,\alpha\wedge\underline{X}} & & & \\
& & \Sigma A \wedge X & & & &
\end{array}$$

The right-hand triangle commutes because $\text{adj}\,\alpha = \Omega\alpha\cdot i_A$, where i_A denotes the inclusion $A \to \Omega\Sigma A$. The left-hand triangle commutes because $\text{adj}\,\beta = \Omega\beta\cdot i_B = \Omega\Sigma g\cdot i_B = i_A\cdot g$ where i_B denotes the inclusion $B \to \Omega\epsilon B$. But it follows immediately from the definition of the Hopf construction associated with α that $J(\gamma)\cdot(\Sigma\text{adj}\,\alpha \wedge \underline{X}) = J(\alpha)$, and the lemma is proved.

If B is not a suspension, then the factorization of the map adj β through the inclusion $A \to \Omega\Sigma A$, which was crucial in the proof of the preceding lemma, is not available to us, and we are forced to reexamine the map Σadj β in (2.1). In fact, it is at precisely this point that the Hopf invariants of β enter into the description of $J(\alpha\cdot\beta)$.

Remark. Before we examine the Hopf invariants of β, an additional remark about Lemma 2 in connection with these invariants is in order. The Hopf invariants which we are about to consider are among (but not necessarily the only) obstructions to a map of suspensions being itself the suspension of a map. Even though they do not necessarily supply a complete set of obstructions, however, we will see that the Hopf invariants of β do provide us with all the information we need to obtain the general description of $J(\alpha\cdot\beta)$ and that Lemma 2 is somewhat crude in that we can relax the hypothesis that β be a suspension and insist only that all its Hopf invariants are zero, and the conclusion will follow.

For the space $\Sigma\Omega\Sigma A$ in (2.1) we have, implicit in work of I. James [6], but as described by Milnor in [8], a homotopy equivalence

$$M\colon \vee_{n=1}^{\infty}\Sigma A^{(n)} \xrightarrow{\simeq} \Sigma\Omega\Sigma A. \tag{2.2}$$

Then, as noted by Barratt [3], if we compose the map Σadj β with M^{-1} followed by the projection onto the $n^{\underline{th}}$ space in the bouquet, we obtain a map $\Sigma B \to \Sigma A^{(n)}$ whose homotopy class depends only on the homotopy class of β. We will denote this class by $M_n(\beta)$ and call it the $n^{\underline{th}}$ Milnor-Hopf invariant of β. (These invariants are closely related to the invariants James obtained in [7] using his reduced product construction and "combinatorial extensions" $A_\infty \to A_\infty^{(n)}$). Justification for the name "Milnor-Hopf invariant" is given by

Lemma 3. Let $\beta: \Sigma B \to \Sigma A$ be a map with A connected. Then $M_n(\beta) = H_{b_{n+1}}(\beta) \in [\Sigma B, \Sigma A_{b_{n+1}}] = [\Sigma B, \Sigma A^{(n)}]$ for $n = 1,2,3,\ldots$ where the invariants $H_{b_{n+1}}(\beta)$ are the Hilton-Hopf invariants singled out by Barcus and Barratt in [1] corresponding to the basic products of Milnor in [8] defined recursively by $A_{b_1} = A_1$; $A_{b_2} = A_2$; $\ldots$; $A_{b_n} = (A_{b_{n-1}} \wedge A_1)$ with $A_1 = A_2 = A$.

Primary sources for this lemma may be found in [2] and [3], so we will not prove it here. For further details the reader is referred to [4] where the lemma appears as Theorem 3.28 and a complete proof is given. A remark on the statement of the lemma is in order. The original "higher" Hopf invariants as defined by Hilton in [5] arose from the application of his theorem on the homotopy groups of a bouquet of simply-connected spheres to the map $S^r \to S^n \xrightarrow{\rho} S^n \vee S^n$. Milnor's generalization to suspensions of Hilton's theorem, applied to the special case of $F(A \vee A)$, made possible the definition of Hilton-Hopf invariants for a map between suspensions. The Hilton-Hopf invariants $H_{b_{n+1}}(\beta)$ of the lemma refer to the Hopf invariants arising from Milnor's analysis of $F(A \vee A)$, and the proof of the lemma is an algebraic exercise involving Milnor's formulae.

The Hilton-Milnor theorem ([5],[8]) describes the homotopy type of a a loop space ΩT, when T is an ordered finite bouquet of suspensions ΣT_n of connected spaces T_n, as an ordered countable product of loop spaces $\Omega\Sigma S_t$, where some S_t's are the original T_n's and the rest are recursively defined smash products of the T_n's indexed by symbolic basic products in symbols T_n (corresponding to a Witt basis for the free Lie algebra on these symbols). The ΣS_t are mapped to T either by an inclusion map, or a generalized Whitehead product of the inclusion maps $\Sigma T_n \to T$ corresponding to a symbolic basic product. The theorem in fact applies also when T is an ordered countable bouquet. It follows that the group $[\Sigma B, T]$ is, as a set, isomorphic to a strong product of groups $[\Sigma B, \Sigma S_t]$. While there are various ways of organizing this isomorphism, we will use the one described by Milnor in [8]. Let $I_t: [\Sigma B, T] \to [\Sigma B, \Sigma S_t]$ denote the projection, which is a homomorphism. If B is a finite complex, and so compact, for any element $\xi \in [\Sigma B, T]$ only a finite number of the $I_t(\xi)$ will be non-zero, and ξ can be written as an ordered track sum $\xi = \sum_t \omega_t \cdot I_t(\xi)$ taken over the non-zero $I_t(\xi)$, where $\omega_t: \Sigma S_t \to T$ is an inclusion or a Whitehead product of inclusions. The isomorphism described in [8] is such that the terms in the sum corresponding to the S_t's which are T_n's precede the other terms and occur in the same order as in the presentation of T as a bouquet.

Under the assumption that B is finite-dimensional and taking as T the bouquet $\bigvee_{n=1}^{\infty} \Sigma A^{(n)}$, (so that $T_n = A^{(n)}$), we can apply these results to the map $\Sigma \mathrm{adj}\ \beta$ in (2.1) composed with M^{-1} from (2.2) and obtain

<u>Lemma 4</u>. Given complexes X, A, B with A connected and B finite-dimensional, and a map $\beta: \Sigma B \to \Sigma A$, then

$$(M^{-1} \wedge \underline{X}) \cdot (\Sigma \mathrm{adj}\ \beta \wedge \underline{X}) \in [\Sigma B \wedge X, \bigvee_{n=1}^{\infty} \Sigma A^{(n)} \wedge X]$$

is a finite sum

$$\sum_{1}^{\infty} M_n(\beta) \wedge \underline{X} + \sum_t (\omega_t \cdot I_t(\beta)) \wedge \underline{X}$$

where $M_n(\beta) \wedge \underline{X}$ maps into $\Sigma A^{(n)} \wedge X$, the ω_t are generalized Whitehead products, and the $I_t(\beta)$ denote the generalized Hilton-Hopf invariants of β corresponding to basic products in the symbols $T_n = A^{(n)}$ of length greater than 1.

The lemma is a direct application of the Hilton-Milnor theorem together with the definition of the Milnor-Hopf invariants. We note that the hypotheses that A is connected and B is finite-dimensional ensure that each of the sums is finite. The formula given by the lemma is simplified considerably if X has the homotopy type of a co-H space, as we see in the following

Corollary. If X is a suspension, or more generally a co-H-space, then $(M^{-1} \wedge \underline{X})\cdot(\Sigma \mathrm{adj}\ \beta \wedge \underline{X}) = \sum_n M_n(\beta) \wedge \underline{X}$.

Proof. For such a space X, $\omega_t \wedge \underline{X}$ is zero if ω_t is a Whitehead product.

If we now let $\theta_n\colon \Sigma A^{(n)} \wedge X \to \Sigma X$ denote the composition $\Sigma A^{(n)} \wedge X \xrightarrow{M|\Sigma A^{(n)} \wedge \underline{X}} \Sigma\Omega\Sigma A \wedge X \xrightarrow{\Sigma\Omega\alpha\wedge\underline{X}} EG \wedge X \xrightarrow{J(\gamma)} \Sigma X$ and apply Lemma 4 and its corollary to the description of $J(\alpha\cdot\beta)$ given by (2.1), we obtain

Lemma 5. For a connected space A and a finite-dimensional space B, if the fibre X has the homotopy type of a co-H-space, then the Hopf construction $J(\alpha\cdot\beta)$ associated with a composition $\Sigma B \xrightarrow{\beta} \Sigma A \xrightarrow{\alpha} BG$ may be expressed as a sum in $[\Sigma B \wedge X, \Sigma X]$ as follows:

$$J(\alpha\cdot\beta) = \sum_n \theta_n(M_n(\beta) \wedge \underline{X}).$$

We proceed in the next section to analyze the maps θ_n. We will note here that $M_1(\beta) = H_{b_2}(\beta) = \beta$; that
$\theta_1 = J(\gamma)\cdot(\Sigma\Omega\alpha \wedge \underline{X})\cdot(\Sigma i_A \wedge \underline{X}) = J(\alpha)$, (because in (2.2) $M|\Sigma A = \Sigma a_A$);

and that if β is a suspension then $M_n(\beta) = H_{b_{n+1}}(\beta) = 0$ for $n \geq 2$, so that in the case β is a suspension Lemma 5 reduces to Lemma 2.

§3. The iterated Hopf construction

The maps $\theta_n: \Sigma A^{(n)} \wedge X \to \Sigma X$ depend upon the homotopy equivalence (2.2) which in turn is based on a homotopy equivalence $F(V_{n=1}^{\infty} A^{(n)}) \xrightarrow{\simeq} FFA$ ([8], Theorem 3) for which Milnor gives explicit formulae. Using those formulae, together with the description given by Lemma 1 and its corollary for the adjoint of the Hopf construction, we can describe the maps

$$\text{adj}\ \theta_n: A^{(n)} \wedge X \xrightarrow{\text{adj}(M|\Sigma A^{(n)} \wedge \underline{X})} F(FA \wedge X)$$

$$\xrightarrow{F(\Omega\alpha \wedge \underline{X})} F(G \wedge X) \xrightarrow{\Omega J(\gamma)} FX.$$

for each n. For instance, the map adj $\theta_2: A \wedge A \wedge X \to FX$ is defined by the correspondence:

$$\text{adj}\ \theta_2: (a_1, a_2, x) \to (a_1, x)(\overline{a_2 a_1, x})(a_2, x) \to$$

$$(\overline{a_1 x_0})(a_1 x)(\overline{a_0 x})(a_0 x)(\overline{a_2 a_1 x})(a_2 a_1 x)(\overline{a_2 x_0})(a_2 x)(\overline{a_0 x}) \in FX \qquad (3.1)$$

where for notational convenience we have suppressed the effects of the maps $\Omega\alpha$: $FA \to G$ and γ: $G \times X \to X$ and represented the action of FA on X by juxtaposition, so that $yx = \gamma((\Omega\alpha)(y), x)$ for $y \in FA$. The similarity between the formula given by Lemma 1 for the adjoint of the Hopf construction and Milnor's formulae (which are based on the commutator identity for groups $[[b,a],\emptyset] = [b,a][b,\emptyset a]^{-1}[b,\emptyset]$) is very strong, and indeed one can view Milnor's maps as "twisted" Hopf constructions on the obvious map from $A \times A \to FA$. To be more precise about this we first define the iterated Hopf construction and the "twisting" map, which can be viewed as a generalized sign. Then our knowledge of the formula for the adjoint of the Hopf construction will enable us to recognize the maps θ_n as iterated Hopf constructions modulo the generalized sign.

Definition. Given a map $\alpha\colon \Sigma A \to BG$, the iterated Hopf construction $J(\alpha)_n\colon \Sigma A^{(n)} \wedge X \to \Sigma X$ will refer to the composition:

$$\Sigma A^{(n)} \wedge X \xrightarrow{A_1\wedge\ldots\wedge A_{n-1}\wedge J(\alpha)} \Sigma A^{(n-1)} \wedge X \xrightarrow{J(\alpha)_{n-1}} \Sigma X$$

where $J(\alpha)_1 = J(\alpha)\colon \Sigma A \wedge X \to \Sigma X$.

Definition. Let $A_1 = A_2 = \cdots = A_n = A$. The generalized sign

$$s_n\colon \Sigma A^{(n)} \wedge X \to \Sigma A^{(n)} \wedge X \text{ for } n = 1,2,3,\ldots$$

shall refer to the map

$$\Sigma A^{(n)} \wedge X = A_1 \wedge A_2 \wedge \ldots \wedge A_n \wedge X \wedge S^1 \xrightarrow{\tau_n\wedge \underline{X}\wedge\phi_n}$$

$$A_n \wedge A_{n-1} \wedge \ldots \wedge A_1 \wedge X \wedge S^1 = \Sigma A^{(n)} \wedge X$$

where τ_n permutes the A-factors as indicated and $\phi_n\colon S^1\ S^1$ maps $e^{i\theta} \to e^{-i\theta}$ if n is even and $e^{i\theta} \to e^{i\theta}$ if n is odd.

The point of the generalized sign is that it is just the correction factor needed to obtain

Lemma 6. For each integer $n \geq 2$ the images (words in FX) of an element $(a_1,a_2,\ldots,a_n,x)\in A^{(n)} \wedge X$ under $\mathrm{adj}(J(\alpha)_n\cdot s_n)$ and $\mathrm{adj}\ \theta_n\colon A^{(n)} \wedge X \to FX$ are anagrams.

The proof is by induction on n; details will not be given, but the case for $n = 2$ will be illustrated. For the adjoint of $J(\alpha)_2 \cdot S_2$ we have, by Lemma 1 and its corollary, that

$$\operatorname{adj}(J(\alpha)_2 \cdot s_2)\colon (a_1,a_2,x) \to (a_2,a_1,x) \to$$
$$(a_2,a_0,)(a_2,a_1x)(a_2,a_1x_0) \to$$
$$(\overline{a_2x_0})(a_2a_0x)(\overline{a_0a_0x})(a_0a_1x)(\overline{a_2a_1x})(a_2x_0)(\overline{a_2x_0})(a_2a_1x_0)(a_0a_1x_0) \in FX,$$

and we note that this is indeed an anagram of the word $\operatorname{adj} \theta_2(a_1,a_2,x)$ described above at (3.1).

Now, if, in addition to the requirement that X be a suspension, we add the hypothesis that the action of G on X is basepoint-preserving, then we obtain

<u>Lemma 7</u>. If X is a suspension (co-H-space suffices) and the action of G on X is basepoint-preserving, then $J(\alpha)_n \cdot s_n = \theta_n$ for each n.

<u>Proof</u>. Because the action is basepoint-preserving, all the letters involving x_0 in the image words of $\operatorname{adj}(J(\alpha)_n \cdot s_n)$ and $\operatorname{adj} \theta_n$ become the identity in FX. The remaining words are still anagrams, of course, but now, because all the letters contain x, and again because the action is basepoint-preserving, it follows that the difference $\operatorname{adj}(J(\alpha)_n \cdot s_n)(\operatorname{adj} \theta_n)^{-1} \in [A^{(n)} \wedge X,FX]$ is the pre-image under $\eta^*\colon [A^{(n)} \wedge X,FX] \to [A^n \ltimes X,FX]$ of a product of commutators in the group on the right, where A^n denotes the n-fold cartesian product, $\ltimes$ denotes half-smash $((a_1,\ldots,a_n,x_0) \sim \text{basepoint})$, and $\eta\colon A^n \ltimes X \to A^{(n)} \wedge X$ denotes the standard collapsing map. But if X is a suspension the group $[A^n \ltimes X,FX]$ is abelian, from which it follows that $\operatorname{adj}(J(\alpha)_n \cdot s_n) = \operatorname{adj} \theta_n$, (the map η^* is a monomorphism),

and therefore that $J(\alpha)_n \cdot s_n = \theta_n$.

Combining Lemma 5 with Lemma 7 we obtain the

Theorem. For a connected space A and a finite-dimensional space B, if the fibre space X has the homotopy type of a co-H-space and the action of G on X is basepoint-preserving, then the Hopf construction $J(\alpha\cdot\beta)$ associated with a composition $\Sigma B \xrightarrow{\beta} \Sigma A \xrightarrow{\alpha} BG$ may be expressed as a sum in $[\Sigma B \wedge X, \Sigma X]$ as

$$J(\alpha\cdot\beta) = \sum_{1}^{\infty} J(\alpha)_n \cdot s_n \cdot (M_n(\beta) \wedge \underline{X})$$

where $J(\alpha)_n$ denotes the nth iterated Hopf construction associated with α, s_n denotes a generalized sign, $M_n(\beta)$ denotes the nth Milnor-Hopf invariant of β, and where at most a finite number of terms are non-zero.

Remark. The effect of the generalized signs s_n that occur here, and those that occur in reconciling different definitions of the generalized Hopf invariants (in [2], [3], and [4]), will be examined elsewhere. It appears that their effect can be considerably simplified.

References

1. W. D. Barcus and M. G. Barratt, "On the homotopy classification of the extensions of a fixed map", Trans. Amer. Math. Soc. 88 (1958), 57-74.

2. M. G. Barratt, "Higher Hopf Invariants", (mimeographed notes), University of Chicago (Summer 1957).

3. M. G. Barratt, "Remarks on James' invariants", Colloquium on Algebraic Topology, Matematisk Institute, Aarhus Universitet,

4. C. H. Hanks, "The Hopf Construction Associated to a Composition", Thesis, Northwestern University, 1975.

5. P. J. Hilton, "On the homotopy groups of the union of spheres", J. London Math. Soc. 30 (1955), 154-172.

6. I. M. James, "Reduced product spaces", Ann. of Math. (2) 62 (1955), 170-197.

7. I. M. James, "On the suspension triad", Ann. of Math. (2) 63 (1956), 191-247.

8. J. Milnor, "On the construction FK", (mimeographed notes), Princeton University, (1956).

On the stable decomposition of $\Omega^\infty S^\infty A$

by

Daniel S. Kahn*

Section 1.

In [S], Snaith gives a stable decomposition of $\Omega^n S^n A$, a connected, into a wedge of spaces which, when $n = \infty$, have the homotopy type of the m-adic constructions $D_m A = W\Sigma_m \ltimes_{\Sigma_m} A^{(m)}$. Both the result and Snaith's proof may be thought of as extending Milnor's proof of the decomposition of $\Omega S \Omega SA$ [Mil]. Proofs of the stable decomposition of $\Omega^\infty S^\infty A$ have also been given by F. Cohen and L. Taylor [CT] and, for $n = \infty$, by Barratt and Eccles [BE1, BE2, E] which are similar in spirit to the proof given in [S] but which use different models for $\Omega^n S^n A$.

The decomposition of ΩSA after a single suspension may also be given a proof derived from the facts that $S(K \times L) \simeq SK \vee SL \vee S(K \wedge L)$ and that the James reduced product $A_\infty \simeq \Omega SA$ is filtered by subspaces A_n which are quotients of A^n. Since the stable decomposition of $\Omega^n S^n A$ and its ingredients have proven to be of continued use (see, for example, [Ka, Ki, Ma]), it seems worthwhile to record here a proof of the stable decomposition of $\Omega S A$ which is modeled on the proof of the decomposition of $S\Omega SA$ which uses $S(K \times L) \simeq SK \vee SL \vee S(K \wedge L)$.

For simplicity, we assume throughout that spaces are countable CW complexes. Our basic reference for stable homotopy theory is [A]. The main result, Theorem 2.3, is proved in Section 2, except that the proof of Lemma 2.1 is deferred to Section 3.

Section 2.

In this section we state the decomposition theorem and give a

*Partially supported by NSF Grant No. MCS76-07051 A01

proof based on a lemma whose proof is deferred until Section 3.

We begin by recalling the Barratt-Eccles model Γ^+A for $\Omega^\infty S^\infty A$ (when A is connected), or rather, those features of the model which are need in our discussion. See [BE2] for fuller details. Γ^+A is defined to be an identification space of $\coprod_{n=1}^{\infty} W\Sigma_n \times_{\Sigma_n} A^n$, where $W\Sigma_n$ is one of the usual models of the universal bundle for the symmetric group Σ_n (we don't need to know which model). The only fact about the identifications we will use is that a point $[w; a_1,\ldots,a_n] \in W\Sigma_n \times_{\Sigma_n} A^n$ is identified with a point of $W\Sigma_{n-1} \times_{\Sigma_{n-1}} A^{n-1}$ if any of $a_1,\ldots,a_n$ is the base-point $a_0 \in A$.

Denote by Γ_n^+A the image in Γ^+A of $W\Sigma_n \times_{\Sigma_n} A^n$. It follows that $\Gamma_n^+A/\Gamma_{n-1}^+A = D_mA = W\Sigma_n \ltimes_{\Sigma_n} A^{(n)}$, the n-adic construction on A. Barratt and Eccles prove that if A is connected, $\Gamma^+A \simeq \Omega^\infty S^\infty A$. If A is not connected, Γ^+A is altered to obtain $\Gamma A \simeq \Omega^\infty S^\infty A$. Since we deal only with A connected, we only use Γ^+A. However, to simplify notation we will henceforth write ΓA for Γ^+A and Γ_nA for Γ_n^+A.

<u>Lemma (2.1)</u>. There exist spectra Y_n and stable maps $\phi_n\colon Y_n \to \Gamma_nA$ such that the composites

$$Y_n \xrightarrow{\phi_n} \Gamma_nA \xrightarrow{\psi_n} D_nA$$

are stable homotopy equivalences. (ψ_n is the natural projection $\Gamma_nA \to \Gamma_nA/\Gamma_{n-1}A = D_nA$).

<u>Proof</u>: The proof is deferred to Section 3.

<u>Theorem 2.2</u>. Γ_nA has the same stable homotopy type as $D_1A \vee D_2A \vee \cdots \vee D_nA$ for $n < \infty$.

Proof: The proof will be by induction. For $n = 1$, $\Gamma_1 A$ is actually equal to $D_1 A$.

Now assume that $\Gamma_{n-1}A$ is stably equivalent to $D_1A \vee \cdots \vee D_{n-1}A$. The proof of 2.2 will be complete if we show that $\Gamma_n A$ is stably equivalent to $\Gamma_{n-1}A \vee D_n A$. Let $i_{n-1}: \Gamma_{n-1}A \to \Gamma_n A$ be the inclusion and let $\alpha_n: D_n A \to Y_n$ be the stable homotopy inverse that exists by virtue of (2.1). Then the stable map $i_{n-1} \vee (\emptyset_n \cdot \alpha_n): \Gamma_{n-1}A \vee D_n A \to \Gamma_n A$ induces an isomorphism of homotopy groups. This may be seen by applying the 5-lemma to the induced map of pairs $(\Gamma_{n-1}A \vee D_n A, \Gamma_{n-1}A) \to (\Gamma_n A, \Gamma_{n-1}A)$ and using (2.1). It follows that $i_{n-1} \vee (\emptyset_n \cdot \alpha_n)$ is a stable equivalence [A; 35] and the proof of (2.2) is complete.

Theorem 2.3. ΓA is stably equivalent with $\bigvee_{n=1}^{\infty} D_n A$.

Proof: Consider the stable map

$$\chi = \bigvee_{n=1}^{\infty} (j_n \cdot \emptyset_n \cdot \alpha_n): \bigvee_{n=1}^{\infty} D_n A \to \Gamma A,$$

where $j_n: \Gamma_n A \to \Gamma A$ is the natural inclusion. Since the $D_n A$ are $(n-1)$-connected, it follows from (2.2) that χ_* maps homotopy groups isomorphically. The result now follows from [A; 3.5].

Section 3.

This section is devoted to the proof of Lemma 2.1. In the proof we will use various spaces with given actions of Σ_n. It will be convenient to list them at the outset.

Notation (3.1)

a) S^t = t-sphere with the trivial action of Σ_n.

b) Δ^{n-1} = standard (n-1)-simplex with the action of Σ_n induced by the permutation of its vertices.

c) $A^n = A \times \cdots \times A$ with the action of Σ_n given by permuting the coordinates. A^n has $p = (a_0, \ldots, a_0)$ as basepoint, where a_0 is the basepoint of A.

d) $A^{(n)} = A \wedge \cdots \wedge A$ with the action and basepoint induced from c).

e) Σ^{n-1} = unit $(n-1)$-sphere in R^n with the action of Σ_n induced by the permutation representation $\rho\colon \Sigma_n \to O(n)$. We take $\alpha = (1/\sqrt{n})(1, \ldots, 1)$ to be the basepoint of Σ^{n-1}.

f) Denote by $\rho'\colon \Sigma_n \to O(n-1)$ the restriction of ρ to the plane $\Sigma x_i = 0$. Thus ρ is equivalent to $\rho' + \epsilon$, where ϵ is the trivial one dimensional representation of Σ_n.

g) Σ^{qn-1} = unit $(qn-1)$-sphere in R^{qn} with the action of Σ_n induced by the direct sum representation $q\rho\colon \Sigma_n \to O(nq)$. The basepoint is $(1/\sqrt{qn})(1, \ldots, 1)$.

h) Cartesian products shall have diagonal actions of Σ_n; quotient spaces shall have induced actions of Σ_n.

i) An important case of h) is $\tilde{J}^n A$, the n-fold join of A with itself, which is an identification space of $A^n \times \Delta^{n-1}$ [Mi2]. In this section, we will use $J^n A$, the space obtained from $\tilde{J}^n A$ by collapsing to a point (= basepoint) the contractible subcomplex $\tilde{J}^n\{a_0\} \subset \tilde{J}^n A$. This does not change the homotopy type.

j) Σ^{qn} denotes the one point compactification of R^{qn} with the action of Σ_n induced from $q\rho\colon \Sigma_n \to O(qn)$. We take ∞ as the basepoint of Σ^{qn}.

The following two lemmas are evident.

<u>Lemma 3.2</u>. Σ^{n-1} is equivariantly homeomorphic with the one point compactification of ρ' (see 3.1e,f) with α corresponding to the point at infinity.

<u>Lemma 3.3</u>. We have the following equivariant homeomorphisms:

(i) $\Sigma^n \equiv \Sigma^{n-1} \wedge S^1$,

(ii) $\Sigma^{(p+q)n} \equiv \Sigma^{pn} \wedge \Sigma^{qn}$, and

(iii) $\Sigma^{(q+1)n} \equiv \Sigma^{n-1} \wedge S^1 \wedge \Sigma^{qn}$.

Now let $f\colon A^n \to A^{(n)}$ be the identification map.

Lemma 3.4. (i) f is Σ_n-equivariant.

(ii) The exists an equivariant relative homeomorphism

$$g\colon (\Delta^{n-1}, \dot{\Delta}^{n-1}) \to (\Sigma^{n-1}, \alpha).$$

(iii) $f \times g\colon A^n \times \Delta^{n-1} \to A^{(n)} \times \Sigma^{n-1}$ induces a homotopy equivalence $h\colon J^n A \to A^{(n)} \wedge \Sigma^{n-1}$, h being Σ_n-equivariant.

(iv) The map h factors as the composite of two equivariant maps

$$J^n A \xrightarrow{h_1} A^n \wedge \Sigma^{n-1} \xrightarrow{h_2} A^{(n)} \wedge \Sigma^{n-1}.$$

Proof: The only part of (3.4) whose proof may not be immediately evident is (iii). Letting a_i denote the basepoint $a_0 \in A$, the map h collapses the union U of the subcomplexes $A * \cdots * A * a_i * A * \cdots * A$ ($*$ denotes the join operation). Since the finite intersection of these are of the form $A * \cdots * A * a_{i_1} * A * \cdots * A * a_{i_t} * A * \cdots * A$, the finite intersections are all acyclic. A simple Mayer-Vietoris argument shows that U is acyclic. Hence $h_*\colon H_*(J^n A) \xrightarrow{\cong} H_*(A^{(n)} \wedge \Sigma^{n-1})$. Since A is connected, $J^n A$ and $A^{(n)} \wedge \Sigma^{n-1}$ are both simply connected, hence h is a homotopy equivalence. ($J^n A$ is simply-connected when $n \geq 2$. But for $n = 1$, h is a homeomorphism, so (2.3iii) is true in that case also.)

Lemma (3.5). If $\beta\colon B\Sigma_n \to BO(r)$ is an r-plane bundle over $B\Sigma_n$, there exist integers K_i so that

(i) $K_i\beta = \beta \oplus \cdots \oplus \beta$ is trivial over $B^i\Sigma_n$,

(ii) K_i divides K_{i+1}

(iii) the trivializations γ_i can be chosen so that γ_{i+1} extends $(K_{i+1}/K_i)\cdot\gamma_i$.

Proof: (3.5) follows by simple obstruction theory from the finiteness of the groups $\tilde{H}^i(B\Sigma_n; \pi_j(O))$.

The following is evident by inspection.

Lemma (3.6). Let B be a space on which Σ_n acts leaving the base-point b_0 fixed. Then we have the following equivariant homomorphism which is natural with respect to equivariant maps of (B,b_0):

$$(W\Sigma_n \ltimes B) \wedge S^N \equiv W\Sigma_n \ltimes (B \wedge S^N).$$

Lemma 3.7. Let B be as in (3.6). Then we have the following equivariant homeomorphism which is natural with respect to equivariant maps of (B,b_0):

$$W^i\Sigma_n \ltimes (B \wedge E^{K_i n}) \equiv W^i\Sigma_n \ltimes (B \wedge S^{K_i n}).$$

(K_i is as in (3.5) with $\beta = B\rho\colon B\Sigma_n \to BO(n)$.)

Proof: This follows from (3.5). (Refer to (3.1) for notation.)

We are now ready to give the proof of Lemma (2.1). Consider Diagram 3.8.

In Diagram 3.8, the homeomorphism ① is derived from (3.3iii). The homeomorphism ② is derived from (3.5i). The homeomorphism ③ is derived from (3.6). The homeomorphism ⑥ is analogous to the composite of the homeomorphisms ①, ②, ③. The maps ④ and ⑤ give the evident factorization of the map induced by the projection

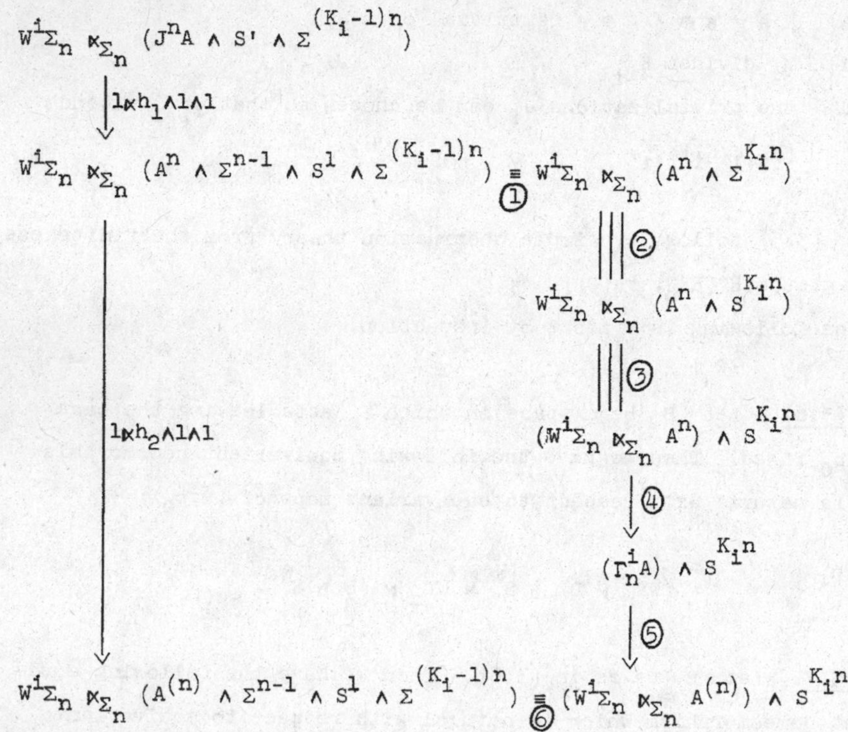

Diagram 3.8

$A^n \to A^{(n)}$. The diagram clearly commutes.

By (3.4iii), $h_2 \cdot h_1$ is a homotopy equivalence. A simple spectral sequence argument shows that $(1 \times h_2 \wedge 1 \wedge 1)\cdot(1 \times h_1 \wedge 1 \wedge 1)$ induces an isomorphism in homology. Since the spaces are high suspensions, this composite is a homotopy equivalence.

We next examine the question of letting $i \to \infty$. By (3.5) and (3.6), we have a homeomorphism

$$W^i\Sigma_n \ltimes_{\Sigma_n} (J^nA \wedge S^1 \wedge \Sigma^{(K_i-1)n}) \wedge S^{[(K_{i+1}-K_i)n]}$$

$$\equiv W^i\Sigma_n \ltimes_{\Sigma_n} (J^nA \wedge S^1 \wedge \Sigma^{(K_{i+1}-1)n}).$$

Combining this with the inclusion

$$W^i\Sigma_n \ltimes_{\Sigma_n} (J^n A \wedge S^1 \wedge \Sigma^{(K_{i+1}-1)n}) \subset (W^{i+1} \ltimes_{\Sigma_n} (J^n \wedge S^1 \wedge E^{(K_{i+1}-1)n})$$

we obtain a spectrum Y_n. (3.5ii,iii) guarantee that the maps of Diagram (3.8) define a spectrum map

$$\emptyset_n\colon Y_n \to \Sigma^\infty \Gamma_n A.$$

It follows from the above discussion that $\psi_n \cdot \emptyset_n$ induces an isomorphism of homotopy groups and hence is an equivalence. This completes the proof of Lemma 2.1.

Bibliography

[A] J. F. Adams, "Stable homotopy and generalized homology" Mimeographed notes, University of Chicago, 1971.

[BE1] M. G. Barratt and P. J. Eccles, "On a theorem of D. S. Kahn's," Duplicated notes, Manchester University, 1971.

[BE2] M. G. Barratt and P. J. Eccles, "Γ^+-structures I, II, III", Topology 13 (1974), 25-45, 113-126, 199-207.

[CT] F. Cohen and L. Taylor, "A stable decomposition for certain spaces", to appear.

[E] P. J. Eccles, The Γ Construction for Stable Homotopy Theory, Thesis, University of Manchester, 1972.

[Ka] D. S. Kahn, "Homology of the Barratt-Eccles decomposition maps", Notas de Mathematics y Simposia No. 1, Soc. Mat. Mexicana 1975.

[Ki] P. O. Kirley, On the Indecomposability of Iterated Loop Spaces, Thesis, Northwestern University, 1975.

[Ma] M. Mahowald, "A new infinite family in ${}_2\Pi_*^S$", Topology 16 (1977), 249-256.

[Mi1] J. Milnor, "On the construction FK" in J. F. Adams, Algebraic Topology: A Student's Guide, Cambridge University Press, 1972.

[Mi2] J. Milnor, "Construction of Universal Bundles I,II," Ann. Math. (2), 63 (1956), 272-284, 430-436.

[S] V. P. Snaith, "A stable decomposition for $\Omega^n S^n X$," Jour. London Math. Soc. (2), 7 (1974), 577-583.

Multiplications in two-cell spectra

by Wojciech Komornicki

Let $C = S^n \cup_\alpha e^{n+k+1}$ where $\alpha: S^{n+k} \longrightarrow S^n$ and let $\underline{E}$ be a ring spectrum. One can then form the spectrum $C \wedge \underline{E}$ where $(C \wedge \underline{E})_m = C \wedge E_{m-n}$. The spectrum $C \wedge \underline{E}$ is both a left and a right module spectrum over the spectrum $\underline{E}$ and the inclusion map $S^n \longrightarrow C$ defines a map of spectra $\underline{E} \longrightarrow C \wedge \underline{E}$. We describe the obstruction to making $C \wedge \underline{E}$ a ring spectrum with multiplication extending the above module structure. Call such a ring structure on $C \wedge \underline{E}$ admissible. We also describe obstructions to making this ring structure commutative and/or associative.

The obstruction to the existence of an admissible ring structure on $C \wedge \underline{E}$ is described in Section 3. The commutativity and associativity obstructions are described in Sections 4 and 5 respectively. A generalized James construction due to Barratt and Mahowald (unpublished) is described in Section 1; the above obstructions are constructed as homotopy operations on homotopy groups on spheres taking values in the homotopy groups of this James construction. A generalized coextension construction is described in Section 2. In order to avoid signs (which would otherwise appear in most of the formulas) and so that the homotopies involved will be "nice", it will be assumed that $n \equiv 0 \pmod 4$.

Although the construction of the obstructions can be done unstably, the application will be in the stable range. Hence we will not differentiate between a space and its suspension spectrum. Likewise a map from a space (i.e., its suspension spectrum) to a spectrum will be identified with the cohomology class the map determines.

Section 1: A Generalized James Construction

We will denote the n-fold suspension, $S^n \wedge X$, of a space X by S^nX.

Since $1 \simeq T\colon S^nS^n \longrightarrow S^nS^n$ (where T is the twisting map, $T(x,y) = (y,x)$) there exists a map $\delta\colon S^nC \longrightarrow C \wedge S^n$ such that

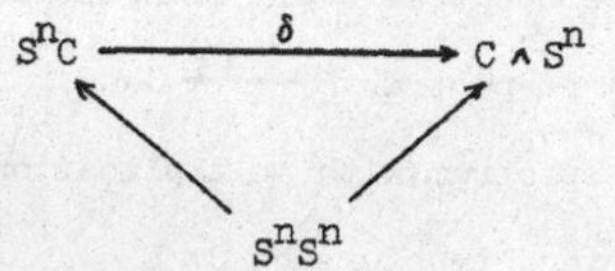

commutes.

Let $W_2 = S^nC \cup C \wedge S^n$ and let $\delta_2 = \delta \cup 1\colon W_2 \longrightarrow C \wedge S^n$. Define C_2 as the pushout

$$\begin{array}{ccc} C \wedge C & \overset{\pi_2}{\dashrightarrow} & C_2 \\ \uparrow & & \uparrow \\ W_2 & \longrightarrow & C \wedge S^n . \end{array}$$

Similarily, if $W_3 = C \wedge C \wedge S^n \cup C \wedge S^nC \cup S^nC \wedge C$ is the $3n+2k+2$-skeleton of $C \wedge C \wedge C$ and if $\delta_3\colon W_3 \longrightarrow C \wedge C \wedge S^n \longrightarrow C_2 \wedge S^n$ is the map induced by δ, then again we can form the pushout

$$\begin{array}{ccc} C \wedge C \wedge C & \overset{\pi_3}{\dashrightarrow} & C_3 \\ \uparrow & & \uparrow \\ W_3 & \longrightarrow & C_2 \wedge S^n . \end{array}$$

Continuing this process one forms a spectrum $\{C_m\}$ with maps $C_m \wedge S^n \longrightarrow C_{m+1}$ and $C_1 = C$. From the universal property of a pushout it is easily seen that $\{C_m\}$ is an associative ring spectrum with multiplication $\pi_{ij}\colon C_i \wedge C_j \longrightarrow C_{i+j}$. This generalization of the James construction is due to M. Barratt and M. Mahowald.

Section 2: The Difference Construction

Let $f,g\colon A \longrightarrow B$, $h\colon B \longrightarrow C$ be maps such that $hf \simeq hg$ by a homotopy K. Then the difference construction on the data (DCD)

$A \underset{g}{\overset{f}{\rightrightarrows}} B \xrightarrow{h} C\colon K$ is the map $SA \longrightarrow C \cup_h IB$ (where $IB = I \wedge B$)

$$\text{given by:}\quad (t,a) \longmapsto \begin{cases} 1-3t,\ f(a) & 0 \le t \le 1/3 \\ K(3t-1,\ a) & 1/3 \le t \le 2/3 \\ 3t-2,\ g(a) & 2/3 \le t \le 1 \end{cases}.$$

The virtue of this construction is its naturality; i.e., maps between data sets induce maps between the difference constructions. If f is the trivial map, this is just the construction of a coextension.

Section 3: The Existence Obstruction

Let $K\colon 1 \simeq T\colon S^nS^n \longrightarrow S^nS^n$ be any homotopy. Consideration of the space C_2 constructed in Section 1 yields:

Proposition 3.1 a) The mapping cone of the DCD

$$S^{n+k}S^{n+k} \underset{(1\wedge\alpha)T}{\overset{1}{\rightrightarrows}} S^{n+k}S^n \xrightarrow{\alpha\wedge 1} S^nS^n\colon K(\alpha\wedge\alpha)$$

is homotopy equivalent to C_2. Call this difference construction $\tilde{\alpha}\colon S^{2n+2k+1} \longrightarrow C\,S^n$.

b) $\tilde{\alpha}$ is independent (up to homotopy) of the homotopy K.

An admissible ring structure on $C \wedge \underline{E}$ is given by maps $\bar{\rho}\colon C\wedge C\wedge E_m \longrightarrow C\wedge E_{m+n}$ where

$$\begin{array}{ccc} C\wedge C\wedge E_m & \xrightarrow{\tilde{\rho}} & C\wedge E_{m+n} \\ \uparrow & & \uparrow \\ W_2\wedge E_m & \longrightarrow & C\wedge S^n\wedge E_m \end{array}$$

homotopy commutes.

By the universality of pushouts we get:

Proposition 3.2 $C \wedge \underline{E}$ is an admissible ring spectrum iff $\tilde{\alpha}_* = 0 \in E_{n+2k+1}(C)$.

Notation: Let the $\underline{E}$-homology (respectively $\underline{E}$-cohomology) with α-coefficients be defined as the homology (respectively cohomology) theory given by the spectrum $C \wedge \underline{E}$. Then $E_*(C) \cong E_*(S^0; \alpha)$ up to dimension shift.

Proposition 3.3 i) If k is even, $\tilde{\alpha}$ factors as $S^{2n+2k+1} \longrightarrow S^{2n} \longrightarrow C \wedge S^n$.

ii) If k is odd,

$$\begin{array}{ccc} S^{2n+2k+1} & \xrightarrow{\tilde{\alpha}} & C \wedge S^n \\ & \searrow^{2\alpha} & \downarrow \pi \\ & & S^{2n+k+1} \end{array}$$

is homotopy commutative.

If

$$\begin{array}{ccc} A & \xrightarrow{f} & B \\ h\downarrow & & \downarrow k \\ C & \xrightarrow{g} & D \end{array}$$

is a commutative diagram, we denote by $(k,h): B \cup_f IA \longrightarrow D \cup_g IC$ the induced map.

Proposition 3.4 $\widetilde{\alpha \wedge \beta} \simeq (\alpha,1)(\alpha \wedge \tilde{\beta})(T \wedge 1) + (\beta,1)(\tilde{\alpha} \wedge \beta)$

Proof: Let $\beta: S^{n+1} \longrightarrow S^n$. Then $\widetilde{\alpha \wedge \beta}$ is the DCD

$$S^{n+k}S^{n+k}S^{n+1}S^{n+1} \underset{(1\wedge\alpha)T\wedge(1\wedge\beta)T}{\overset{1\wedge\alpha\wedge 1\wedge\beta}{\rightrightarrows}} S^{n+k}S^{n}S^{n+1}S^{n} \xrightarrow{\alpha\wedge 1\wedge\beta\wedge 1} S^{4n}: K(\alpha\wedge\alpha) \wedge K(\beta\wedge\beta)$$

Using the homotopy $K \wedge K \simeq (T \wedge K) + (K \wedge 1)$ (here we consider a homotopy as a path) the result follows.//

Corollary 3.5 If $E^*(\ ;\alpha)$ and $E^*(\ ;\beta)$ are multiplicative theories, then so is $E^*(\ ;\alpha\beta)$.

Corollary 3.6 For all α, $\widetilde{\alpha \wedge \alpha} \simeq *$.

Corollary 3.7 If $\alpha = \beta^2$ for some $\beta \in \pi_*^S$, $C \wedge \underline{E}$ is an admissible ring spectrum, or equivalently, $E^*(\ ;\alpha)$ is a multiplicative theory.

If $C \wedge \underline{E}$ admits an admissible ring structure, then there exists a map $\rho: C_2 \longrightarrow C \wedge E_n$ such that

$$\begin{array}{ccc} C_2 & \xrightarrow{\rho} & C \wedge E_n \\ \uparrow & \nearrow_{1} & \\ C \wedge S^n & & \end{array}$$

commutes. And any such map

defines an admissible ring structure on $C \wedge \underline{E}$. This defines a one-to-one correspondence between different admissible ring structures on $C \wedge \underline{E}$ and elements $\rho \in E^*(C_2; \alpha)$. The commutativity of the above diagram yields:

Proposition 3.8 If $C \wedge \underline{E}$ is an admissible ring spectrum, then the homotopy classes of admissible ring structures on $C \wedge \underline{E}$ are in one-to-one correspondence with $E^{2k+2}(S^0; \alpha)$.

Section 4: The Commutativity Obstruction

The spectrum $C \wedge \underline{E}$ is a commutative ring spectrum if the maps $\mu: (C \wedge \underline{E}) \wedge (C \wedge \underline{E}) \longrightarrow C \wedge \underline{E}$ defining an admissible ring structure on $C \wedge \underline{E}$ make the diagram

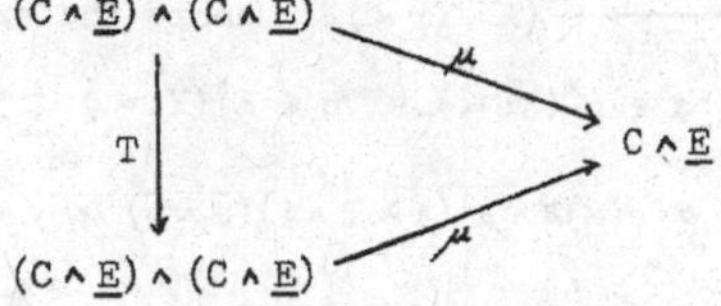

homotopy commute up to the appropriate sign. Assuming that $\underline{E}$ itself is a commutative ring spectrum, this is equivalent to the requirement that

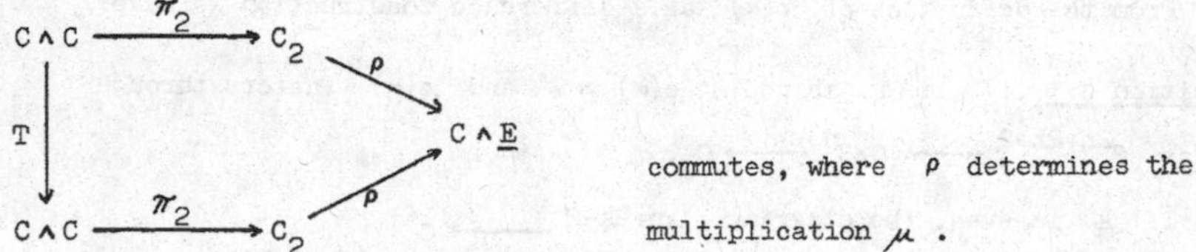

commutes, where ρ determines the multiplication μ.

For the remainder of this section it will be assumed that $\underline{E}$ is a commutative ring spectrum.

Since $n \equiv 0 \pmod 4$, for any homotopy $K: 1 \simeq T: S^nS^n \longrightarrow S^nS^n$, there exists a homotopy $K \simeq K(-1 \wedge T)$. This homotopy then defines a homotopy $B: \tilde{\alpha} \simeq \tilde{\alpha}(-1 \wedge T)$ and hence the DCD

$$S^1S^{n+k}S^{n+k} \underset{-1\wedge T}{\overset{1}{\rightrightarrows}} S^1S^{n+k}S^{n+k} \xrightarrow{\tilde{\alpha}} C\wedge S^n : B$$

is defined. Call it $c(\alpha): S^{2n+2k+2} \longrightarrow C_2$.

Proposition 4.1 i) $\pi_2(1-T) \simeq c(\alpha)\pi : C\wedge C \longrightarrow S^{2n+2k+2} \longrightarrow C_2$.

ii) $c(\alpha)$ is independent (up to homotopy) of the homotopy B.

Proposition 4.2 If $C\wedge \underline{E}$ is an admissible ring spectrum with ring structure induced by $\rho \in E^{2n}(C_2;\alpha)$, then this ring structure is commutative if and only if $c(\alpha)^*\rho = 0 \in E^{2k+2}(S^0;\alpha)$.

Not only does $c(\alpha)$ tell us whether or not the spectrum $C\wedge\underline{E}$ is commutative, it gives us an explicit formula for the commutator: Let $\bar{\wedge}: E^*(X;\alpha) \otimes E^*(Y;\alpha) \longrightarrow E^*(X\wedge Y;\alpha)$ denote the external product determined by ρ. Then for $a \in E^*(X;\alpha)$, $b \in E^*(Y;\alpha)$

$$a\,\bar{\wedge}\,b \;-\; T^*(b\,\bar{\wedge}\,a) \;=\; \rho(c(\alpha)\pi\wedge\mu)(1\wedge T\wedge 1)(\bar{a}\wedge\bar{b})$$

where $\bar{a}: X \longrightarrow C\wedge\underline{E}$, $\bar{b}: Y \longrightarrow C\wedge\underline{E}$, $\pi: C\wedge C \longrightarrow S^{2n+2k+2}$ and μ is the multiplication on $\underline{E}$.

From the definition of $c(\alpha)$ as a difference construction we have:

Proposition 4.3 i) If k is odd, $2c(\alpha) \simeq *$ and $c(\alpha)$ factors through $C\wedge S^n$ as $S^{2n+2k+2} \longrightarrow C\wedge S^n \longrightarrow C_2$.

ii) If k is even, the diagram

$$\begin{array}{ccc} S^{2n+2k+2} & \longrightarrow & C_2 \\ & \searrow^{2} & \downarrow \pi \\ & & S^{2n+2k+2} \end{array}$$

homotopy commutes.

Corollary 4.4 i) If k is even, $2\tilde{\alpha} \simeq *$.

ii) If k is even and α is of odd order, $\tilde{\alpha} \simeq *$.

iii) If k is even and either α is of odd order or $\mathrm{Tor}(E^*, Z_2) = 0$, then $C\wedge\underline{E}$ is an admissible ring spectrum.

<u>Corollary 4.5</u> If $C \wedge \underline{E}$ is an admissible ring spectrum, then

i) if k is odd and there exists an admissible commutative ring structure on $C \wedge \underline{E}$, then all admissible ring structures on $C \wedge \underline{E}$ are commutative;

ii) if k is even and there exists an admissible commutative ring structure on $C \wedge \underline{E}$, then the set of all admissible commutative ring structures on $C \wedge \underline{E}$ is in one-to-one correspondence with $\mathrm{Tor}(E^{2k+2}(S^0;\alpha), Z_2)$.

If $\beta: S^{n+1} \longrightarrow S^n$, by the naturality of the difference construction, the following diagram is commutative:

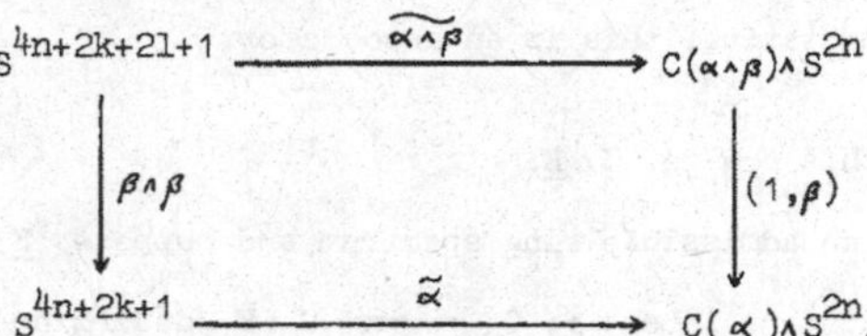

Again using the naturality of the difference construction we get the commutative diagram:

$$\begin{array}{ccc} S^{4n+2k+2l+2} & \xrightarrow{c(\alpha\wedge\beta)} & C_2(\alpha\wedge\beta) \\ \downarrow{\scriptstyle \beta\wedge\beta} & & \downarrow{\scriptstyle (1,\beta,\beta\wedge\beta)} \\ S^{4n+2k+2} & \xrightarrow{c(\alpha)} & C_2(\alpha). \end{array}$$

Let now α be a map in an odd stem and of odd order q. Letting β be a map of degree q in the 0-stem we have

<u>Proposition 4.6</u> If k is odd and the order of α is odd, then $c(\alpha) \simeq *$.

Calculations: 1) $c(2\iota): S^6 \longrightarrow S^4 \cup_\eta e^6$ is the unique coextension of 2ι. Hence $BU^*(\ ;2\iota) = K^*(\ ;Z_2)$ is not a commutative theory for any admissible ring structure.

2) $c(\eta): S^8 \longrightarrow S^4 \cup_\eta e^6 \cup_\nu e^8$ is null-homotopic since $\pi_{n+4}(S^n \cup_\eta e^{n+2}) = 0$.

Section 5: The Associativity Obstruction

As in the study of any multiplicative object, the problem of determining the associativity obstruction for $C \wedge \underline{E}$ is more complex than that of determining the existence or commutativity obstructions. If x, y, and z are elements of some multiplicative theory, we can look at the element $(xy)z - (zx)y$. If this is zero for all x, y, and z then the multiplication is both commutative and associative. If the multiplication is associative, we may take y to be the identity and then the above element is the commutator of x and z. If the multiplication is commutative, this is an associator.

We can apply this now to $C \wedge \underline{E}$:

Let $C \wedge \underline{E}$ be an admissible ring spectrum and suppose $\underline{E}$ is both commutative and associative. Let $\rho: C_2 \longrightarrow C \wedge \underline{E}$ determine the ring structure on $C \wedge \underline{E}$. Let $T': C \wedge C \wedge C \longrightarrow C \wedge C \wedge C$ be defined by $T'(x,y,z) = (z,x,y)$. Then $\pi_3 - \pi_3 T' \simeq (\pi_{2,1} - \pi_{1,2}T)(\pi_2 \wedge 1): C \wedge C \wedge C \longrightarrow C_2 \wedge C \longrightarrow C_3$ is the joint associator-commutator describe above. If we let $\overline{C} = C_2 \wedge S^n \cup S^{2n} C \subseteq C_2\, C$ then $\pi_3 - \pi_3 T'$ factors through a map

$$S^{3n+2k+2} \cup e^{3n+3k+3} = C_2 \wedge C_{/\overline{C}} \xrightarrow{a(\alpha)} C_2 \wedge S^n \longrightarrow C_3$$

and $c(\alpha) \wedge 1 \simeq a(\alpha)\iota: S^{3n+2k+2} \longrightarrow C_2 \wedge S^n$.

<u>Proposition 5.1</u> If k is even or if $2\alpha \simeq *$ or if $c(\alpha) \simeq *$, then there exists a map $\bar{a}(\alpha): S^{3n+3k+3} \longrightarrow C_2 \wedge S^n$ such that if $\rho: C_2 \longrightarrow C \wedge \underline{E}$ defines an admissible multiplication on $C \wedge \underline{E}$ then this multiplication is associative if an only if $\bar{a}(\alpha)^* \rho = 0 \in E^{3k+3}(S^0; \alpha)$.

<u>Proof</u>: If k is even or if $2\alpha \simeq *$ then the domain of $a(\alpha)$ is homotopy equivalent to the wedge of two spheres; so define $\bar{a}(\alpha)$ as the component of $a(\alpha)$ on the 3n+3k+3-sphere. If $c(\alpha) \simeq *$, then $a(\alpha)$ factors through the 3n+3k+3 sphere; define $\bar{a}(\alpha)$ as this factorization. //

<u>Proposition 5.2</u> $\iota a(\alpha)\colon S^{3n+2k+2} \cup e^{3n+3k+3} \longrightarrow C_2 \wedge S^n \longrightarrow C_3$ is uniquely defined up to homotopy by the formula $\pi_3(1 - T') \simeq \iota a(\alpha)\pi$.

<u>Proof</u>: This follows from examination of the attaching maps in $C \wedge C \wedge C$ and in C_3.//

By Proposition 5.2 it is now trivial to calculate the indeterminacy of $\bar{a}(\alpha)$ when it is defined. In the cases k even or $2\alpha \simeq *$, $\bar{a}(\alpha)$ is defined modulo $c(\alpha)\pi^S_{k+1}$; if $c(\alpha) \simeq *$, then $\bar{a}(\alpha)$ is well-defined. Since T' is of degree 1 on the $3n+3k+3$-cell of $C \wedge C \wedge C$, we have:

<u>Proposition 5.3</u> i) If α is in an odd stem, $6\iota a(\alpha) \simeq *$.

ii) If $\bar{a}(\alpha)$ is defined, $3\iota\bar{a}(\alpha) = 0$ modulo its indeterminacy.

<u>Example</u>: $\bar{a}(\eta) \simeq *$ since the 3-component of $\pi_{14}(S^8 \cup_\eta e^{10} \cup_\nu e^{12}) = \pi_{14}(C_2(\eta))$ is zero. Combining this with previous calculations we see that any admissible ring structure on $E^*(\ ;\eta)$ is both commutative and associative.

Department of Pure Mathematics
University of Waterloo

Current address:
Department of Mathematics
Hamline University
St. Paul, Minnesota

LINE BUNDLES, COHOMOLOGY AUTOMORPHISMS, AND HOMOTOPY RIGIDITY OF LINEAR ACTIONS

Arunas Liulevicius *

Let U be a unitary group, $H \subset U$ a closed subgroup, then the homogeneous space U/H is a smooth manifold. We will investigate the G-homotopy types of linear actions of a compact group G on U/H and will prove a striking rigidity result.

An action of G on U/H is said to be <u>linear</u> if there exists a representation $\alpha : G \longrightarrow U$ such that the action is given by $g.uH = \alpha(g)uH$. We write $(U/H, \alpha)$ for this linear action.

<u>Theorem 1</u>. If $U = U(m+n+1)$, H a subgroup of U conjugate to $U(1)xU(m)xU(n)$, G a compact group with representations $\alpha, \beta : G \longrightarrow U$, then a G-map $f: (U/H, \alpha) \longrightarrow (U/H, \beta)$ with $f: U/H \longrightarrow U/H$ a homotopy equivalence exists if and only if there is a homomorphism $\chi : G \longrightarrow S^1$ such that either β or $\bar{\beta}$ is similar to $\chi\alpha$.

The implication one way is trivial: if β or $\bar{\beta}$ is similar to $\chi\alpha$ then we can construct a self-map of C^{m+n+1} as a vector space over the real numbers which induces a G-equivalence $(U/H, \alpha) \longrightarrow (U/H, \beta)$. The point of the theorem is this surprising <u>homotopy rigidity</u> of linear actions on U/H: two linear actions on U/H are G-homotopy equivalent if and only if the associated linear representations are similar as <u>real</u> projective representations of G.

It is not surprising that projective representations of G come in since if H contains the center C of U (for example this is true if H is of maximal rank) then under the left action of U on U/H the center C of U fixes every element of U/H, hence the left action of U induces a left action of the projective

---

* Research partially supported by NSF grant MCS 75-08280.

unitary group $P(U) = U/C$ on U/H. We can sharpen the theorem to

Theorem 2. If U, H are as in Theorem 1, $\alpha, \beta : G \longrightarrow P(U)$ are projective representations of G then a G-map $f: (U/H, \alpha) \longrightarrow (U/H, \beta)$ with $f: U/H \longrightarrow U/H$ a homotopy equivalence exists if and only if β or its complex conjugate $\bar{\beta}$ is similar to α as representations of G to P(U).

The homotopy rigidity result holds for other conjugacy classes of subgroups H of U. Our first result on the homotopy rigidity of linear actions on CP^n (see [4],[5]) is the case $U = U(n+1)$, $H = U(1)xU(n)$. This was followed by the result for $U = U(m+k)$ and $H = T^k xU(m)$, where T^k is the standard k-torus and $m \geqslant k$ (see [6] for details and discussion of conjectures). The following conjecture no longer seems wild.

Conjecture 3. If H is a connected subgroup of U of maximal rank then linear actions of a compact group G on U/H are rigid under homotopy.

Work is in progress on this conjecture with Wu-Yi Hsiang.

The key technique of this paper depends on a result about cohomology automorphisms of U/H induced by homotopy equivalences and their action on the Chern class of a certain complex line bundle on U/H. Notice that the group of U-maps $U/H \longrightarrow U/H$ is given by $N_U(H)/H$ where $N_U(H)$ is the normalizer of H in U. Let $c: U \rightarrow U$ be conjugation ($c(A) = \bar{A}$, the complex conjugate of the matrix A). We have chosen H so that $c(H) = H$, so c induces an equivalence $c: U/H \longrightarrow U/H$.

To state the result: consider $U = U(m+n+1)$, $H=U(1)xU(m)xU(n)$ and let $\pi: U/H \longrightarrow CP^{m+n}$ be the map $\pi(uH) = [u\varepsilon_1]$, where ε_1 is the first vector of the standard basis for C^{m+n+1}. Of course π is the projection of a smooth fiber bundle with fiber $CG_{m,n}$, the Grassmann manifold of complex m-planes in C^{m+n} . Let $h: S^{2m+2n+1} \longrightarrow CP^{m+n}$ be the Hopf bundle and $y = c_1(h) \in H^2(CP^{m+n};Z)$ its first Chern class.

Theorem 4. If H, U, $\pi: U/H \longrightarrow CP^{m+n}$ are as above, $u = \pi^*(y)$, φ an algebra automorphism of $H^*(U/H;Z)$ then there exists a $k \in N_U(H)/H$ such that $\varphi(u) = k^*(u)$ or $\varphi(u) = -k^*(u)$.

Conjecture 5. If H is a connected subgroup of maximal rank in $U = U(n)$ and $n \geq 3$ then the group of algebra automorphisms of $H^*(U/H;Z)$ is isomorphic to $N_U(H)/H \times Z/2Z$, where the generator of $Z/2Z$ acts as c^*.

We shall prove Conjecture 5 for $U = U(n+2)$, $H=U(1)xU(1)xU(n)$ (compare [6] for other results and a more restrained conjecture).

The paper is organized as follows: section 1 shows the equivalence of Theorems 1 and 2; section 2 shows how Theorem 4 implies Theorem 1; section 3 proves Conjecture 5 (hence also Theorem 4) for $U = U(3)$, $H= U(1)xU(1)xU(1)$; section 4 proves Conjecture 5 for $U = U(n+2)$, $H = U(1)xU(1)xU(n)$ for $n \geq 2$; section 5 proves Theorem 4 for $U = U(5)$, $H = U(1)xU(2)xU(2)$; section 6 proves Theorem 4 for $U = U(m+n+1)$, $H = U(1)xU(m)xU(n)$ in case $mn \geq m+n+1$. If the reader is just interested in a sample argument, section 6 is to be recommended.

The whole project of studying homotopy rigidity of linear actions on U/H began through numerous consultations with Ted Petrie while we both were enjoying the hospitality of the Matematisk Institut, Aarhus Universitet. This paper was presented at the homotopy conference at Northwestern University - thanks for helpful comments go to J.F.Adams, H.Glover, D.Gottlieb, P.Landweber, I.Madsen, R.Narasimhan, T.Petrie, M.Rothenberg, R.Schultz, R.Stong.

1. Theorem 1 is equivalent to Theorem 2.

Theorem 2 implies Theorem 1 trivially, so let us show the converse. Let $\alpha, \beta : G \longrightarrow P(U)$ be representations of a compact group, $p: U \longrightarrow P(U)$ the quotient map, and consider the pullback of pxp under the map $k = (\alpha \times \beta)\Delta$ where Δ is the diagonal map of G. We have the commutative diagram

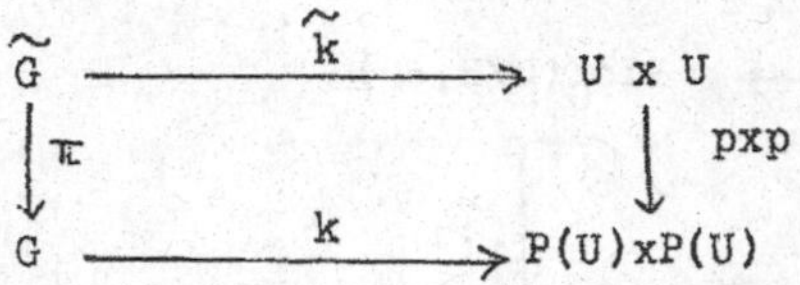

where G is a compact group (since the fiber of π is $S^1 x S^1$), $\tilde{k}(x) = (\tilde{\alpha}(x), \tilde{\beta}(x))$ and $\tilde{\alpha}, \tilde{\beta} : \tilde{G} \longrightarrow U$ are representations, $\alpha\pi = p\tilde{\alpha}$, $\beta\pi = p\tilde{\beta}$, so if $f:(U/H, \alpha) \longrightarrow (U/H, \beta)$ is a G-map then $f: (U/H, \tilde{\alpha}) \longrightarrow (U/H, \tilde{\beta})$ is a $\tilde{G}$-map and if $f: U/H \longrightarrow U/H$ is a homotopy equivalence then Theorem 1 says that there exists a homomorphism $\chi : \tilde{G} \longrightarrow S^1$ such that $\tilde{\beta}$ or its complex conjugate is similar to $\chi\tilde{\alpha}$, and when we apply p we obtain that β or its conjugate $\bar{\beta}$ is similar to α.

2. Theorem 4 implies Theorem 1.

Let $\gamma : G \longrightarrow U$ be a representation, then the Hopf bundle $h(\gamma) : (S^{2m+2n+1}, \gamma) \longrightarrow (CP^{m+n}, \gamma)$ determines an element $h(\gamma) \in Pic_G(CP^{m+n}, \gamma)$, the Picard group of G-equivariant complex line bundles over (CP^{m+n}, γ). The map π is a G-equivariant map $\pi_\gamma : (U/H, \gamma) \longrightarrow (CP^{m+n}, \gamma)$. Consider the equivariant K-theory functor K_G (see [1],[8]), then

$$\pi^!_\gamma : K_G(CP^{m+n}, \gamma) \longrightarrow K_G(U/H, \gamma)$$

is a monomorphism (see [8]), moreover $K_G(CP^{m+n}, \gamma)$ is a free $K_G(\text{point}) = R(G)$ - module on $1, s, \ldots, s^{m+n}$, where $s = h(\gamma)$ and

$$s^{m+n+1} - \gamma s^{m+n} + \Lambda^2\gamma s^{m+n-1} - \ldots + (-1)^{m+n+1} \Lambda^{m+n+1}\gamma = 0.$$

Now suppose $f: (U/H, \alpha) \longrightarrow (U/H, \beta)$ is a G-map such that $f: U/H \longrightarrow U/H$ is a homotopy equivalence, so f^* is an algebra automorphism of $H^*(U/H;Z)$. Let $u = \pi^* c_1(h)$, then according to Theorem 4 there exists a U-equivalence $k: U/H \longrightarrow U/H$ such that $f^*u = k^*u$ or c^*k^*u. If the first holds we replace f by $k^{-1}f$, and if $f^*u = c^*k^*u$ we replace f by $ck^{-1}f$ (and replace β by $\bar{\beta}$) to obtain $f^*u = u$, that is $f^*\pi^*h = \pi^*h$. Stated in another way, if $s = h(\alpha)$, $t = h(\beta)$ then $i^! f^! \pi^!_\beta t = i^! \pi^!_\alpha s$, where $i: E \hookrightarrow G$ is the inclusion of the identity subgroup. We have the commutative diagram

$$\begin{array}{ccc}
Pic_G(U/H,\beta) & \xrightarrow{f^!} & Pic_G(U/H,\alpha) \\
\downarrow i^! & & \downarrow i^! \\
Pic_E(U/H) & \xrightarrow{f^*} & Pic_G(U/H)
\end{array} .$$

Since U/H is simply connected we have an exact sequence

$$Hom(G,S^1) \xrightarrow{c^!} Pic_G(U/H,\alpha) \xrightarrow{i^!} Pic_E(U/H)$$

where $c^!\chi = \chi.1$ (. see [5] or [6] - the proof uses G.Segal's technique of cohomology of topological groups [9]). Using this exact sequence we find that $i^!f^!\pi_\beta^! t = i^!\pi_\alpha^! s$ implies that $f^!\pi_\beta^! t = \chi\pi_\alpha^! s = \pi_\alpha^!(\chi s)$ for some homomorphism $\chi: G \to S^1$. Since $\pi_\alpha^!$ is a monomorphism we can define

$$\psi = (\pi_\alpha^!)^{-1} f^! \ \pi_\beta^! : K_G(CP^{m+n},\beta) \longrightarrow K_G(CP^{m+n},\alpha)$$

which is a homomorphism of R(G)-algebras and $\psi(t) = \chi s$, so using the expansion of s^{m+n} we obtain $\chi^{-1}\beta = \alpha$ in R(G) which is precisely Theorem 1.

For a more leisurely discussion of the method see [6].

3. Conjecture 5 is true for U(1,1,1).

Let U = U(3), $H = T^3 = U(1)xU(1)xU(1)$, the standard maximal torus in U. We use the classical results of Borel [2] on the integral cohomology algebra of U/H = U(1,1,1). Let $x_i = c_1(\lambda_i^! h)$, i =1,2,3 where $\lambda_i: U(1,1,1) \longrightarrow CP^2$ are given by $\lambda_i(uH) = [u\varepsilon_i]$, $\{\varepsilon_1, \varepsilon_2, \varepsilon_3\}$ being the standard basis of C^3. The cohomology algebra H*(U(1,1,1);Z) is the quotient of $Z[x_1,x_2]$ by the homogeneous ideal I generated by $h_2 = x_1^2 + x_1x_2 + x_2^2$ and x_1^3. The element x_3 is given by $x_3 = -x_1 - x_2$. The group $N_U(H)/H$ is S_3, the symmetric group on three letters and it acts on H*(U(1,1,1);Z) by permuting x_1,x_2,x_3 - for example if $\sigma = (13)$ then $\sigma(x_1) = -x_1 - x_2$, $\sigma(x_2) = x_2$. Conjugation $c: U/H \to U/H$ is given by $c^*x_i = -x_i$.

Lemma 6. If $u \in H^2(U(1,1,1);Z)$ with $u^3 = 0$ then $u = ax_i$ with $a \in Z$, i = 1,2,3.

Proof. The element $x_1x_2^2$ generates H^6 and $x_1^2x_2 = -x_1x_2^2$.

Suppose $u = ax_1+bx_2$ and both a, b are non-zero, then $0 = u^3 = 3a^2bx_1^2x_2 + 3ab^2x_1x_2^2 = 3(-a^2b + ab^2)x_1x_2^2$, so a=b and $u = -ax_3$.

<u>Proposition 7</u>. Conjecture 5 is true for U(1,1,1).

<u>Proof</u>. Let φ be an automorphism of $H^*(U(1,1,1);Z)$. Lemma 6 shows that φ is up to sign a permutation of x_1,x_2,x_3 (since $x_i^3 = 0$ implies $\varphi(x_i)^3 = 0$), $\varphi x_i = \varepsilon_i x_{\sigma(i)}$, $\sigma \in S_3$ and $\varepsilon_i = 1$ or -1. We have to show that $\varepsilon_1 = \varepsilon_2 = \varepsilon_3$, but $x_3 = -x_1-x_2$, so $\varepsilon_3 x_{\sigma(3)} = -\varepsilon_1 x_{\sigma(1)} - \varepsilon_2 x_{\sigma(2)}$, but $\{x_{\sigma(1)}, x_{\sigma(2)}\}$ is a free basis for H^2, hence $\varepsilon_3 = \varepsilon_1 = \varepsilon_2$, so $\varphi = \sigma$ if $\varepsilon_1 = 1$, or $\varphi = c^*\sigma$ if $\varepsilon_1 = -1$.

<u>Remark</u>. Conjecture 5 is true for U/T where T is a maximal torus in U = U(n), but the proof is considerably more complicated than for n=3, partly because it is not yet known whether the analogue of Lemma 6 holds. For details see [7].

4. <u>Conjecture 5 is true for U(1,1,n), $n \geqslant 2$.</u>

We proceed as in the case of U(1,1,1), but we notice that here there is less symmetry. We have $H^*(U(1,1,n);Z) = Z[x_1,x_2] / I_n$, where as before $x_i = c_1(\lambda_i^! h)$, i = 1,2, $I_n = (h_{n+1}, x_1^{n+2})$, $h_{n+1} = x_1^{n+1} + x_1^n x_2 + \dots + x_1 x_2^n + x_2^{n+1}$. Here $N_U(H)/H = S_2$ (since $n \geqslant 2$) and $(12)x_1 = x_2$, $(12)x_2 = x_1$, $c^*x_i = -x_i$.

<u>Lemma 8</u>. If $u \in H^2(U(1,1,n);Z)$ with $u^{n+2} = 0$ then $u = ax_i$, $a \in Z$, i = 1,2.

<u>Proof</u>. H^{2n+4} has as basis $\{x_1^n x_2^2, \dots, x_1 x_2^{n+1}\}$ and $x_1^{n+1}x_2 = -x_1^n x_2 - \dots - x_1 x_2^{n+1}$. Let $u = ax_1 + bx_2$ and we may assume (using the action of S_2) that $a \neq 0$. We wish to show that b = 0. Now the coefficient of $x_1 x_2^{n+1}$ in $u^{n+2} = 0$ is

$$-\binom{n+2}{n+1} a^{n+1} b + \binom{n+2}{1} ab^{n+1} = 0$$

so $b \neq 0$ implies a = b, but then the coefficient of $x_1^n x_2^2$ is

$$-\binom{n+2}{n+1} a^{n+2} + \binom{n+2}{2} a^{n+2} = 0$$

which is ridiculous since $n \geqslant 2$.

Remark. This lemma is rather weak - but enough for our current purpose. A much stronger result (see [6]) is that if $u \in H^2(U(1,1,n);Z)$ and $u^{2n} = 0$ then $u = ax_i$, $a \in Z$, $i = 1,2$.

Proposition 9. Conjecture 5 is true for $U(1,1,n)$, $n \geq 2$.

Proof. Let φ be a cohomology automorphism of $H^*(U(1,1,n);Z)$, then we may assume that $\varphi(x_1) = x_1$ (by using $(12) \in S_2$ and/or c^*, if necessary), since $\varphi(x_i) = \varepsilon_i x_j$, $\varepsilon_i = 1$ or -1. We have $\varphi(x_2) = \varepsilon x_2$ and want to show that $\varepsilon = 1$. Now φ must preserve h_{n+1}: $x_1^{n+1} + x_1^n x_2 + \ldots + x_1 x_2^n + x_2^{n+1} = 0$, so applying φ we get $x_1^{n+1} + \varepsilon x_1^n x_2 + \ldots + \varepsilon^n x_1 x_2^n + \varepsilon^{n+1} x_2^{n+1} = 0$ hence $\varepsilon = 1$ by comparing coefficients (since $\{x_1^n x_2, \ldots, x_2^{n+1}\}$ is a free basis for H^{2n+2}). Thus φ = identity and we are done.

Remark. $U(1,1,n)$ with increasing n is a better and better approximation to $CP^\infty \times CP^\infty = Y$, and the algebra automorphisms of $H^*(Y;Z)$ are given by the group of 2×2 integer matrices with determinant 1 or -1. This is a huge group, so the result that the algebra automorphisms of $H^*(U(1,1,n);Z)$ for all $n \geq 2$ is the Klein four-group $S_2 \times Z/2Z$ is doubly surprising.

5. Theorem 4 for $U(1,2,2)$.

This surprisingly enough is the most complicated case we have to examine. We let $\lambda^1, \lambda^2, \mu^2$ be the standard complex vector bundles over $U(1,2,2) = U(5)/U(1)\times U(2)\times U(2)$ with complex dimension indicated by the superscripts such that $\lambda^1 + \lambda^2 + \mu^2 = 1^5$, the trivial 5-plane bundle. If we let $x = c_1(\lambda^1)$, $a_1 = c_1(\lambda^2)$, $a_2 = c_2(\lambda^2)$, then we have (see Borel [2])

$$H^*(U(1,2,2);Z) = Z[x, a_1, a_2] / (r_3, r_4, r_5)$$

where $r_i = \bar{c}_i(\lambda^1 + \lambda^2) = c_i(\mu^2)$ are given by

$$r_3 = -a_1^3 + 2a_1a_2 + (-a_1^2 + a_2)x - a_1x^2 - x^3,$$

$$r_4 = a_1^4 - 3a_1^2a_2 + a_2^2 + (a_1^3 - 2a_1a_2)x + (a_1^2 - a_2)x^2 + a_1x^3 + x^4,$$

$$r_5 = -a_1^5 + 4a_1^3a_2 - 3a_1a_2^2 + (-a_1^4 + 3a_1^2a_2 - a_2^2)x + (-a_1^3 + 2a_1a_2)x^2$$
$$+ (-a_1^2 + a_2)x^3 - a_1x^4 - x^5.$$

Consider the fibration

$$\begin{array}{ccc} U(2,2) & \longrightarrow & U(1,2,2) \\ & & \downarrow \pi \\ & & CP^4 \end{array}$$

induced by the standard inclusions $U(1)xU(2)xU(2) \subset U(1)xU(4)$. Since cohomology of the base and fiber is concentrated in even dimensions the Serre spectral sequence for π^* collapses and $H^*(U(1,2,2);Z)$ is a free $H^*(CP^4;Z)$-module with free basis $\{1, a_1, a_1^2, a_2, a_1a_2, a_1^2a_2\}$ since these elements project to a basis of $H^*(U(2,2);Z)$. An easy calculation gives:

$$\begin{aligned} a_1^5 &= -5\,a_1^2a_2x - 5\,a_1a_2x^2 + 5\,a_1x^4, \\ a_1^4x &= 2\,a_1^2a_2x - a_1a_2x^2 - a_2x^3, \\ a_1^3x^2 &= 2\,a_1^2a_2x - a_1^2x^3 + a_2x^3 - a_1x^4. \end{aligned}$$

We now prove

Lemma 10. Let $u = \alpha a_1 + \xi x \in H^2(U(1,1,2);Z)$ with $u^5 = 0$, then $\alpha = 0$.

Proof. We shall first show that one of α and ξ is zero. If not, inspect the coefficient of $a_1^2x^3$ in $u^5 = 0$: we obtain $-10\,\alpha^2\xi^3 + 10\,\alpha^3\xi^2 = 0$, so $\alpha = \xi$ if both are non-zero. Now inspect the coefficient of $a_1^2a_2x$: it is $-5\,\alpha^5 + 10\,\alpha^5 + 20\,\alpha^5 = 25\,\alpha^5$, and is non-zero, which is ridiculous since $u^5 = 0$. To prove the lemma we only have to point out that $a_1^5 \neq 0$ (so $\xi = 0$, $\alpha \neq 0$ is impossible).

Corollary 11. Theorem 4 holds for U(1,2,2).

Proof. If φ is an algebra automorphism then $\varphi(x)^5 = 0$, so according to Lemma 10 $\varphi(x) = x$ or $-x$ which is precisely Theorem 4.

6. Theorem 4 holds for U(1,m,n) if $mn \geqslant m+n+1$.

The argument in this case is very simple and is due to an observation of R.Narasimhan that since $U(m,n)=U(m+n)/U(m)\times U(n)$ is a Kähler manifold, then the class $\omega \in H^2(U(m,n);Z)$ corresponding to the closed 2-form defined by the Kähler metric has its mn-th power nonzero. This means that since $H^2(U(m,n);Z)=Z$ if we choose a generator a_1 its mn-th power is non-zero. We again inspect the familiar fibration

$$\begin{array}{ccc} U(m,n) & \xrightarrow{i} & U(1,m,n) \\ & & \downarrow \pi \\ & & CP^{m+n} \end{array}$$

and let $H^2(U(1,m,n);Z)$ have a basis $\{a, u\}$, where $i^*a = a_1$, $u = \pi^*y$. Suppose φ is an automorphism of $H^*(U(1,m,n);Z)$, then $\varphi(u) = \alpha a + \beta u$. We claim: $\alpha = 0$. Notice that $u^{m+n+1} = 0$, hence $\varphi(u)^{m+n+1} = 0$, so a fortiori

$$i^*\varphi(u)^{m+n+1} = (\alpha a_1)^{m+n+1} = \alpha^{m+n+1} a_1^{m+n+1} = 0,$$

but $m+n+1 \leqslant mn$ and $a_1^{mn} \neq 0$, so $\alpha = 0$, as claimed. This of course means that $\varphi(u) = u$ or $\varphi(u) = -u = c^*u$ which is precisely Theorem 4 for U(1,m,n) with $mn \geqslant m+n+1$.

REFERENCES

1. M.F.Atiyah and G.B.Segal, Lectures on equivariant K-theory, Mimeographed notes, Oxford 1965.

2. A.Borel, Sur la cohomologie des espaces fibrés principaux et des espaces homogènes de groupes de Lie compacts, Annals of Math. 57(1953), 115-207.

3. H.Glover and W.Homer, Automorphisms of the cohomology ring of a finite Grassmann manifold (manuscript, March 1977).

4. A.Liulevicius, Homotopy types of linear G-actions on complex projective spaces. Matematisk Institut, Aarhus Universitet, Preprint Series 1975/76, No. 14.

5. -------------, Characters do not lie. Transformation Groups (ed. Czes Kosniowski), Proceedings of the conference on Transformation Groups, Newcastle upon Tyne, August 1976, Cambridge University Press (1976), 139-146.

6. -------------, Homotopy rigidity of linear actions:

characters tell all (to appear in the Bulletin AMS).

7. ---------------, Homogeneous forms of high degree and homotopy rigidity (to appear in the proceedings of the summer conference on algebraic topology, Vancouver 1977).

8. G.B.Segal, Equivariant K-theory, Publ. Math. I.H.E.S. 34 (1968), 129-151.

9. ----------, Cohomology of topological groups. Symposia Mathematica, vol. IV (INDAM, Rome, 1968/69),377-387.

DEPARTMENT OF MATHEMATICS
THE UNIVERSITY OF CHICAGO
CHICAGO, ILLINOIS 60637

July 14, 1977

The construction of small ring spectra

by

Mark Mahowald

The purpose of this note is to prove the following theorem conjectured by D. Ravenel.

Theorem 1. Let λ_k be an element in π^S_{k-1} of order 4 and in the image of the J-homomorphism. Then there is a ring spectra $X_k = S^0 \cup_{4\iota} e^1 \cup_{\lambda_k} e^{8k+1} \cup_{4\iota} e^{8k+2}$. ($\iota$ represents the identity map $S^0 \to S^0$)

Recall that the J-homomorphism is a map $J\colon \pi_j(S^0) \to \pi^S_j$ where $\pi^S_j = \lim\limits_{n\to\infty} \pi_{j+n}(S^n)$. If $j = 8k - 1$ the image is a cyclic summand of 2-order greater than 4.

The existence of such ring spectra is expected to have application in stable homotopy theory. Note that there are possibly many homotopically distinct CW complexes with the given attaching maps. The theorem asserts that at least one is a ring spectra and describes uniquely the attaching maps of one such spectra.

2. We will prove in detail the case k = 1 and indicate modification necessary for the general case. We will use the results of [4] and other known calculations in stable homotopy theory. Recall that for any spectra X we can construct a resolution

$$\begin{array}{ccccccc} X & \xleftarrow{p_1} & X_1 & \xleftarrow{p_2} \cdots & X_i & \leftarrow \cdots \\ \downarrow f_1 & & \downarrow f_2 & & \downarrow f_{i+1} & \\ X \wedge K(Z_2,0) & & X_1 \wedge K(Z_2,0) & & X_i \wedge K(Z_2,0) & \end{array}$$

where (X_i, X_{i-1}, p_i) is the fiber triple induced by f_i. If we apply homotopy as a functor we get a spectral sequence, the Adams spectral sequence, where

$$E_1^{s,t+s} = \pi_{t+1}(X_i \wedge K(Z_2,0)) = (H_*(X) \otimes I^s A)_{t+s}$$

where $I^s(A)$ is $IA \otimes IA \cdots \otimes IA$ s-times and IA is the augmentation ideal of A, the Steenrod algebra,

$$E_2^{s,t+s} = \mathrm{Ext}_A^{s,t+s}(\tilde{H}^*(X), Z_2) \text{ and } E_\infty = E_0({}_2\pi_*^S(X)).$$

Let $g: S^j \to X$ be a map which lifts to X_i. Then we have

$$\begin{array}{ccccc} X \leftarrow X_1 + \cdots \leftarrow & X_i & \leftarrow & X_{i+1} & \leftarrow \cdots \\ & \uparrow g_i & & \uparrow g_{i+1} & \\ & S^j & \leftarrow & S_1^j & \leftarrow \cdots \end{array}$$

If the map g_i does not lift to X_{i+1} then g_i induces a monomorphism in Z_2 homology and

$$X \cup_g e^{j+1} \leftarrow X_1 \cup_{g_1} e^{j+1} \leftarrow \cdots \leftarrow X_{i-1} \cup_{g_i} e^{j+1} \leftarrow x_i/S^j \leftarrow x_{i+1}/S_1^j \leftarrow \cdots$$

is a tower of fibrations whose fiber at each stage is a product of Eilenberg-MacLane spaces. The choice of notation $X \cup e^{j+1}$ and X_{i+1}/S_k^j is intended to convey the fact that

$H_*(X \cup e^{j+1}) \approx H_*(X) \oplus \tilde{H}_*(S^{j+1})$ while

$H_*(X_{i+k}/S_k^j) \approx H_*(X_{i+k})/(g_{i+k})_* H_*(S_k^j)$. Again using homotopy as the functor we have a spectral sequence and

$E_1^{s,t+s} = (H_{t+1}(X_s/S_{s-1}))$. The E_2 term fits into an exact sequence

$$\cdots \to E_2^{\sigma,t}(X) \to E_2^{\sigma,t}(X \cup e^{j+1}) \to E_2^{\sigma-j-1,t-j}(S) \xrightarrow{g_\#} E_2^{\sigma+1,t}(X)$$

where $g_\#$ is the map induced by g. A resolution built in this manner is called proper.

If we start with S^0 and build our complex cell by cell we get:

a) $S^0 \xrightarrow[4_\iota]{} S^0$ has filtration 2.

b) $S^8 \xrightarrow[4_\sigma \#]{} S^0 \cup_{4_\iota} e^1$ has filtration 4 in the proper resolution of $S^0 \cup_{4_\iota} e^1$.

c) the map $S^9 \xrightarrow[4_\iota \#]{} S^0 \cup_{4_\iota} e^1 \cup_{4_\sigma} e^9$ has filtration 5 in the proper resolution of $S^0 \cup_{4_\iota} e^1 \cup_{4_\sigma} e^9 \cup_{4_\iota} e^{10}$.

Proposition 2. There is a resolution for $S^0 \cup_{4_\iota} \cup_{4_\sigma} e^9 \cup_{4_\iota} e^{10}$ whose E_2 term is given by Table 1.

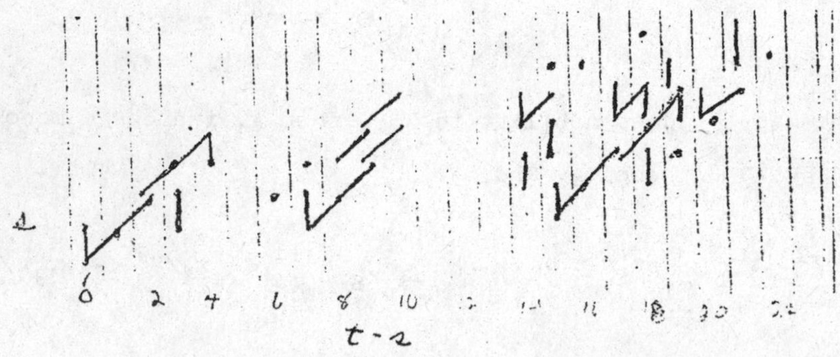

Table 1

<u>Proof</u>. The proof is a direct calculation using the calculation for $Ext_A(Z_2,Z_2)$ of Tangora [4].

Let $Z_0 \leftarrow Z_1 \leftarrow \cdots$ be the resolution of $S^0 \cup_{4\iota} e^1 \cup_{4\sigma} e^9 \cup_{4\iota} e^{10}$ whose E_1 term is its E_2 term and thus is given by Table 1. Associated to such a resolution is a chain complex

$$\underline{Z} \qquad K(E_2^{0,t}) \leftarrow K(E_2^{1,t}) \leftarrow \cdots$$

The homology of this resolution is

$H_{0,0} = Z_2$, $H_{1,2} = Z_2$, $H_{3,12} = Z_2$, $H_{4,14} = Z_2$ and $H_{s,t} = 0$ for

all other (s, t). Associated to the resolution $\underline{Z}$ there is the induced resolution over a point.

$$\begin{array}{cccccccc} & K(E^{1,t}) & & K(E^{2,t-1}) & & K(E^{i,t-i+1}) & & \\ & \uparrow & & \uparrow & & \uparrow & & \\ \overline{\underline{Z}} & \overline{Z}_0 = K(Z_2,0) & \leftarrow & \overline{Z}_1 & \leftarrow \cdots & \overline{Z}_i & & \leftarrow \cdots \end{array}$$

Consider $\overline{\underline{Z}}$ as a resolution by inclusion and let

$$(\overline{\underline{Z}} \otimes \overline{\underline{Z}})_i = \bigcup_{j+k=i} \overline{Z}_j \wedge \overline{Z}_k .$$

Theorem 3 $(\overline{\underline{Z}} \otimes \overline{\underline{Z}})_i$ is a resolution by Eilenberg-MacLane spaces of $X \wedge X$ where $X = S^0 \cup_{4\iota} e^1 \cup_{4\sigma} e^9 \cup_{4\iota} e^{10}$.

The proof is standard using the ideas in [2] 19.01-19.13 Cartan Seminare 58/59.

Theorem 4. There is a mapping of $\overline{Z} \otimes \overline{Z} \to \overline{Z}$ which is the identity for $(\overline{Z} \otimes \overline{Z})_0 = K(Z_2,0) \to K(Z_2,0) = Z_0$.

Proof. Suppose we have lifted to the ith level

$$\begin{array}{ccc} (\overline{Z} \otimes \overline{Z})_{i+1} & \xrightarrow{f_{i+1}} & \overline{Z}_{i+1} \\ p_{i+1}\downarrow & & \downarrow\rho_{i+1} \\ (\overline{Z} \otimes \overline{Z})_{i} & \xrightarrow[f_i]{} & \overline{Z}_{i} \to K(E^{i,T-i+1}) \\ \downarrow & & \downarrow \\ (\overline{Z} \otimes \overline{Z})_{0} & = & \overline{Z}_{0} \end{array}$$

If p^*_{i+1} is zero in each dimension $t - i + 1$ where $E^{i,t-i+1} \neq 0$ then clearly f_{i+1} exists. But p^*_{i+1} is not zero precisely in those dimensions for which the homology, $H_{s,t}$, of the associated resolution is non-zero for $s \leq i$ and all t. This homology is $H_{**}(Z) \otimes H_{**}(Z)$. Thus the only possible obstruction occurs when $i = 3$ in which case

$$H_{3,12}(\overline{Z} \otimes \overline{Z}) = Z_2 \oplus Z_2 \text{ and } E_2^{3,8} = Z_2$$

Consider the composite

$$X \vee X \xrightarrow[g]{} X \wedge X \to (\overline{Z} \otimes \overline{Z})_3 \xrightarrow{f_3} \overline{Z}_3 \to K(Z_2,5).$$

Any class in $H^9((\overline{Z} \otimes \overline{Z})_3)$ which is mapped nonzero by p^*_4 is mapped non-zero by g^*. Since $f_3 \circ g$ lifts to $\overline{Z}_4$ the composite

$X \vee X \to K(Z_2,5)$ is zero and so f_4 exists. There are no other obstructions.

Corollary 5. $X \longrightarrow X \wedge X \xrightarrow{f} X$ is a homotopy equivalence and thus X is a ring spectrum.

Any map of degree 1 on the bottom cell has this property.

3. Using Adams periodicity [1], Tangora's calculation, and the identification of the element of order 4 in the im of J in the $8k - 1$ stem with the class in $\mathrm{Ext}_A^{4k-1,(2k-2}(Z_2,Z_2)$ [3], it is easy to verify that the above argument works <u>mutatis mutandis</u> to cover the general case.

References

1. Adams, J. F. Stable homotopy theory, Lecture notes 3, Berlin, Springer 1964.
2. Cartan Seminar 58/59 19.01-19.13.
3. Mahowald, M. E., Metastable homotopy of S^n, Memoir AMS #72, 1967.
4. Tangora, M.C., The cohomology of the Steenrod Algebra, Math. J. 116 (1970), 18-64.

A_∞ RING SPACES AND ALGEBRAIC K-THEORY

J.P. May

In [22], Waldhausen introduced a certain functor $A(X)$, which he thought of as the algebraic K-theory of spaces X, with a view towards applications to the study of the concordance groups of PL manifolds among other things. Actually, the most conceptual definition of $A(X)$, from which the proofs would presumably flow most smoothly, was not made rigorous in [22] on the grounds that the prerequisite theory of rings up to all higher coherence homotopies was not yet available.

As Waldhausen pointed out, my theory of E_∞ ring spaces [12] gave a successful codification of the stronger notion of commutative ring up to all higher coherence homotopies. We begin this paper by pointing out that all of the details necessary for a comprehensive treatment of the weaker theory appropriate in the absence of commutativity are already implicit in [12]. Thus we define A_∞ ring spaces in section 1, define A_∞ ring spectra in section 2, and show how to pass back and forth between these structures in section 3. The reader is referred to [13] for an intuitive summary of the E_∞ ring theory that the present A_∞ ring theory will imitate.

The general theory does not immediately imply that Waldhausen's proposed definition of $A(X)$ can now be made rigorous. One must first analyze the structure present on the topological space M_nX of $(n \times n)$-matrices with coefficients in an A_∞ ring space X. There is no difficulty in giving M_nX a suitable additive structure, but it is the multiplicative structure that is of interest and its analysis requires considerable work. We prove in section 4 that M_nX is a multiplicative A_∞ space and compare these A_∞ structures as n varies in section 6. Technically, the freedom to use different A_∞ operads is crucial to the definition of these A_∞ structures, and a curious change of operad pairs trick is needed for their comparison as n varies. We study the relationship between the additive and multiplicative structures on M_nX in section 5. It turns out that M_nX is definitely not an A_∞ ring space, although it may satisfy the requirements of an appropriate strong homotopy generalization of this notion.

With this theoretical background in place, we find ourselves in a position to develop a far more general theory than would be needed solely to obtain the algebraic K-theory of spaces. Thus we construct the algebraic K-theory of A_∞ ring spaces X in section 7. The basic idea is simple enough. We take the homotopy groups of the plus construction KX on the telescope of the classifying spaces of the A_∞ spaces of unit components FM_nX. The technical work here involves the construction of the relevant compatible classifying spaces and of a modified telescope necessary for functoriality. In section 8, we analyze the effect of restricting this chain of functors to the sub A_∞ spaces F_nX of monomial matrices. The resulting plus construction turns out to be equivalent to the zero component of $Q(BFX \amalg \{0\})$, where FX is the A_∞ space of unit components of X and QY denotes colim $\Omega^n\Sigma^nY$. The proof involves the generalization of a standard consequence of the Barratt-Quillen theorem for wreath products of monoids to wreath products of A_∞ spaces together with a comparison between the A_∞ spaces F_nX and $\Sigma_n \int FX$.

This completes the development of the technical machinery. Of course, the proofs in sections 3-8 (all of which are relegated to the ends of the sections) are necessarily addressed to those interested in a close look at the machinery. The consumer who wishes to inspect the finished product without taking the tour through the factory is invited to first read section 1 and skim sections 2 and 3 (up to the statement of Proposition 3.7), then skip to section 7 and read as far as Remarks 7.4, next read section 8 as far as Remarks 8.5, and finally turn to sections 9-12.

We begin the homotopical and homological analysis of our functors in section 9, giving general homotopy invariance properties and pointing out a spectral sequence converging to H_*KX and some general formulas relevant to the computation of its E_2-term.

We discuss examples and display various natural maps and diagrams in section 10. On discrete rings, our theory reduces to Quillen's [18, 5]. For general A_∞ ring spaces X, the discretization map $X \to \pi_0 X$ is a map of A_∞ ring spaces. This establishes a natural map from the new algebraic K-groups of X to Quillen's algebraic K-groups of π_0X. To illustrate the force of this assertion, we record the following trifling consequence of the diagram displayed in Theorem 10.7.

Corollary. The usual map from the $q^{\underline{th}}$ stable homotopy group of spheres to the $q^{\underline{th}}$ algebraic K-group of Z factors through the qth algebraic K-group of X for any A_∞ ring space X such that $\pi_0 X = Z$.

On topological rings, our theory reduces to that of Waldhausen [22, § 1], or rather, to the topological analog of his simplicial theory. However, it should be pointed out that virtually all of the proofs in the earlier sections become completely trivial in this special case. The force of the theory is the translation of the obvious intuition, that much that is true for rings remains true for A_∞ ring spaces, into rigorous mathematics. The crucial question is, how much? We have already observed that the matrix "ring" functor M_n does not preserve A_∞ ring spaces, and other examples of phenomena which do not directly generalize are discussed in Remarks 10.3 and 12.4. What I find truly remarkable is how very much does in fact generalize. The point is that there are vast numbers of interesting A_∞ ring spaces which are far removed from our intuition of what a ring looks like.

Perhaps the most fascinating examples come from the fact that E_∞ ring spaces are A_∞ ring spaces by neglect of structure. For instance, the category of finitely generated projective modules (or free modules) over a commutative ring R gives rise to an E_∞ ring space in which the addition comes from the direct sum of modules and the multiplication comes from the tensor product. The homotopy groups of this space are Quillen's algebraic K-groups of R, and the present theory gives rise to a second order algebraic K-theory based on its A_∞ ring structure. I shall say a little bit more about such examples in section 10, but I should add at once that I have not yet had time even to contemplate the problem of making actual calculations.

We specialize to obtain the algebraic K-theory of topological spaces X in section 11, defining A(X) to be $KQ(\Omega X \amalg \{0\})$, as proposed by Waldhausen. The general theory gives an immediate calculation of the rational homotopy type of A(X), and various other properties claimed by Waldhausen also drop out by specialization. In particular, we discuss the algebraic K-theory of X with coefficients in a (discrete) commutative ring and give a complete account of the stabilization of the various algebraic K-theories of X to generalized homology theories, this being based on a general stabilization theorem given in the Appendix.

While the theory discussed above was inspired by Waldhausen's ideas, it is logically independent of his work and should be of independent interest. The connection with his theory is work in progress and is discussed very briefly in section 12. The basic point to be made is that, at this writing, there exist two algebraic K-theories of spaces, the one developed here and the one rigorously defined by Waldhausen, and a key remaining problem is to prove their equivalence.

I would like to thank Mel Rothenberg for insisting that I try to make Waldhausen's ideas rigorous and for many helpful discussions. Conversations with Bob Thomason have also been very useful. I am profoundly indebted to Waldhausen for envisioning the possibility of such a theory as that presented here.

Contents

§1. A_∞ ring spaces

We begin by recalling the definitional framework of [12]; details are in [12,VI §1 and §2] and a more leisurely discussion of some of the main ideas is in [10, §1-3].

The notion of an E_∞ ring space is based on the notion of an operad pair $(\mathcal{C}, \mathcal{G})$, which consists of an "additive" operad $\mathcal{C}$, a "multiplicative" operad $\mathcal{G}$, and an action of $\mathcal{G}$ on $\mathcal{C}$. An operad $\mathcal{C}$ has associated to it a monad (C, μ, η) in $\mathcal{J}$, the category of (nice) based spaces. There is a notion of an action of $\mathcal{C}$ on a space X, and this is equivalent to the standard notion of an action of the monad C on X. These notions apply equally well to $\mathcal{G}$. Actions by $\mathcal{G}$ are thought of as multiplicative, with basepoint 1, and a $\mathcal{G}_0$-space is a $\mathcal{G}$-space with a second basepoint 0 which behaves as zero under the action. When $\mathcal{G}$ acts on $\mathcal{C}$, the monad C restricts to a monad in $\mathcal{G}_0[\mathcal{J}]$, the category of $\mathcal{G}_0$-spaces. That is, CX is a $\mathcal{G}_0$-space if X is a $\mathcal{G}_0$-space and $\mu: CCX \to CX$ and $\eta: X \to CX$ are then maps of $\mathcal{G}_0$-spaces. An action of $(\mathcal{C}, \mathcal{G})$ on a space X is an action of the monad C in $\mathcal{G}_0[\mathcal{J}]$ on X. That is, a $(\mathcal{C}, \mathcal{G})$-space is both a $\mathcal{C}$-space and a $\mathcal{G}_0$-space such that the additive action $CX \to X$ is a map of $\mathcal{G}_0$-spaces. The last condition encodes distributivity homotopies in a simple conceptual way, and the multiplicative theory is to be thought of as obtained from the additive theory by a change of ground categories from spaces to $\mathcal{G}_0$-spaces.

$\mathcal{C}$ is an E_∞ operad if its jth space $\mathcal{C}(j)$ is contractible and is acted on freely by the symmetric group Σ_j. An E_∞ space is a space together with an action by any E_∞ operad $\mathcal{C}$. Examples are the spaces CX for any space X. These are commutative monoids up to all coherence homotopies. $(\mathcal{C}, \mathcal{G})$ is an E_∞ operad pair if $\mathcal{C}$ and $\mathcal{G}$ are both E_∞ operads. An E_∞ ring space is a space together with an action by any E_∞ operad pair $(\mathcal{C}, \mathcal{G})$. Examples are the spaces CX for any $\mathcal{G}_0$-space X. These are commutative semi-rings up to all coherence homotopies (semi-ring because additive inverses are not built in).

A theory of A_∞ spaces is developed in [10, §3 and 13]. An operad $\mathcal{G}$ is an A_∞ operad if $\pi_0 \mathcal{G}(j)$ is Σ_j and each component of $\mathcal{G}(j)$ is contractible. An A_∞ space is a space together with an action by any A_∞ operad $\mathcal{G}$. These are monoids up to all coherence homotopies. As explained in [10, p. 134], A_∞ spaces are equivalent to monoids, hence admit classifying spaces; there is also a direct delooping construction independent of the use of monoids.

<u>Definition 1.1.</u> An operad pair $(\mathcal{C}, \mathcal{G})$ is an A_∞ operad pair if $\mathcal{C}$ is an E_∞ operad and $\mathcal{G}$ is an A_∞ operad. An A_∞ ring space is a space together with an action by any A_∞ operad pair $(\mathcal{C}, \mathcal{G})$. Examples are the spaces CX for any $\mathcal{G}_0$-space X.

This is the desired notion of a ring (or rather, semi-ring) up to all coherence homotopies. There is also a notion with both $\mathcal{C}$ and $\mathcal{G}$ A_∞ operads, but it seems unprofitable to study rings up to homotopy for which not even addition is homotopy commutative.

Note that the product of an E_∞ operad and an A_∞ operad is an A_∞ operad. Thus if $(\mathcal{C}, \mathcal{G})$ is an E_∞ operad pair and $\mathcal{G}'$ is an A_∞ operad, then $(\mathcal{C}, \mathcal{G}' \times \mathcal{G})$ is an A_∞ operad pair, the action of $\mathcal{G}' \times \mathcal{G}$ on $\mathcal{C}$ being obtained by pullback from the action of $\mathcal{G}$ on $\mathcal{C}$. Since a $(\mathcal{C}, \mathcal{G})$-space is a $(\mathcal{C}, \mathcal{G}' \times \mathcal{G})$-space, again by pullback, E_∞ ring spaces are A_∞ ring spaces.

Some discussion of discrete operads may clarify ideas. There are operads $\mathcal{M}$ and $\mathcal{N}$ such that $\mathcal{M}(j) = \Sigma_j$ and $\mathcal{N}(j)$ is a point. An $\mathcal{M}$-space is a monoid and an $\mathcal{N}$-space is a commutative monoid. Both $\mathcal{M}$ and $\mathcal{N}$ act on $\mathcal{N}$. An $(\mathcal{N}, \mathcal{M})$-space is a semi-ring and an $(\mathcal{N}, \mathcal{N})$-space is a commutative semi-ring. Say that a map of operads is an equivalence if the underlying map of jth spaces is a homotopy equivalence for each j. An E_∞ operad $\mathcal{C}$ admits an evident equivalence $\mathcal{C} \to \mathcal{N}$. An operad $\mathcal{G}$ is an A_∞ operad if and only if it admits an equivalence $\mathcal{G} \to \mathcal{M}$. Thus A_∞ operad pairs and E_∞ operad pairs map by equivalences onto the respective operad pairs $(\mathcal{N}, \mathcal{M})$ and $(\mathcal{N}, \mathcal{N})$.

Remark 1.2. I would like to correct an annoying misprint in the crucial definition, [12,VI.1.6], of an action of $\mathcal{G}$ on $\mathcal{C}$. In (a') of the cited definition, the displayed formula is missing some symbols, d_I being written for $\lambda(g, d_I)$, and should read

$$\gamma(\lambda(g; c_1, \ldots, c_k); \underset{I \in S(j_1, \ldots, j_k)}{\times} \lambda(g; d_I))\nu = \lambda(g; e_1, \ldots, e_k) .$$

§2 A_∞ ring spectra

An E_∞ space determines a spectrum and thus a cohomology theory. The notion of E_∞ ring spectrum encodes the additional multiplicative structure on the spectra derived from the underlying additive E_∞ spaces of E_∞ ring spaces. A_∞ ring spaces also have underlying additive E_∞ spaces, and we have an analogous notion of A_∞ ring spectrum. Only the multiplicative operad $\mathcal{G}$ appears in these definitions. Let $\mathcal{L}$ denote the linear isometries E_∞ operad of [12,I.1.2]. For good and sufficient reasons explained in [12,IV §1], we assume given a map of operads $\mathcal{G} \to \mathcal{L}$. In the cited section, $\mathcal{G}$ was assumed to be an E_∞ operad. We may instead assume that $\mathcal{G}$ is an A_∞ operad. For example, $\mathcal{G}$ might be the product of an A_∞ operad and an E_∞ operad which maps to $\mathcal{L}$. Now the assumption that $\mathcal{G}$ was an E_∞ operad played no mathematical role whatever in the definition,

[12,IV.1.1], of a $\mathcal{G}$-spectrum. This notion of an action by $\mathcal{G}$ only required the map $\mathcal{G} \to \mathcal{L}$ and is thus already on hand in our A_∞ context.

Definition 2.1. An A_∞ ring spectrum is a $\mathcal{G}$-spectrum over any A_∞ operad $\mathcal{G}$ with a given morphism of operads $\mathcal{G} \to \mathcal{L}$.

As explained in [12, p. 68-70], an A_∞ ring spectrum is a (not necessarily commutative) ring spectrum with additional structure. In particular, its zero[th] space is a $\mathcal{G}_0$-space. The formal lemmas [12, IV 1.4-1.9] apply verbatim to $\mathcal{G}$-spectra for any operad $\mathcal{G}$ which maps to $\mathcal{L}$. We summarize the conclusions they yield.

Recall that a (coordinate-free) spectrum E consists of a space EV for each finite-dimensional sub inner product space V of R^∞ together with an associative and unital system of homeomorphisms $EV \to \Omega^W E(V+W)$ for V orthogonal to W; here $E_0 = E\{0\}$. The stabilization functor Q_∞ from spaces to spectra is defined by

$$Q_\infty X = \{Q\Sigma^V X \mid V \subset R^\infty\}, \text{ where } QX = \operatorname{colim} \Omega^V \Sigma^V X;$$

here the loop and suspension functors Ω^V and Σ^V are defined in terms of the sphere tV, the one-point compactification of V. The inclusion $\eta: X \to QX$ and colimit of loops on evaluation maps $\mu: QQX \to QX$ give a monad (Q, μ, η) in $\mathcal{T}$, and the analogous colimit map $\xi: QE_0 \to E_0$ gives an action of Q on E_0. The sphere spectrum S is defined to be $Q_\infty S^0$, and S^0 is a $\mathcal{G}_0$-space for any operad $\mathcal{G}$. Use of these notions is vital for rigor, but the reader may prefer to think of spectra in more classical terms, restricting attention to $V = R^i$ for $i \geq 0$.

Proposition 2.2. Let X be a $\mathcal{G}_0$-space and E a $\mathcal{G}$-spectrum.

(i) $Q_\infty X$ is a $\mathcal{G}$-spectrum and is the free $\mathcal{G}$-spectrum generated by X in the sense that a map $f: X \to E_0$ of $\mathcal{G}_0$-spaces extends uniquely to a map $\tilde{f}: Q_\infty X \to E$ of $\mathcal{G}$-spectra such that $\tilde{f}_0 \eta = f$.

(ii) S is a $\mathcal{G}$-spectrum and the unit $e: S \to E$ is a $\mathcal{G}$-map.

(iii) The monad Q in $\mathcal{J}$ restricts to a monad in $\mathcal{G}_0[\mathcal{J}]$ and $\xi: QE_0 \to E$ is a $\mathcal{G}_0$-map, so that E_0 is a Q-algebra in $\mathcal{G}_0[\mathcal{J}]$.

The following analog of [12,IV.1.10] is central to the definitions proposed by Waldhausen.

Example 2.3. For a $\mathcal{G}$-space X without zero, construct a $\mathcal{G}$-space X^+ with zero by adjoining a disjoint basepoint 0 to X and extending the action in the evident way. $Q_\infty X^+$ is then a $\mathcal{G}$-spectrum and $\eta: X^+ \to QX^+$ is a map of $\mathcal{G}_0$-spaces. If $\mathcal{G}'$ is any A_∞ operad and $\mathcal{G} = \mathcal{G}' \times \mathcal{L}$, then a $\mathcal{G}'$-space is a $\mathcal{G}$-space via the projection $\mathcal{G} \to \mathcal{G}'$, while the projection $\mathcal{G} \to \mathcal{L}$ allows $\mathcal{G}$ to be used in the present theory. Therefore $Q_\infty X^+$ is an A_∞ ring spectrum for any A_∞ space X.

Remarks 2.4. For what it is worth, we note that much of the discussion of orientation theory given in [12,IV §3] remains valid for A_∞ ring spectra. One first checks that commutativity of the underlying ring spectra is not essential to the general theory in [12,III]. Independently of this, one finds that the assertion of [12,IV.3.1] is valid for $\mathcal{G}$-spectra E for A_∞ operads $\mathcal{G}$ as well as for E_∞ operads $\mathcal{G}$. The cited result gives a certain commutative diagram of $\mathcal{G}$-spaces and $\mathcal{G}$-maps, the middle row of which yields a fibration sequence

$$G \xrightarrow{e} FE \xrightarrow{\tau} B(G;E) \xrightarrow{q} BG \xrightarrow{Be} BFE$$

after passage one step to the right by use of the classifying space functor on $\mathcal{G}$-spaces. Here FE is the union of those components of E_0 which are units in the ring $\pi_0 E_0$, G is the infinite group or monoid corresponding to some theory of bundles or fibrations, such as O, U, Top, or F, and B(G; E) is the classifying space for E-oriented G-bundles or fibrations. The map q corresponds to neglect of orientation and the maps e and τ are interpreted in [12,III. 2.5]. The point is that the notion of A_∞ ring spectrum is just strong enough to yield Be: BG → BFE, which is the universal obstruction to the E-orientability of G-bundles; compare [12, IV.3.2].

§3. The recognition principle

We first show that the zeroth space of an A_∞ ring spectrum is an A_∞ ring space and then show that the spectrum determined by the additive E_∞ structure of an A_∞ ring space is an A_∞ ring spectrum. We also obtain comparisons between the two evident composite functors and give an A_∞ ring level version of the Barratt-Quillen theorem. All of this is in precise analogy with the corresponding development for E_∞ ring spaces and spectra in [12, VII], and we need only point out the trivial changes of definition involved.

Let $(\mathcal{C}, \mathcal{G})$ be an A_∞ operad pair and suppose given a map of operad pairs $(\pi, \rho): (\mathcal{C}, \mathcal{G}) \to (\mathcal{K}_\infty, \mathcal{L})$, where $\mathcal{K}_\infty$ is the infinite little convex bodies E_∞ operad. Here $\mathcal{K}_\infty$ and its action by $\mathcal{L}$ are defined in [12, VII §1 and §2] (and we are suppressing technical problems handled there). $\mathcal{K}_\infty$ acts naturally on the zeroth spaces of spectra, and there is a morphism $\alpha_\infty: K_\infty \to Q$ of monads in $\mathcal{J}$ Similarly, π induces a morphism $C \to K_\infty$ of monads in $\mathcal{J}$. With these notations, the proof of [12, VII. 2.4] applies verbatim to prove the following result, in which the second part follows from the first via part (iii) of Proposition 2.2.

Theorem 3.1. (i) The morphisms $\pi: C \to K_\infty$ and $\alpha_\infty: K_\infty \to Q$ of monads in $\mathcal{J}$ restrict to morphisms of monads in $\mathcal{G}_0[\mathcal{J}]$.
(ii) If E is a $\mathcal{G}$-spectrum, then its zeroth space E_0 is a $(\mathcal{C}, \mathcal{G})$-space by pullback of its Q-action $\xi: QE_0 \to E_0$ along $\alpha_\infty \pi$.

An E_∞ space X determines a spectrum $\underline{B}X$. Thanks to recent work by Thomason and myself [16], we now know that all infinite loop space machines yield equivalent spectra when applied to X, but it is essential to the present multiplicatively enriched theory that we use the construction presented in [12, VII §3]. We assume now that the A_∞ operad pair $(\mathcal{C}, \mathcal{G})$ is $(\mathcal{C}' \times \mathcal{K}_\infty, \mathcal{G}' \times \mathcal{L})$ and that (π, ρ) is given by the projections, where $(\mathcal{C}', \mathcal{G}')$ is an operad pair such that each $\mathcal{C}'(j)$ is contractible (but not necessarily Σ_j-free) and $\mathcal{G}'$ is an A_∞ operad. For definiteness, one might think of the example $(\mathcal{C}', \mathcal{G}') = (\mathcal{N}, \mathcal{M})$. The proofs of [12, VII.4.1 and 4.2] apply verbatim to yield the following results.

Theorem 3.2. If X is a $(\mathcal{C}, \mathcal{G})$-space, then $\underline{B}X$ (formed with respect to the $\mathcal{C}$-space structure) is a $\mathcal{G}$-spectrum.

The relationship between X and the zeroth space B_0X is summarized by a natural diagram

$$X \xleftarrow{\simeq} B(C, C, X) \longrightarrow B(Q, C, X) \xrightarrow{\simeq} B_0X.$$

(with a dotted arrow ι from X to B_0X)

The first and third solid arrows are equivalences, and ι is obtained by use of a canonical homotopy inverse to the first arrow. The middle solid arrow, and therefore also ι, is a group completion (see [12, p.168] or, for a full discussion, [11, §1]).

Theorem 3.3. The solid arrows in this diagram are maps of $(\mathcal{C}, \mathcal{G})$-spaces. The dotted arrow ι is a map of $\mathcal{G}_0$-spaces.

The canonical homotopy inverse, and ι, are not $\mathcal{C}$-maps, but this is of little significance. The basic idea is that we have group completed the additive structure of X while carrying along the multiplicative structure.

We have the following consistency statements in special cases, the proofs being identical to those in [12, p.191-192]. For a spectrum E, there is a natural map of spectra $\tilde{\omega}: \underline{B}E_0 \to E$, and $\tilde{\omega}$ is an equivalence if E is connective (that is, if $\pi_i E = 0$ for $i < 0$).

Proposition 3.4. If E is a $\mathcal{G}$-spectrum, then $\tilde{\omega}: \underline{B}E_0 \to E$ is a map of $\mathcal{G}$-spectra.

For a $\mathcal{G}_0$-space Y, CY is a $(\mathcal{C}, \mathcal{G})$-space by Definition 1.1 and QY is a $(\mathcal{C}, \mathcal{G})$-space by Proposition 2.2(i) and Theorem 3.1(ii). Moreover, $\alpha_\infty \pi: CY \to QY$ is a map of $(\mathcal{C}, \mathcal{G})$-spaces by Theorem 3.1(i).

Proposition 3.5. For a $\mathcal{G}_0$-space Y, the composite map of $\mathcal{G}$-spectra

$$\underline{B}CY \xrightarrow{\underline{B}(\alpha_\infty \pi)} \underline{B}QY \xrightarrow{\tilde{\omega}} Q_\infty Y$$

is a strong deformation retraction. Its inverse inclusion $\nu: Q_\infty Y \to \underline{B}CY$ is induced by the freeness of $Q_\infty Y$ from the $\mathcal{G}_0$-map $Y \xrightarrow{\eta} CY \xrightarrow{\iota} B_0 CY$ and is thus a map of $\mathcal{G}$-spectra.

When $Y = CS^0$, $CY = \coprod \mathcal{C}(j)/\Sigma_j = \coprod K(\Sigma_j, 1)$. Here the last result is a multiplicatively enriched form of the Barratt-Quillen theorem, the strongest form of which appears on the E_∞ ring level. Interesting A_∞ ring level applications come from A_∞ spaces, such as monoids, via Example 2.3.

The previous result can be related to the Hurewicz homomorphism. The monad N associated to $\mathcal{N}$ assigns to a space Y its infinite symmetric product, or free commutative topological monoid, and any operad $\mathcal{G}$ acts on $\mathcal{N}$. Therefore N restricts to a monad in $\mathcal{G}_0[\mathcal{T}]$. If $\varepsilon: \mathcal{C} \to \mathcal{N}$ is the augmentation, then $(\varepsilon, 1): (\mathcal{C}, \mathcal{G}) \to (\mathcal{N}, \mathcal{G})$ is a map of operad pairs. These observations imply the following result.

Lemma 3.6. For a $\mathcal{G}_0$-space Y, $\varepsilon: CY \to NY$ is a map of $(\mathcal{C}, \mathcal{G})$-spaces, hence $\underline{B}\varepsilon: \underline{B}CY \to \underline{B}NY$ is a map of $\mathcal{G}$-spectra.

Now forget all about the multiplicative structure on Y. By an oversight, the following result was omitted from my earlier works in this area.

Proposition 3.7. For a based space Y, $\pi_*\underline{B}NY$ is naturally isomorphic to $\tilde{H}_*Y$ and $h = \underline{B}\varepsilon \circ \nu : Q_\infty Y \to \underline{B}NY$ induces the stable Hurewicz homomorphism on passage to homotopy groups. In particular, h is a rational equivalence.

Proof. The zeroth map $h: QY \to B_0NY$ is obtained by passage to direct limits from the top composite in the commutative diagram

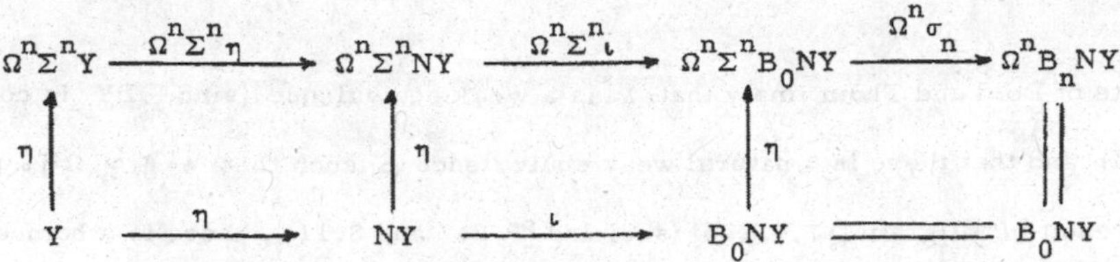

$$\begin{array}{ccccccc} \Omega^n\Sigma^n Y & \xrightarrow{\Omega^n\Sigma^n\eta} & \Omega^n\Sigma^n NY & \xrightarrow{\Omega^n\Sigma^n\iota} & \Omega^n\Sigma^n B_0NY & \xrightarrow{\Omega^n\sigma_n} & \Omega^n B_n NY \\ \uparrow\eta & & \uparrow\eta & & \uparrow\eta & & \| \\ Y & \xrightarrow{\eta} & NY & \xrightarrow{\iota} & B_0NY & = & B_0NY \end{array}$$

where σ_n is the iterated structure map of the spectrum $\underline{B}NY$ (and η is written for the unit of both monads N and $\Omega^n\Sigma^n$). Therefore h will induce the stable Hurewicz homomorphism on homotopy groups if $\iota\eta: Y \to B_0NY$ induces the ordinary Hurewicz homomorphism. If Y is connected, then ι is a natural equivalence while $\pi_*NY = \tilde{H}_*Y$ and η induces the Hurewicz homomorphism on π_* by results of Dold and Thom [3]. Thus the problem is to account for non-connected spaces. Let $\tilde{N}Y$ denote the free commutative topological group generated by Y and let $\tilde{\eta}: Y \to \tilde{N}Y$ and $\lambda : NY \to \tilde{N}Y$ denote the natural inclusions, so that $\lambda\eta = \tilde{\eta}$. Dold and Thom give that $\pi_*\tilde{N}Y = \tilde{H}_*Y$ and $\tilde{\eta}$ induces the Hurewicz homomorphism on π_*. One could prove that λ is a group completion by direct homological calculation and then deduce an equivalence $\tilde{N}Y \simeq B_0NY$, but we shall reverse this idea. Consider the following diagram.

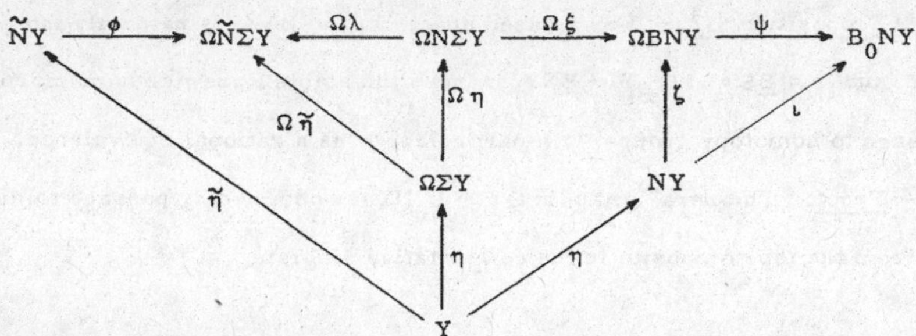

Results of Dold and Thom imply that λ is a weak equivalence (since ΣY is connected) and that there is a natural weak equivalence ϕ such that $\phi \circ \tilde{\eta} \simeq \Omega \tilde{\eta} \circ \eta$. By a result of Milgram [17, p. 245] (see also [9, 8.7 and 8.11]), there is a homeomorphism $\xi : N\Sigma Y \to BNY$ such that $\Omega \xi$ restricts on Y to $\zeta\eta$, where ζ is the standard map [9, 8.7]. Finally, since the product on NY is a map of monoids and thus of N-spaces, [12, 3.4] together with the proof of [11, 3.7 (p. 75)] give a natural weak equivalence ψ such that $\psi\zeta \simeq \iota$ (modulo the use of weak equivalences with arrows going the wrong way). Since $\tilde{\eta}_*$ on the left is the Hurewicz homomorphism, so is $(\iota\eta)_*$ on the right. It is a standard consequence of the finiteness of the stable homotopy groups of spheres that, upon tensoring with the rationals, the stable Hurewicz homomorphism becomes an isomorphism of homology theories.

We record the following corollary of the proof and an elaboration of the diagram above which shows that $\phi\lambda$ is homotopic to $\Omega(\lambda\xi^{-1})\zeta : NY \to \Omega\tilde{N}\Sigma Y$.

<u>Corollary 3.8</u>. For any space Y, $\lambda : NY \to \tilde{N}Y$ is a group completion.

§4. Matrices with entries in A_∞ ring spaces

Let X be a $(\mathcal{C}, \mathcal{G})$-space, where $(\mathcal{C}, \mathcal{G})$ is any operad pair, and consider the set M_nX of $n\times n$ matrices with entries in X. Clearly M_nX is a $\mathcal{C}$-space, namely the n^2-fold Cartesion product of the $\mathcal{C}$-space X with itself. We wish to show that if $(\mathcal{C}, \mathcal{G})$ is an A_∞ operad pair, then, while M_nX is not an A_∞ ring space, it is at least a (multiplicative) A_∞ space. Even this much is non-trivial, since M_nX is not actually a $\mathcal{G}$-space.

Since M_nX is certainly not commutative, it is convenient to first eliminate the extraneous actions by symmetric groups on the spaces $\mathcal{G}(j)$; these serve only to handle commutativity homotopies in the general theory of operads. Thus assume now that $\mathcal{G}$ is a non-Σ operad, in the sense of [10,3.12]. By [10,3.13], a typical A_∞ operad has the form $\mathcal{G}\times\mathcal{M}$ (with j^{th} space $\mathcal{G}(j)\times\Sigma_j$) for some non-$\Sigma$ operad $\mathcal{G}$ with each $\mathcal{G}(j)$ contractible. An action of $\mathcal{G}$ on the operad $\mathcal{C}$ is given by maps

$$\lambda : \mathcal{G}(k)\times\mathcal{C}(j_1)\times\ldots\times\mathcal{C}(j_k) \to \mathcal{C}(j_1\cdots j_k)$$

with the properties specified in [12,VI.1.6] (see Remark 1.2), except that its extraneous equivariance condition (c) must be deleted. A slight elaboration of [10,3.13] shows that the non-Σ operad $\mathcal{G}$ acts on $\mathcal{C}$ if and only if the operad $\mathcal{G}\times\mathcal{M}$ acts on $\mathcal{C}$. A $(\mathcal{C}, \mathcal{G})$-space is defined to be a $\mathcal{C}$-space and $\mathcal{G}_0$-space X such that the additive action $CX \to X$ is a map of $\mathcal{G}_0$-spaces, and the notions of $(\mathcal{C},\mathcal{G})$-space and $(\mathcal{C},\mathcal{G}\times\mathcal{M})$-space are then equivalent. Therefore the theory of A_∞ ring spaces may as well be developed in terms of $(\mathcal{C},\mathcal{G})$-spaces for an E_∞ operad $\mathcal{C}$ and a non-Σ operad $\mathcal{G}$ with each $\mathcal{G}(j)$ contractible. However, the work in the rest of this section requires only that $\mathcal{C}$ be an operad acted upon by a non-Σ operad $\mathcal{G}$ and that X be a $(\mathcal{C},\mathcal{G})$-space. Let

$$\theta_j : \mathcal{C}(j)\times X^j \to X \quad \text{and} \quad \xi_j : \mathcal{G}(j)\times X^j \to X$$

denote the actions of $\mathcal{C}$ and of $\mathcal{G}$ on X.

We want to use matrix multiplication to define an A_∞ space structure on M_nX. Obviously this entails use of both the multiplication and the addition on X. We are quite happy to use arbitrary j-fold products, that is, the products given by arbitrary elements of $\mathcal{G}(j)$. We are less happy to use arbitrary n^{j-1}-fold additions, but, since a canonical addition is present only in the trivial case of actual commutativity, with $\mathcal{C} = \mathcal{N}$, we have no choice. Thus define

$$\mathcal{H}_n(j) = \mathcal{C}(n^{j-1}) \times \mathcal{G}(j).$$

Let $T(j)$ denote the set of all sequences $U = (u_0, \ldots, u_j)$ with $1 \leq u_i \leq n$ and order $T(j)$ lexicographically. Let $T(r,s,j)$ denote the subset of those U such that $u_0 = r$ and $u_j = s$. Define

$$\psi_j\colon \mathcal{H}_n(j) \times (M_nX)^j \to M_nX$$

by the following formula, where $x(r,s)$ denotes the $(r,s)^{th}$ entry of a matrix x.

$$(1) \qquad \psi_j(c,g;x_1,\ldots,x_j)(r,s) = \theta_{n^{j-1}}(c; \mathop{\times}_{U \in T(r,s,j)} \xi_j(g; \mathop{\times}_{q=1}^{j} x_q(u_{q-1},u_q))).$$

All we have done is to write down ordinary iterated matrix multiplication, allowing for parametrized families of both multiplications and additions on the underlying "ring". The rest of this section will be devoted to the proof of the following result.

<u>Theorem 4.1</u>. The $\mathcal{H}_n(j)$ are the jth spaces of a non-Σ operad $\mathcal{H}_n$, and the maps ψ_j specify an action of $\mathcal{H}_n$ on M_nX.

<u>Proof</u>. By convention, $\mathcal{H}_n(0) = \{*\}$ and ψ_0 is the inclusion of the identity matrix I_n in M_nX. Let $1 = (1,1) \in \mathcal{C}(1) \times \mathcal{G}(1) = \mathcal{H}_n(1)$. Clearly $\psi_1(1)$ is the identity map of M_nX. We must specify maps

$$\gamma\colon \mathcal{H}_n(k) \times \mathcal{H}_n(j_1) \times \ldots \times \mathcal{H}_n(j_k) \to \mathcal{H}_n(j), \qquad j = j_1 + \ldots + j_k,$$

with respect to which $\mathcal{H}$ is an operad and which make the following diagram commute, where μ is the evident shuffle homeomorphism:

$$
\begin{array}{ccc}
\mathcal{H}_n(k)\times\mathcal{H}_n(j_1)\times\ldots\times\mathcal{H}_n(j_k)\times(M_nX)^j & \xrightarrow{\gamma\times 1} \mathcal{H}_n(j)\times(M_nX)^j \xrightarrow{\psi_j} & M_nX \\
\downarrow{\scriptstyle 1\times\mu} & (*) & \uparrow{\scriptstyle \psi_k} \\
\mathcal{H}_n(k)\times\mathcal{H}_n(j_1)\times(M_nX)^{j_1}\times\ldots\times\mathcal{H}_n(j_k)\times(M_nX)^{j_k} & \xrightarrow{1\times\psi_{j_1}\times\ldots\times\psi_{j_k}} & \mathcal{H}_n(k)\times(M_nX)^k
\end{array}
$$

(Compare [10,1.5].) We first chase the diagram to see how γ must be defined and then verify that, with this definition, $\mathcal{H}_n$ is an operad. In principle, the details are perfectly straightforward: one does what one has to do and it works. However, since I omitted all such routine verifications from [12] and since this one is much less intuitively obvious than most, I will try to give some idea of the combinatorics involved.

We first calculate the composite around the bottom of the diagram (*). By (1), we have

(2) $$\psi_k(1\times\psi_{j_1}\times\ldots\times\psi_{j_k})(1\times\mu)(c,g;c_1,g_1,\ldots,c_k,g_k;x_1,\ldots,x_j)(r,s)$$

$$= \theta_{n^{k-1}}(c;\ \underset{U\in T(r,s,k)}{\times}\ \xi_k(g;\ \underset{q=1}{\overset{k}{\times}} z_q(u_{q-1},u_q))),$$

where, with $e_q = j_1+\ldots+j_q$,

(3) $$z_q = \psi_{j_q}(c_q,g_q;x_{e_{q-1}+1},x_{e_{q-1}+2},\ldots,x_{e_q}).$$

If $j_q = 0$, then $z_q = I_n$. Since X is a $\mathcal{G}_0$-space,

(4) $$\xi_k(g;\ \underset{q=1}{\overset{k}{\times}} z_q(u_{q-1},u_q)) = 0 \text{ if } j_q = 0 \text{ and } u_{q-1}\neq u_q \text{ for any } q.$$

Let $\{q_1,\ldots,q_m\}$, $m\leq k$, denote those q, in order, such that $j_q > 0$ and set

(5) $$i_p = j_{q_p},\ d_p = e_{q_p},\ b_p = c_{q_p},\ f_p = g_{q_p},\ \text{and } y_p = z_{q_p} \text{ for } 1\leq p\leq m.$$

Let $s(j_1,\ldots,j_k) \in \mathcal{G}(\varepsilon_1) \times \ldots \times \mathcal{G}(\varepsilon_k)$, $\varepsilon_q = 0$ or 1, have q^{th} coordinate $1 \in \mathcal{G}(1)$ if $j_q > 0$ and q^{th} coordinate $* \in \mathcal{G}(0)$ if $j_q = 0$. Set

(6) $$f = \gamma(g; s(j_1,\ldots,j_k)) \in \mathcal{G}(m).$$

If $U \in T(k)$ satisfies $u_{q-1} = u_q$ whenever $j_q = 0$, define $V = (v_0,\ldots,v_m) \in T(m)$ by deletion of the duplicated entries u_q. Then, since $1 \in X$ is the basepoint for the $\mathcal{G}$-action and $z_q(u_{q-1},u_q) = 1$ if $j_q = 0$,

(7) $$\xi_k(g; \mathop{\times}_{q=1}^{k} z_q(u_{q-1},u_q)) = \xi_m(f; \mathop{\times}_{p=1}^{m} y_p(v_{p-1},v_p))$$

Note that each $V \in T(m)$ arises uniquely from such a $U \in T(k)$ and let $t(j_1,\ldots,j_k) \in \mathcal{C}(\varepsilon_1) \times \ldots \times \mathcal{C}(\varepsilon_{n^{m-1}})$, $\varepsilon_r = 0$ or 1, have r^{th} coordinate $1 \in \mathcal{C}(1)$ if $j_q = 0$ implies $u_{q-1} = u_q$ for the rth element $U \in T(k)$ and r^{th} coordinate $* \in \mathcal{C}(0)$ otherwise. Set

(8) $$b = \gamma(c; t(j_1,\ldots,j_k) \in \mathcal{C}(n^{m-1}).$$

Since $0 \in X$ is the basepoint for the $\mathcal{C}$-action, (4)-(8) imply

(9) $$\theta_{n^{k-1}}(c; \mathop{\times}_{U \in T(r,s,k)} \xi_k(g; \mathop{\times}_{q=1}^{k} z_q(u_{q-1},u_q)))$$
$$= \theta_{n^{m-1}}(b; \mathop{\times}_{V \in T(r,s,m)} \xi_m(f; \mathop{\times}_{p=1}^{m} y_p(v_{p-1},v_p))).$$

Evaluating y_p by (1), (3), and (5) and then using the definition, [12, VI. 1.8 and VI. 1.10], of a $(\mathcal{C}, \mathcal{G})$-space, we find

(10) $$\xi_m(f; \mathop{\times}_{p=1}^{m} y_p(v_{p-1},v_p)) = \xi_m(f; \mathop{\times}_{p=1}^{m} \theta_{n^{i_p-1}}(b_p; \mathop{\times}_{W_p \in T(v_{p-1},v_p,i_p)} y(W_p)))$$
$$= \theta_{n^{j-m}}(\lambda(f; b_1,\ldots,b_m); \mathop{\times}_{H \in S(n^{i_1-1},\ldots,n^{i_m-1})} \xi_m(f; y_H(V))).$$

where if $W_p = (w_{p,0},\ldots,w_{p,i_p}) \in T(i_p)$, then

(11) $$y(W_p) = \xi_{i_p}(f_p; \mathop{\times}_{t=1}^{p} x_{d_{p-1}+t}(w_{p,t-1},w_{p,t}))$$

and if $H = (h_1, \ldots, h_m)$ with $1 \leq h_p \leq n^{i_p - 1}$, then

(12) $y_H(V) \in X^m$ has p^{th} coordinate the h_p^{th} coordinate of $\underset{W_p \in T(v_{p-1}, v_p, i_p)}{\times} y(W_p)$.

By (10) and the definition of a $\mathcal{C}$-space [10, 1.3 or 12, VI. 1.3],

$$(13) \quad \theta_{n^{m-1}}(b; \underset{V \in T(r,s,m)}{\times} \xi_m(f; \overset{m}{\underset{p=1}{\times}} y_p(v_{p-1}, v_p)))$$

$$= \theta_{n^{j-1}}(\gamma(b; \lambda(f; b_1, \ldots, b_m)^{n^{m-1}}); \underset{V \in T(r,s,m)}{\times} \quad \underset{H \in S(n^{i_1 - 1}, \ldots, n^{i_m - 1})}{\times} \xi_m(f; y_H(V))).$$

Here $\lambda(f; b_1, \ldots, b_m) \in \mathcal{C}(n^{j-m})$, since $i_1 + \ldots + i_m = j$, hence application of $\gamma(b; -)$ to its n^{m-1} st power yields an element of $\mathcal{C}(n^{j-1})$. Further, by (11), (12), and the definition of a $\mathcal{B}$-space,

$$(14) \quad \xi_m(f; y_H(V)) = \xi_j(\gamma(f; f_1, \ldots, f_m); \overset{j}{\underset{q=1}{\times}} x_q(u_{q-1}, u_q)),$$

where $U_H(V) = (u_0, \ldots, u_j) \in T(j)$ is the sequence

$$(w_{1,0}, \ldots, w_{1,i_1}, w_{2,1}, \ldots, w_{2,i_2}, \ldots, w_{m,1}, \ldots, w_{m,i_m})$$

obtained by splicing together the h_p^{th} elements W_p of the ordered sets $T(v_{p-1}, v_p, i_p)$ for $1 \leq p \leq m$. As V runs through $T(r, s, m)$ and H runs through $S(n^{i_1 - 1}, \ldots, n^{i_m - 1})$, $U_H(V)$ runs through $T(r, s, j)$. Let $\zeta \in \Sigma_{n^{j-1}}$ be that permutation which changes the given lexicographic ordering of $T(r, s, j)$ to the ordering specified by $U_H(V) < U_{H'}(V')$ if $V < V'$ or if $V = V'$ and $H < H'$ (in the lexicographic ordering; see [12, VI. 1.4]). Substituting (14) into (13), (13) into (9), and (9) into (2) and using the evident equivariance identification (to rearrange "addends"), we arrive at the formula

$$(15)\quad \psi_k(1\times\psi_{j_1}\times\ldots\times\psi_{j_k})(1\times\mu)(c,g;c_1,g_1,\ldots,c_k,g_k;x_1,\ldots,x_j)(r,s)$$

$$=\theta_{n^{j-1}}(\gamma(b;\lambda(f;b_1,\ldots,b_m)^{n^{m-1}})\zeta;\ \underset{U\in T(r,s,j)}{\times}\ \xi_j(\gamma(f;f_1,\ldots,f_m);\ \overset{j}{\underset{q=1}{\times}} x_q(u_{q-1},u_q))).$$

Comparing (1) and (15), we see that the diagram (*) will commute provided that we define

$$(16)\quad \gamma(c,g;c_1,g_1,\ldots,c_k,g_k)=(\gamma(b;\lambda(f;b_1,\ldots,b_m)^{n^{m-1}})\zeta,\ \gamma(f;f_1,\ldots,f_m))\ .$$

Here, when $j_q=0$, $(c_q,g_q)=*$ and we may think of g_q as $*\in\mathcal{G}(0)$; (5), (6), and the definition of an operad then imply

$$(17)\quad \gamma(f;f_1,\ldots,f_m)\ =\ \gamma(g;g_1,\ldots,g_k).$$

No such reinterpretation of the first factor of (16) is possible (as we would have to interpret c_q as an element of $\mathcal{C}(n^{-1})$ to make the numbers work out).

We must show that, with this definition, $\mathcal{H}_n$ is a non-Σ operad. Certainly $\gamma(1;c,g)=(c,g)$ and $\gamma(c,g;1^k)=(c,g)$ for $(c,g)\in\mathcal{H}_n(k)$, by [12,VI.1.6(b) and (b')]. It remains to check the associativity formula [12,VI.1.2(a)] for iteration of the maps γ, and the reader who has followed the combinatorics so far should not have too much trouble carrying out the requisite verification for himself. The details involve use of the corresponding associativity formulas for $\mathcal{C}$ and $\mathcal{G}$, the equivariance formulas [12,VI.1.2(c) and VI.1.6(c')], the interaction formulas [12,VI.1.6(a) and (a')] as corrected in Remark 1.2, and a rather horrendous check that the permutations come out right.

§5. Strong homotopy $(\mathcal{C}, \mathcal{G})$-spaces and matrix rings

In this rather speculative section (which will play little part in our later work), we make an initial definition in the direction of an up to homotopy elaboration of the theories of A_∞ and E_∞ ring spaces and explain its likely relevance to the matrix "rings" studied in the previous section.

Lada, in [2, V], has developed an up to homotopy generalization of the theory of $\mathcal{C}$-spaces. (See [13, §6] for a sketch.) His theory is based on use of the associated monad C, and the essential starting point of the analogous up to homotopy generalization of the theory of $(\mathcal{C}, \mathcal{G})$-spaces surely must be the fact that C restricts to a monad in the category of $\mathcal{G}_0$-spaces. The following definition should be appropriate.

Definition 5.1. Let $(\mathcal{C}, \mathcal{G})$ be an operad pair (where $\mathcal{G}$ might be a non-Σ operad). A strong homotopy, or sh, $(\mathcal{C}, \mathcal{G})$-space X is a $\mathcal{C}$-space (X, θ) with basepoint 0 and a $\mathcal{G}_0$-space (X, ξ) with basepoint 1 such that $\theta: CX \rightarrow X$ is an sh G-map.

The notion of an sh G-map is defined and discussed in [2, V §3]. It is required that the basic distributivity diagram

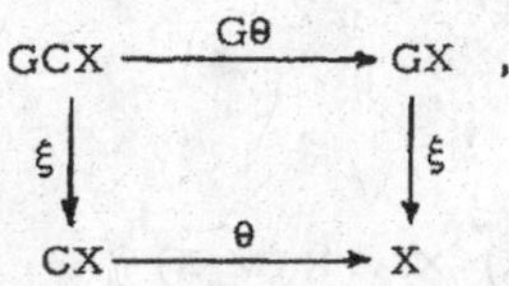

the commutativity of which is the defining property of a $(\mathcal{C}, \mathcal{G})$-space, should homotopy commute and that this homotopy should be the first of an infinite sequence of higher coherence homotopies.

More general notions of sh $(\mathcal{C}, \mathcal{G})$-spaces, with X only an sh $\mathcal{C}$-space or only an sh $\mathcal{G}$-space, surely also exist but would be much more complicated.

Unfortunately, the most general notion would presumably be essential to a fully homotopy invariant theory.

I would hope that if X is an sh $(\mathcal{C}, \mathcal{G})$-space, then there is an actual $(\mathcal{C}, \mathcal{G})$-space UX equivalent to X in an appropriately structured way, so that the passage from A_∞ ring spaces to A_∞ ring spectra and from E_∞ ring spaces to E_∞ ring spectra directly generalizes to sh $(\mathcal{C}, \mathcal{G})$-spaces for suitable pairs $(\mathcal{C}, \mathcal{G})$. This would be in analogy to Lada's one operad theory, and his cubical bar construction $UX = \widetilde{B}(C, C, X)$ of [2, V §2] would be the obvious candidate for UX. However, I have not attempted to pursue these ideas.

We return to consideration of M_nX for a $(\mathcal{C}, \mathcal{G})$-space X, with $\mathcal{G}$ being a non-Σ operad for convenience. As formulae (4.16) and (4.17) make clear, projection on the second factor gives a morphism of non-Σ operads $\mathcal{H}_n \to \mathcal{G}$. Therefore $\mathcal{H}_n$ acts on $\mathcal{C}$ by pullback of the action of $\mathcal{G}$ on $\mathcal{C}$. However, M_nX is not a $(\mathcal{C}, \mathcal{H}_n)$-space because the distributivity diagram fails to commute (for this or any other action of $\mathcal{H}_n$ on $\mathcal{C}$). As explained in [12, VI §1 and 2, p. 77], the diagram in question results by passage to disjoint unions and then to quotients from the following diagrams:

$$\begin{array}{ccc}
\mathcal{H}_n(k) \times \mathcal{C}(j_1) \times (M_nX)^{j_1} \times \cdots \times \mathcal{C}(j_k) \times (M_nX)^{j_k} & \xrightarrow{1 \times \theta_{j_1} \times \cdots \times \theta_{j_k}} & \mathcal{H}_n(k) \times (M_nX)^k \\
\downarrow{\scriptstyle 1 \times \mu} & & \\
\mathcal{H}_n(k) \times \mathcal{C}(j_1) \times \cdots \times \mathcal{C}(j_k) \times (M_nX)^{j_1} \times \cdots \times (M_nX)^{j_k} & & \downarrow{\scriptstyle \psi_k} \\
(*) \qquad \downarrow{\scriptstyle \Delta \times 1 \times \delta} & & M_nX \\
\mathcal{H}_n(k)^{j+1} \times \mathcal{C}(j_1) \times \cdots \times \mathcal{C}(j_k) \times ((M_nX)^k)^j & & \uparrow{\scriptstyle \theta_j} \\
\downarrow{\scriptstyle \mu} & & \\
\mathcal{H}_n(k) \times \mathcal{C}(j_1) \times \cdots \times \mathcal{C}(j_k) \times (\mathcal{H}_n(k) \times (M_nX)^k)^j & \xrightarrow{\lambda \times (\psi_k)^j} & \mathcal{C}(j) \times (M_nX)^j
\end{array}$$

Here $j = j_1 \cdots j_k$, the μ are shuffle homeomorphisms, Δ is the iterated diagonal, ψ_k is as defined in formula (4.1), the θ_j give the additive action on $(M_nX) = X^{n^2}$, and

$$\delta: (M_nX)^{j_1} \times \ldots \times (M_nX)^{j_k} \to ((M_nX)^k)^j$$

is specified by

$$\delta(y_1, \ldots, y_k) = \underset{I \in S(j_1, \ldots, j_k)}{\times} y_I, \quad \text{with}$$

$$y_I = (x_{1i_1}, \ldots, x_{ki_k}) \text{ if } y_q = (x_{q1}, \ldots, x_{qj_q}).$$

The following result shows that, as far as the relationships between any homological and homotopical invariants of the maps θ_j and ψ_k go, any results valid for A_∞ ring spaces are also valid for matrix rings with entries in A_∞ ring spaces. (Compare the analysis of the homology of E_∞ ring spaces in [2, II]; we shall return to this point in section 9.)

Proposition 5.2. If X is a $(\mathcal{C}, \mathcal{G})$-space, where $\mathcal{C}$ is an E_∞ operad and $\mathcal{G}$ is any non-Σ operad which acts on $\mathcal{C}$, then the diagram (*) is $\Sigma_{j_1} \times \ldots \times \Sigma_{j_k}$-equivariantly homotopy commutative.

Proof. On the one hand, the corresponding diagram for the $(\mathcal{C}, \mathcal{G})$-space X, formula (4.1), and the diagram which expresses that X is a $\mathcal{C}$-space [10,1.5] imply

(1) $\quad \psi_k(1 \times \theta_{j_1} \times \ldots \times \theta_{j_k})(c, g; c_1, y_1, \ldots, c_k, y_k)$

$$= \theta_{jn^{k-1}}(\gamma(c; \lambda(g; c_1, \ldots, c_k)^{n^{k-1}}); \underset{U \in T(r,s,k)}{\times} \underset{I \in S(j_1, \ldots, j_k)}{\times} \xi_k(g; y_I(U))),$$

where $y_I(U) = (x_{1i_1}(u_0, u_1), \ldots, x_{ki_k}(u_{k-1}, u_k))$. On the other hand, formula (4.1) and the fact that X is a $\mathcal{C}$-space imply

(2) $\quad \theta_j(\lambda \times \psi_k^j)\mu(\Delta \times 1 \times \delta)(1 \times \mu)(c, g; c_1, y_1, \ldots, c_k, y_k)$

$$= \theta_{jn^{k-1}}(\gamma(\lambda(c, g; c_1, \ldots, c_k); c^j); \underset{I \in S(j_1, \ldots, j_k)}{\times} \underset{U \in T(r,s,k)}{\times} \xi_k(g; y_I(U))).$$

No definition of $\lambda(c, g; c_1, \ldots, c_k)$ will make the right sides of (1) and (2) agree, and we take the pullback definition $\lambda(g; c_1, \ldots, c_k)$. In view of the difference in order of appearance of the indexing sets $S(j_1, \ldots, j_k)$ and $T(r, s, k)$, the addends in (1) and (2) differ by a permutation $\nu \in \Sigma_{jn^{k-1}}$. The maps

$$f, g: \mathcal{H}_n(k) \times \mathcal{C}(j_1) \times \ldots \times \mathcal{C}(j_k) \to \mathcal{C}(jn^{k-1})$$

specified by

$$f(c, g; c_1, \ldots, c_k) = \gamma(c; \lambda(g; c_1, \ldots, c_k)^{n^{k-1}})$$

and

$$g(c, g; c_1, \ldots, c_k) = \gamma(\lambda(g; c_1, \ldots, c_k); c^j)\nu$$

are $\Sigma_{j_1} \times \ldots \times \Sigma_{j_k}$ -equivariant, where the action of this group on $\mathcal{C}(jn^{k-1})$ is determined by its tensorial embedding in Σ_j [10,VI.1.4], the diagonal embedding of Σ_j in $(\Sigma_j)^{n^{k-1}}$, and the block sum embedding of the latter in $\Sigma_{jn^{k-1}}$. Since $\mathcal{C}$ is an E_∞ operad, the domain and codomain of f and g are $\Sigma_{j_1} \times \ldots \times \Sigma_{j_k}$ -free and contractible, hence f and g are equivariantly homotopic. The conclusion follows.

Of course, if we had chosen to work with permutations in our multiplicative operads, then the diagram (*) would be $\Sigma_k \times \Sigma_{j_1} \times \ldots \times \Sigma_{j_k}$ -equivariantly homotopy commutative.

If the homotopies of the proposition can be chosen with suitable compatibility as k and the j_q vary, they will together yield the first of the infinite sequence of homotopies needed to verify the following assertion.

<u>Conjecture 5.3</u>. If X is a $(\mathcal{C}, \mathcal{G})$-space, where $\mathcal{C}$ is an E_∞ operad and $\mathcal{G}$ is any non-Σ operad which acts on $\mathcal{C}$, then $M_n X$ is an sh $(\mathcal{C}, \mathcal{H}_n)$-space.

With $\mathcal{C}$ an E_∞ operad and each $\mathcal{G}(j)$ contractible, the notion of an sh $(\mathcal{C}, \mathcal{G})$-space is an up to homotopy generalization of the notion of an A_∞ ring space.

The conjecture gives the appropriate sense in which it might be true that M_nX is an A_∞ ring space if X is an A_∞ ring space.

The proof of Proposition 5.2 is precisely analogous to that of [10, 1.9(ii)], which gives a similar result about the product ϕ on a $\mathcal{C}$-space when $\mathcal{C}$ is an E_∞ operad. Lada [6] has studied the passage from that result to the assertion that ϕ is an sh C-map, and the problems he encountered there illustrate what would be involved in a proof of Conjecture 5.3.

§6. The comparison between M_nX and $M_{n+1}X$

As in section 4, let X be a $(\mathcal{C}, \mathcal{G})$-space, where $\mathcal{C}$ is any operad and $\mathcal{G}$ is any non-Σ operad which acts on X. We have exhibited a $\mathcal{C}$-space structure and an $\mathcal{H}_n$-space structure on M_nX, where $\mathcal{H}_n(j) = \mathcal{C}(n^{j-1}) \times \mathcal{G}(j)$, and have studied the relationship between these actions. We here study the relationship between M_nX and $M_{n+1}X$. We adopt the notations of section 4, but with an identifying subscript n where necessary for clarity. Let $\nu_n : M_nX \to M_{n+1}X$ denote the natural inclusion. First consider the diagram

$$\begin{array}{ccc}\mathcal{C}(j) \times (M_nX)^j & \xrightarrow{\theta_{nj}} & M_nX \\ \downarrow{\scriptstyle 1 \times \nu_n^j} & & \downarrow{\scriptstyle \nu_n} \\ \mathcal{C}(j) \times (M_{n+1}X)^j & \xrightarrow{\theta_{n+1,j}} & M_{n+1}X\end{array}$$

Here θ_n and θ_{n+1} are determined entrywise from the action of $\mathcal{C}$ on X, and this diagram certainly commutes on the $(r,s)^{th}$ matrix entries for $r \leq n$ and $s \leq n$. Similarly, both composites always give $(r,s)^{th}$ matrix entry 0 if either but not both of r and s is $n+1$. However, for $c \in \mathcal{C}(j)$ and $x_i \in M_nX$, we have

$$\nu_n\theta_{nj}(c, x_1, \ldots, x_j)(n+1, n+1) = 1$$

but

$$\theta_{n+1,j}(1 \times \nu_n^j)(c, x_1, \ldots, x_j)(n+1, n+1) = \theta_j(c; 1^j) \quad .$$

Thus ν_n fails to be a $\mathcal{C}$-map. Indeed, as for discrete rings, 1 and $\theta_j(c, 1^j)$ lie in different components in all non-trivial cases since, by [12, p.140], X is contractible if 0 and 1 lie in the same component. For the multiplicative structures, we have the following result. Its hypothesis $\mathcal{C}(1) = \{1\}$ will be discussed after the proof.

Theorem 6.1. Assume that $\mathcal{C}(1)$ is the point 1. For each $n \geq 1$, there is a map $\tau_n: \mathcal{H}_{n+1} \to \mathcal{H}_n$ of non-Σ operads such that $\nu_n: M_nX \to M_{n+1}X$ is an $\mathcal{H}_{n+1}$-map, where M_nX is an $\mathcal{H}_{n+1}$-space by pullback along ν_n.

Proof. Let $t_{nj} \in \mathcal{C}(\varepsilon_1) \times \ldots \times \mathcal{C}(\varepsilon_{(n+1)^{j-1}})$, $\varepsilon_i = 0$ or 1, have i^{th} coordinate $* \in \mathcal{C}(0)$ if the ith element $U \in T_{n+1}(j-2)$ has any $u_q = n+1$ and have i^{th} coordinate $1 \in \mathcal{C}(1)$ otherwise. Observe that if elements of $T_{n+1}(j-2)$ are written in the form $U = (u_1, \ldots, u_{j-1})$, then $U \longleftrightarrow (r, u_1, \ldots, u_{j-1}, s)$ gives a bijective correspondence between the ordered sets $T_{n+1}(j-2)$ and $T_{n+1}(r, s, j)$ for each r and s between 1 and n+1. Define

$$\tau_{nj}: \mathcal{H}_{n+1}(j) = \mathcal{C}((n+1)^{j-1}) \times \mathcal{G}(j) \to \mathcal{C}(n^{j-1}) \times \mathcal{G}(j) = \mathcal{H}_n(j)$$

by

$$\tau_{nj}(c, g) = (\gamma(c; t_{nj}), g).$$

By convention, $\tau_{n0}(*) = *$ and $t_{n1} = 1 \in \mathcal{C}(1)$ so that τ_{n1} is the identity map. Another laborious combinatorial argument, which uses formula (4.16), the associativity and equivariance formulas for the operad $\mathcal{C}$ [12, VI. 1.2(a) and (c)], the interaction, unit, and equivariance formulas [12, VI. 1.6(a'), (b'), (c')] for the action of $\mathcal{G}$ on $\mathcal{C}$ (see Remark 1.2), and a careful consideration of permutations based on the description of ζ given after formula (4.14) shows that the following diagrams commute.

$$\begin{array}{ccc}
\mathcal{H}_{n+1}(k) \times \mathcal{H}_{n+1}(j_1) \times \ldots \times \mathcal{H}_{n+1}(j_k) & \xrightarrow{\gamma} & \mathcal{H}_{n+1}(j_1 + \ldots + j_k) \\
\downarrow{\scriptstyle \tau_{nk} \times \tau_{nj_1} \times \ldots \times \tau_{nj_k}} & & \downarrow{\scriptstyle \tau_{n, j_1 + \ldots + j_k}} \\
\mathcal{H}_n(k) \times \mathcal{H}_n(j_1) \times \ldots \times \mathcal{H}_n(j_k) & \xrightarrow{\gamma} & \mathcal{H}_n(j_1 + \ldots + j_k)
\end{array}$$

Thus τ_n is a map of non-Σ operads. Now consider the following diagrams.

$$\begin{array}{ccccc}
\mathcal{H}_{n+1}(j) \times (M_n X)^j & \xrightarrow{\tau_{nj} \times 1} & \mathcal{H}_n(j) \times (M_n X)^j & \xrightarrow{\psi_{nj}} & M_n X \\
\downarrow {\scriptstyle 1 \times \nu_n^j} & & & & \downarrow {\scriptstyle \nu_n} \\
\mathcal{H}_{n+1}(j) \times (M_{n+1} X)^j & & \xrightarrow{\psi_{n+1,j}} & & M_{n+1} X
\end{array}$$

Our claim is that these diagrams commute, and it is for this that we require $\mathcal{C}(1)$ to be a point. Consider

$$\psi_{n+1,j}(c, g; y_1, \ldots, y_j)(r, s)$$

$$= \theta_{(n+1)^{j-1}}(c; \underset{U \in T_{n+1}(r,s,j)}{\times} \xi_j(g; \underset{q=1}{\overset{j}{\times}} y_q(u_{q-1}, u_q))) ,$$

where $c \in \mathcal{C}((n+1)^{j-1})$, $g \in \mathcal{G}(j)$, and $y_i = \nu_n(x_i)$. The Uth factor on the right is $0 \in X$ if either but not both of u_{q-1} and u_q is n+1 for any q. If $r \leq n$ and $s \leq n$, it follows that the right side is equal to

$$\theta_{n^{j-1}}(\gamma(c; t_{nj}); \underset{V \in T_n(r,s,j)}{\times} \xi_j(g; \underset{q=1}{\overset{j}{\times}} x_q(u_{q-1}, u_q)))$$

$$= \psi_j(\tau_{nj}(c, g); x_1, \ldots, x_j)(r, s) .$$

If either but not both of r and s is n+1, then, for any $U \in T_{n+1}(r,s,j)$, there exists q such that either but not both of u_{q-1} and u_q is n+1, hence $\psi_j(c, g; y_1, \ldots, y_j)(r, s) = 0$. We therefore have

$$\nu_n \psi_{nj}(\tau_{nj} \times 1)(c, g; x_1, \ldots, x_j)(r, s) = \psi_{n+1,j}(1 \times \nu_n^j)(c, g; x_1, \ldots, x_j)(r, s) ,$$

as desired, unless $r = s = n+1$. Here we find that the left side is $1 \in X$ whereas the right side reduces to

$$\theta_{(n+1)^{j-1}}(c; 0, \ldots, 0, 1) = \theta_1(\gamma(c; s_{nj}), 1) \in X ,$$

where

$$s_{nj} = (*, \ldots, *, 1) \in \mathcal{C}(0)^{(n+1)^{j-1} - 1} \times \mathcal{C}(1) .$$

Indeed, all elements of $T_{n+1}(n+1, n+1, j)$ except the last have either but not both of u_{q-1} and u_q equal to n+1 for some q, whereas the last U is $(n+1, \ldots, n+1)$ for which $y_q(n+1, n+1) = 1$ and $\xi_j(g; \times_{q=1}^{j} 1) = 1$. The assumption $\mathcal{C}(1) = \{1\}$ ensures that $\gamma(c; s_{nj}) = 1$ and therefore $\theta_1(\gamma(c; s_{nj}); 1) = 1$.

Unfortunately, it is not in general the case that $\mathcal{C}(1) = \{1\}$; for example, this fails for the canonical E_∞ operad $\mathcal{K}_\infty$ used in section 3. We could avoid this assumption by appealing to Lada's theory. The contractibility of $\mathcal{C}(1)$ can be used to prove that ν_n is an sh H_{n+1}-map. However, this solution (which I worked out in detail in an earlier draft) leads to further complications in later sections. Our preferred solution is to prove that A_∞ ring spaces can be functorially replaced by equivalent A_∞ ring spaces with respect to a different A_∞ operad pair for which $\mathcal{C}(1)$ is a point. This replacement process works equally well in the E_∞ ring context.

We exploit the fact that the particular E_∞ operad $\mathcal{Q}$ of [11, §4] has $\mathcal{Q}(1) = \{1\}$. Moreover, as explained in [12, VI §2 and 4], $\mathcal{Q}$ acts on itself and thus $(\mathcal{Q}, \mathcal{Q})$ is an E_∞ operad pair. Let $(\mathcal{C}, \mathcal{G})$ be any operad pair such that $\mathcal{C}$ is an E_∞ operad. $\mathcal{G}$ might be either an A_∞ or an E_∞ operad. By use of products and projections, we then have operad pairs, and maps thereof,

$$(\mathcal{C}, \mathcal{G}) \xleftarrow{(\pi_1, \pi_1)} (\mathcal{C} \times \mathcal{Q}, \mathcal{G} \times \mathcal{Q}) \xrightarrow{(\pi_2, 1)} (\mathcal{Q}, \mathcal{G} \times \mathcal{Q}).$$

Therefore $(\mathcal{C}, \mathcal{G})$-spaces are $(\mathcal{C} \times \mathcal{Q}, \mathcal{G} \times \mathcal{Q})$-spaces by pullback, while both D and $C \times D$ are monads in the category of $(\mathcal{G} \times \mathcal{Q})_0$-spaces and $\pi_2 : C \times D \to D$ is a morphism of monads in this category.

We proceed to construct a functor W from $(\mathcal{C}, \mathcal{G})$-spaces to $(\mathcal{Q}, \mathcal{G} \times \mathcal{Q})$-spaces. As explained in [12, VI. 2.7(iii)], there is a functor W from $\mathcal{C}$-spaces to $\mathcal{Q}$-spaces specified in terms of the two-sided bar construction of [10, §9] by $WX = B(D, C \times D, X)$ and there is a natural diagram of $(\mathcal{C} \times \mathcal{Q})$-spaces

$$X \xleftarrow{\varepsilon} B(C\times D, C\times D, X) \xrightarrow{B\pi_2} B(D, C\times D, X) = WX.$$

Here ε is a homotopy equivalence with a natural homotopy inverse and $B\pi_2$ is also an equivalence. Technically, we should assume or arrange (without loss of structure by an elaboration of the arguments in [10,A.8 and A.11]) that $1 \in \mathcal{C}(1)$ and $0 \in X$ are non-degenerate basepoints, so that the simplicial spaces used in our constructions are proper [10,11.2 and 11,A.5]. We have the following result.

Proposition 6.2. If X is a $(\mathcal{C}, \mathcal{G})$-space, then WX is a $(\mathcal{D}, \mathcal{G}\times\mathcal{D})$-space, $B(C\times D, C\times D, X)$ is a $(\mathcal{C}\times\mathcal{D}, \mathcal{G}\times\mathcal{D})$-space, and ε and $B\pi_2$ are maps of $(\mathcal{C}\times\mathcal{D}, \mathcal{G}\times\mathcal{D})$-spaces.

Proof. By formal verifications from [10,9.6 and 9.9], the action $DWX \to WX$ of D on WX, the action of $C\times D$ on $B(C\times D, C\times D, X)$, and ε and $B\pi_2$ are all geometric realizations of maps of simplicial $(\mathcal{G}\times\mathcal{D})_0$-spaces and are therefore maps of $(\mathcal{G}\times\mathcal{D})_0$-spaces by [10,12.2].

Clearly, we may as well start our analysis of matrix rings of A_∞ ring spaces by first replacing X by WX. In particular, our assumption that $\mathcal{C}(1) = \{1\}$ results in no real loss of generality. This construction also handles a different technical problem, one that we have heretofore ignored.

Remarks 6.3. As explained in [12,VII § 1 and 2], $\mathcal{K}_\infty$ and the product operads $\mathcal{C} = \mathcal{C}'\times\mathcal{K}_\infty$ necessary to the proof of the recognition principle in section 3 are only partial operads and their associated monads are only partial monads. However, it is not hard to see that our replacement argument above works perfectly well for such $\mathcal{C}$. Thus, by use of the functor W, we may assume without loss of generality that all operads in sight are honest operads in the development of the present theory since $\mathcal{D}$ and all multiplicative operads (see [12,p.178]) are honest.

Remarks 6.4. By the method of proof of Theorem 6.1, one can construct maps $\tau_{pq}: \mathcal{H}_{p+q} \to \mathcal{H}_p$ and $\tau'_{pq}: \mathcal{H}_{p+q} \to \mathcal{H}_q$ of non-Σ operads such that the usual block sum of matrices $\oplus: M_pX \times M_qX \to M_{p+q}X$ specifies an $\mathcal{H}_{p+q}$-map, where M_pX and M_qX are regarded as $\mathcal{H}_{p+q}$-spaces by pullback along τ_{pq} and τ'_{pq} respectively and where $M_pX \times M_qX$ is given the product $\mathcal{H}_{p+q}$-structure [10,1.7]. From this point of view, the problem with $\mathcal{C}(1)$ above simply reflects the fact that the inclusion $\{1\} \to X = M_1X$ is not an $\mathcal{H}_1$-map unless $\mathcal{C}(1) = \{1\}$. In order for these sum maps to be useful, one would have to understand their stabilization, that is, to analyze the diagrams

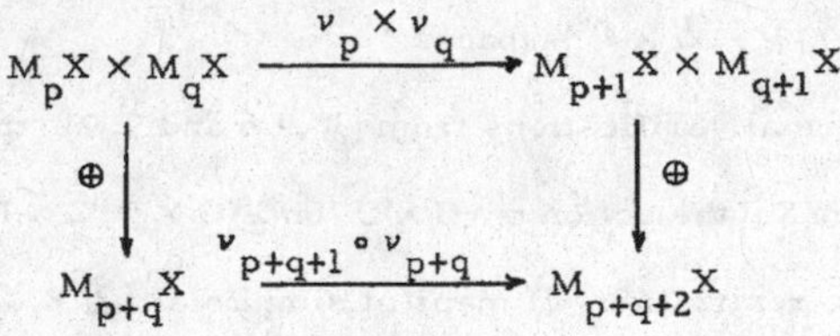

$$\begin{array}{ccc} M_pX \times M_qX & \xrightarrow{\nu_p \times \nu_q} & M_{p+1}X \times M_{q+1}X \\ \downarrow \oplus & & \downarrow \oplus \\ M_{p+q}X & \xrightarrow{\nu_{p+q+1} \circ \nu_{p+q}} & M_{p+q+2}X \end{array}$$

If X is a ring, these composites obviously differ only by conjugation by a permutation matrix. In general, the definition of such a conjugation entails an arbitrary choice of product and yields only a homotopy commutative diagram. A full analysis of the situation would presumably entail application of Lada's theory of strong homotopy maps.

§7. The Algebraic K-theory of A_∞ ring spaces

Let $(\mathcal{C}, \mathcal{G})$ be an A_∞ operad pair. A $(\mathcal{C}, \mathcal{G})$-space X will be said to be grouplike if $\pi_0 X$ is a group under addition and therefore a ring rather than just a semi-ring. Up to weak equivalence then, X is a grouplike A_∞ ring space if and only if it is the zeroth space of an A_∞ ring spectrum.

For a grouplike $(\mathcal{C}, \mathcal{G})$-space X, define $FM_n X$ to be the pullback in the following diagram, where d denotes the discretization map.

$$\begin{array}{ccc} FM_n X & \longrightarrow & M_n X \\ \downarrow & & \downarrow d \\ GL(n, \pi_0 X) & \stackrel{\subset}{\longrightarrow} & M_n(\pi_0 X) \end{array}$$

That is, $FM_n X$ is the space of unit components in $M_n X$. The notation is chosen in analogy with that in Remarks 2.4; Waldhausen would write $\widehat{GL}_n(X)$ for $FM_n X$. If X is a discrete ring then $FM_n X = GL(n, X)$; for general topological rings, $FM_n X$ is larger than $GL(n, X)$. We reiterate that 0 and 1 must be in different components for non-triviality. Clearly ν_n maps $FM_n X$ into $FM_{n+1} X$. If each ν_n is a cofibration, we let $FM_\infty X$ denote the union of the $FM_n X$; otherwise we let $FM_\infty X$ be their telescope. It is the purpose of this section to prove that $FM_\infty X$ has a functorial delooping.

<u>Theorem 7.1</u>. There is a functor T from grouplike $(\mathcal{C}, \mathcal{G})$-spaces to connected based spaces together with a natural weak equivalence between ΩTX and $FM_\infty X$.

In particular, $\pi_1 TX = \pi_0 FM_\infty X = GL(\infty, \pi_0 X)$ has a perfect commutator subgroup. Replacing TX by a naturally weakly equivalent CW-complex if necessary, we can take its plus construction in the sense of Quillen (see e.g. [21, §1]). We have the following definition of the algebraic K-theory of A_∞ ring spaces.

Definition 7.2. Let $i: TX \to KX$ be the plus construction of TX, so that i induces an isomorphism on homology and KX is a simple space. KX is called the connected algebraic K-space of X. For $q > 0$, let $K_q X = \pi_q(KX)$. $K_q X$ is called the qth algebraic K-group of X.

Remarks 7.3. As a space, $FM_n X$ is just the union of some of the components of $M_n X = X^{n^2}$ and each of these components is equivalent to the component $SFM_n X$ of the identity matrix. Indeed, $FM_n X$ is equivalent, although not in general as an H-space, to $SFM_n X \times GL(n, \pi_0 X)$ [2, I.4.6]. Further, $SFM_n X$ is equivalent to the component of the zero matrix, and the latter is just $M_n X_0$ where X_0 is the component of zero in X (and $M_n X$ and $M_n X_0$ are additive infinite loop spaces). It follows that, for $q > 0$, $\pi_q FM_\infty X$ is the direct sum of infinitely many copies of $\pi_q X$; of course, i_* maps this group naturally to $K_{q+1} X$.

Remarks 7.4. By restriction to the spaces $SFM_n X$, the proof of Theorem 7.1 will yield a functor UT from grouplike $(\mathcal{C}, \mathcal{G})$-spaces to simply connected based spaces together with a natural weak equivalence between ΩUTX and $SFM_\infty X = \text{Tel}\, SFM_n X$ and a natural map $UTX \to TX$ compatible with the weak equivalences. Thus, homotopically, UTX will be the universal cover of TX.

The rest of this section will be devoted to the proof of Theorem 7.1. We begin by reviewing the basic theory of classifying spaces of A_∞ spaces. Let $\mathcal{G}$ be an A_∞ operad. There is a functor V from $\mathcal{G}$-spaces to topological monoids specified in terms of the two-sided bar construction of [10, §9] by $VX = B(M, G, X)$. Here M denotes the free monoid, or James construction, monad. The augmentation $\delta: \mathcal{G} \to \mathcal{M}$ induces a map of monads $\delta: G \to M$, and we obtain a natural diagram of $\mathcal{G}$-maps

$$(1) \qquad X \xleftarrow{\varepsilon} B(G, G, X) \xrightarrow{B\delta} B(M, G, X) = VX .$$

Here ε is a homotopy equivalence with a natural homotopy inverse. The map $B\delta$ is also a homotopy equivalence (by [11, A.2(ii)], in which the connectivity assumption of [10,13.5] is removed). If X is itself a monoid considered as a $\mathcal{G}$-space by pullback along δ, there is a natural composite

$$(2) \qquad VX = B(M,G,X) \xrightarrow{B\delta} B(M,M,X) \xrightarrow{\varepsilon} X$$

which is both a map of monoids and a homotopy equivalence. If $X = GY$ for a based space Y, there is a natural equivalence of monoids $VX \to MY$. Moreover, all of this is natural with respect to maps of A_∞ operads. See [10, 13.5] for details.

We can therefore deloop $\mathcal{G}$-spaces by applying the standard product-preserving classifying space functor B of [17 or 9, §7-8] to VX. For our purposes, the crucial property of B is that there is a natural map $\zeta : X \to \Omega BX$ for monoids X such that ζ is a weak equivalence if $\pi_0 X$ is a group (e.g. [9,8.7]).

Turning to the proof of Theorem 7.1, fix an A_∞ operad pair $(\mathcal{C}, \mathcal{G})$ and construct A_∞ operads $\mathcal{H}_n$ as in section 4 (either crossing with $\mathcal{M}$ to obtain actual operads, with permutations, or rephrasing the arguments above in terms of non-Σ operads). By use of the argument at the end of the previous section, replacing X by WX if necessary, we may assume without loss of generality that $\mathcal{C}(1) = \{1\}$.

Clearly Theorems 4.1 and 6.1 imply that FM_nX is a sub $\mathcal{H}_n$-space of M_nX and that ν_n restricts to an $\mathcal{H}_{n+1}$-map $FM_nX \to FM_{n+1}X$, where FM_nX is an $\mathcal{H}_{n+1}$-space by pullback along $\tau_n : \mathcal{H}_{n+1} \to \mathcal{H}_n$. Write V_n for the functor V above defined with respect to the operad $\mathcal{H}_n$. Consider the following maps, where the notation on the left abbreviates that on the right.

$$\begin{array}{ccc}
V_n FM_n X & = & B(M, H_n, FM_n X) \\
\uparrow \tau_n & & \uparrow B(1,\tau_n,1) \\
V_{n+1} FM_n X & = & B(M, H_{n+1}, FM_n X) \\
\downarrow V_{n+1}\nu_n & & \downarrow B(1,1,\nu_n) \\
V_{n+1} FM_{n+1} X & = & B(M, H_{n+1}, FM_{n+1} X)
\end{array}$$

Again, in order to ensure that all simplicial spaces involved in our constructions are proper [10, 11.2 and 11, A.5], we should assume or arrange that $I_n \in FM_n X$ and $1 \in \mathcal{H}_n(1)$ are nondegenerate basepoints before applying the functors V_n and B; see [10, p. 127 and 167-171]. With this precaution, we have the following result.

Lemma 7.5. $\tau_n: V_{n+1} FM_n X \to V_n FM_n X$ is a homotopy equivalence. Therefore $B\tau_n$ is also a homotopy equivalence.

Proof. By [11, A.2(ii)], $\tau_n: H_{n+1} Y \to H_n Y$ is a homotopy equivalence for any $\mathcal{H}_n$-space Y. By an argument just like the proof of [11, A.4 (see 16, 5.5 and 5.6)], $Mf: MX \to MX'$ is a homotopy equivalence for any homotopy equivalence $f: X \to X'$. Therefore $B(1, \tau_n, 1)$ is a homotopy equivalence by [11, A.4(ii)], and the conclusion for $B\tau_n$ follows from the same result; see [9, p. 32].

At this point, we could choose a homotopy inverse $(B\tau_n)^{-1}$ to $B\tau_n$ and let TX be the telescope of the spaces $BV_n FM_n X$. However, we would then run into a naturality problem. Certainly $(B\tau_n)^{-1}$ is natural up to homotopy since $B\tau_n$ is natural (by a trivial formal argument), but functoriality up to homotopy of the telescope would require $(B\tau_n)^{-1}$ to be natural up to natural homotopy. In fact, tracing very carefully through the proofs cited above, one may check that $(B\tau_n)^{-1}$ can be so chosen, but there is a much simpler and more precise solution to the problem.

Definition 7.6. Construct TX as follows. Let T_nX be the reduced double mapping cylinder

$$BV_nFM_nX \bigcup_{B\tau_n} (BV_{n+1}FM_nX \wedge I^+) \bigcup_{BV_{n+1}\nu_n} BV_{n+1}FM_{n+1}X$$

and let TX be obtained from the disjoint union of the T_nX by identifying the top, BV_nFM_nX, of $T_{n-1}X$ with the bottom, BV_nFM_nX, of T_nX for $n > 1$. Clearly T is then the object function of a functor from $(\mathcal{C}, \mathcal{G})$-spaces to connected based spaces.

Visibly, TX is homotopy equivalent to the telescope of the BV_nFM_nX with respect to composites $(BV_{n+1}\nu_n)(B\tau_n)^{-1}$, any questions of naturality being thrown irrelevantly onto the choice of equivalence. The properties of T stated in Theorem 7.1 are immediate from the definition and the general theory discussed above.

Remarks 7.7. One would like to construct a product $TX \times TX \to TX$ (not an H-space structure of course) by use of block sum of matrices so as to be able to obtain an H-space structure on KX by mimicry of Wagoner's proof [21, §1] in the case when X is a discrete ring. The main obstruction is explained in Remarks 6.4.

§8. Monomial matrices and $Q_0(BFX \amalg \{0\})$

As before, let $(\mathcal{C}, \mathcal{G})$ be an A_∞ operad pair and let X be a grouplike $(\mathcal{C}, \mathcal{G})$-space. Let $FX = FM_1X$ be the space of unit components of X and let SFX be the component of $1 \in X$. Then FX and SFX are sub $\mathcal{G}$-spaces of X. Let V be the functor of the previous section, defined with respect to $\mathcal{G}$, and abbreviate $BFX = BVFX$ and similarly for SFX. If FX is a topological monoid regarded as a $\mathcal{G}$-space by pullback, then this classifying space agrees up to natural equivalence with the standard one. Write $Z \amalg \{0\}$ or Z^+ interchangeably for the union of a space Z and a disjoint basepoint 0. (We agree never to use the + notation for the plus construction.) Let Q_0Y denote the component of zero in the space QY. The purpose of this section is to construct a natural map

$$\mu: Q_0(BFX \amalg \{0\}) \to KX$$

and thus a natural transformation from the stable homotopy groups of BFX^+ to the algebraic K-groups of X. Of course, $\pi_q^s(BFX^+)$ is the direct sum of $\pi_q^s BFX$ and the stable stem $\pi_q^s = \pi_q^s S^0$. When $X = QS^0$, such a map was asserted to exist by Waldhausen [22, §2].

The construction is based on the use of monomial matrices.

Definition 8.1. Let F_nX denote the subspace of FM_nX which consists of the monomial matrices with entries in FX, namely those matrices with precisely one non-zero entry in each row and column and all non-zero entries in FX. Let $F_\infty X$ denote the telescope of the F_nX with respect to the restrictions of the maps ν_n. Similarly, let SF_nX denote the space of monomial matrices with entries in SFX and let $SF_\infty X = \text{Tel}\, SF_nX$.

As will become clear below, it is immediate from formula (4.1) that F_nX and SF_nX are sub $\mathcal{H}_n$-spaces of FM_nX. The arguments of the previous section can be carried out word for word with FM_nX replaced by F_nX. There results the following analog of Theorem 7.1.

Theorem 8.2. There is a functor P from grouplike $(\mathcal{C}, \mathcal{H})$-spaces to connected based spaces together with a natural weak equivalence between ΩPX and $F_\infty X$. Moreover, there is a natural map $PX \to TX$ the loop of which agrees under the weak equivalences with the inclusion $F_\infty X \to FM_\infty X$.

In particular, $\pi_1 PX = \pi_0 F_\infty X = \operatorname{colim} \pi_0 F_n X$. It will soon become apparent that $\pi_0 F_n X$ is the wreath product $\Sigma_n \int \pi_0 FX$. It follows easily that $\pi_1 PX$ has a perfect commutator subgroup (see [7, 1.2]), and this will also drop out of our arguments below since they will give a homology isomorphism from PX to the simple space $Q_0(BFX \amalg \{0\})$. This homology isomorphism will immediately imply the following theorem.

Theorem 8.3. The plus construction on PX is naturally equivalent to $Q_0(BFX \amalg \{0\})$, hence the map $PX \to TX$ induces a natural (up to homotopy) map $\mu: Q_0(BFX \amalg \{0\}) \to KX$.

Remarks 8.4. Replacing $F_n X$ by $SF_n X$ throughout, we obtain SPX, a natural weak equivalence between ΩSPX and $SF_\infty X$, and a natural map $SPX \to PX$ compatible with the weak equivalences. Moreover, the plus construction on SPX is naturally equivalent to $Q_0(BSFX \amalg \{0\})$ and the resulting map from this space to KX agrees with the restriction of μ.

Remarks 8.5. There is a natural inclusion $\Sigma FX \to BFX$ (adjoint to $\zeta: FX \to \Omega BFX$), hence μ restricts to a natural map $Q_0(\Sigma FX \amalg \{0\}) \to KX$. Via the basepoint of ΣFX, there is a further natural restriction $QS^0 \to KX$. Similar remarks hold with F replaced by SF.

The rest of this section will be devoted to the proof of Theorem 8.3. We begin with a well-known observation about the classifying spaces of wreath products and an equally well-known consequence of the Barratt-Quillen theorem.

Lemma 8.6. For a topological monoid X, $B(\Sigma_n \int X)$ is naturally homeomorphic to $E\Sigma_n \times_{\Sigma_n} (BX)^n$.

Proof. Our conventions on wreath products are in [2, p.51]. B and E are obtained as geometric realizations of certain simplicial spaces [9, p. 31]. We define a $\Sigma_n \int X$-equivariant homeomorphism of simplicial spaces

$$\psi_* : E_*(\Sigma_n \int X) \to E_*\Sigma_n \times (E_*X)^n$$

by the formula

$$\psi_q[(\sigma_1, y_1), \ldots, (\sigma_q, y_q)](\sigma_{q+1}, y_{q+1}) = ([\sigma_1, \ldots, \sigma_q]\sigma_{q+1}, z),$$

where $\sigma_i \in \Sigma_n$, $y_i = (x_{i1}, \ldots, x_{in}) \in X^n$, and $z \in (E_q X)^n$ has i^{th} coordinate $[x_{1,\sigma_2 \cdots \sigma_{q+1}(i)}, \ldots, x_{q,\sigma_{q+1}(i)}]x_{q+1,i}$; ψ_*^{-1} is given by

$$\psi_q^{-1}([\sigma_1, \ldots, \sigma_q]\sigma_{q+1}, [x_{11}, \ldots, x_{q1}]x_{q+1,1}, \ldots, [x_{1n}, \ldots, x_{qn}]x_{q+1,n})$$
$$= [(\sigma_1, y_1), \ldots, (\sigma_q, y_q)](\sigma_{q+1}, y_{q+1}),$$

where $y_i = (x_{i,\sigma_{q+1}^{-1}\sigma_q^{-1}\cdots\sigma_{i+1}^{-1}(1)}, \ldots, x_{i,\sigma_{q+1}^{-1}\sigma_q^{-1}\cdots\sigma_{i+1}^{-1}(n)})$. The conclusion follows since realization commutes with products.

For a based space Y, the inclusion of Σ_n in Σ_{n+1} as the subgroup fixing the last letter and the inclusion of Y^n in Y^{n+1} as $Y^n \times \{*\}$ induce an inclusion $E\Sigma_n \times_{\Sigma_n} Y^n \to E\Sigma_{n+1} \times_{\Sigma_{n+1}} Y^{n+1}$.

Proposition 8.7. For connected spaces Y, there is a natural homology isomorphism $\mathrm{Tel}\, E\Sigma_n \times_{\Sigma_n} Y^n \to Q_0(Y \amalg \{0\})$.

Proof. For the E_∞ operad $\mathcal{Q}$ of [10, §4 and 12, §2 and 4], the space DY^+ is precisely $\coprod_{n \geq 0} E\Sigma_n \times_{\Sigma_n} Y^n$. Let $\mathcal{C} = \mathcal{Q} \times \mathcal{K}_\infty$ (or $\mathcal{Q} \times \mathcal{C}_\infty$ if one prefers to avoid partial operads) as in section 3, and note that the projection $CY^+ \to DY^+$ is a homotopy equivalence by [11, A.2(ii)]. By Proposition 3.5, with multiplicative structure ignored, QY^+ is naturally a group completion of CY^+. The conclusion follows as in [2, I. 5. 10].

We shall reduce Theorem 8.3 to an application of the previous two results. For this purpose, we require an understanding of $\Sigma_n \int X$ when X is a $\mathcal{C}$-space rather than a monoid.

<u>Definition 8.8.</u> For a $\mathcal{C}$-space X, define a $\mathcal{C}$-space $\Sigma_n \int X$ as follows. As a space, $\Sigma_n \int X = \Sigma_n \times X^n$. The action ξ_n of $\mathcal{C}$ on $\Sigma_n \int X$ is given by the maps

$$\xi_{nj} : \mathcal{C}(j) \times (\Sigma_n \times X^n)^j \to \Sigma_n \times X^n$$

specified for $g \in \mathcal{C}(j)$, $\sigma_q \in \Sigma_n$, and $x_q = (x_{q,1}, \ldots, x_{q,n}) \in X^n$ by

$$\xi_{nj}(g; \sigma_1, x_1, \ldots, \sigma_j, x_j) = (\sigma_1 \cdots \sigma_j, \mathop{\times}_{i=1}^{n} \xi_j(g; \mathop{\times}_{q=1}^{j} x_{q,\sigma_{q+1}\cdots\sigma_j(i)})).$$

(Technically, this formula is appropriate when $\mathcal{C}$ is taken as a non-Σ operad; compare section 4.) This is just the ordinary iterated wreath product, but with a parametrized family of multiplications on X. Let $\nu_n : \Sigma_n \int X \to \Sigma_{n+1} \int X$ denote the natural inclusion, and observe that ν_n is a map of $\mathcal{C}$-spaces.

We wish to commute the functor V past wreath products. The following rather elaborate formal argument based on the maps displayed in (7.1) and (7.2) suffices.

<u>Lemma 8.9.</u> For $\mathcal{C}$-spaces X, the horizontal arrows are homotopy equivalences in the following commutative diagram of $\mathcal{C}$-spaces and $\mathcal{C}$-maps, where $UX = B(G, G, X)$.

$$\begin{array}{ccccccccc}
V(\Sigma_n \int X) & \xleftarrow{B\delta} & U(\Sigma_n \int X) & \xrightarrow{\varepsilon} & \Sigma_n \int X & \xleftarrow{\Sigma_n \int \varepsilon} & \Sigma_n \int UX & \xrightarrow{\Sigma_n \int B\delta} & \Sigma_n \int VX \\
\downarrow{\scriptstyle V\nu_n} & & \downarrow{\scriptstyle W\nu_n} & & \downarrow{\scriptstyle \nu_n} & & \downarrow{\scriptstyle \nu_n} & & \downarrow{\scriptstyle \nu_n} \\
V(\Sigma_{n+1} \int X) & \xleftarrow{B\delta} & U(\Sigma_{n+1} \int X) & \xrightarrow{\varepsilon} & \Sigma_{n+1} \int X & \xleftarrow{\Sigma_{n+1} \int \varepsilon} & \Sigma_{n+1} \int UX & \xrightarrow{\Sigma_{n+1} \int B\delta} & \Sigma_{n+1} \int VX
\end{array}$$

Application of the functor V to this diagram gives a commutative diagram which can be extended to the left and right by the commutative diagrams

$$\begin{array}{ccc} V(\Sigma_n \int X) & \xleftarrow{\varepsilon \circ B\delta} & VV(\Sigma_n \int X) \\ \downarrow V\nu_n & & \downarrow VV\nu_n \\ V(\Sigma_{n+1} \int X) & \xleftarrow{\varepsilon \circ B\delta} & VV(\Sigma_{n+1} \int X) \end{array} \quad \text{and} \quad \begin{array}{ccc} V(\Sigma_n \int VX) & \xrightarrow{\varepsilon \circ B\delta} & \Sigma_n \int VX \\ \downarrow V\nu_n & & \downarrow \nu_n \\ V(\Sigma_{n+1} \int VX) & \xrightarrow{\varepsilon \circ B\delta} & \Sigma_{n+1} \int VX \end{array}$$

The resulting composite diagram is a commutative diagram of maps of topological monoids in which all horizontal arrows are homotopy equivalences.

Via the homotopy invariance properties of the classifying space functor B [9, 7.3(ii)] and the telescope, the previous result implies a chain of natural equivalences of telescopes which, together with the first two results above, leads to the following conclusion.

<u>Theorem 8.10.</u> For $\mathcal{G}$-spaces X, there is a natural homotopy equivalence

$$\text{Tel } BV(\Sigma_n \int X) \simeq \text{Tel } B(\Sigma_n \int VX) \cong \text{Tel } E\Sigma_n \times_{\Sigma_n} (BVX)^n .$$

Therefore there is a natural homology isomorphism

$$\text{Tel } BV(\Sigma_n \int X) \to Q_0(BVX \amalg \{0\})$$

We can now prove Theorem 8.3. By application of the previous theorem to the $\mathcal{G}$-space FX, with BFX = BVFX by notational convention, it suffices to prove the following result. The proof again makes strong use of the assumption $\mathcal{C}(1) = \{1\}$ justified at the end of section 6.

<u>Theorem 8.11.</u> For grouplike $(\mathcal{C}, \mathcal{G})$-spaces X, there is a natural homotopy equivalence $PX \to \text{Tel } BV(\Sigma_n \int FX)$.

<u>Proof.</u> Define a homeomorphism $\alpha_n : F_n X \to \Sigma_n \int FX$ by sending a monomial matrix x to $(\sigma, x(1), \ldots, x(n))$, where $\sigma \in \Sigma_n$ is specified by $\sigma(i) = j$ if $x(i,j) \neq 0$ and where $x(j) = x(i,j)$ for this i. Let $\pi_n : \mathcal{H}_n \to \mathcal{G}$ be the evident projection of operads and regard $\Sigma_n \int FX$ as an $\mathcal{H}_n$-space by pullback along π_n. We claim first that α_n is an $\mathcal{H}_n$-map or, equivalently, that the actions

$\alpha_n^{-1}\xi_n\pi_n H_n\alpha_n$ and ψ_n coincide on F_nX. Recall formula (4.1). Given $x_i \in F_nX$ for $1 \leq i \leq j$, there is for each r between 1 and n a unique s between 1 and n and a unique $U \in T_n(r,s,j)$ such that $x_q(u_{q-1}, u_q) \neq 0$. If this U is the kth element of $T_n(r,s,j)$ and if

$$s_{n,j,k} = (*, \ldots, *, 1, *, \ldots, *) \in \mathcal{C}(0)^{k-1} \times \mathcal{C}(1) \times \mathcal{C}(0)^{n^{j-1}-k},$$

then, for $c \in \mathcal{C}(n^{j-1})$ and $g \in \mathcal{G}(j)$,

$$\psi_{n,j}(c, g; x_1, \ldots, x_j)(r,s) = \theta_1(\gamma(c; s_{n,j,k}); \xi_j(g; \times_{q=1}^{j} x_q(u_{q-1}, u_q)))$$
$$= \xi_j(g; \times_{q=1}^{j} x_q(u_{q-1}, u_q)),$$

the last equality holding since $\gamma(c; s_{n,j,k}) = 1 \in \mathcal{C}(1)$ by assumption. The claim follows by comparison with Definition 8.8. Now consider the following diagram.

$$\begin{array}{ccccc}
BV_nF_nX & \xrightarrow{BV_n\alpha_n} & BV_n(\Sigma_n\int FX) & \xrightarrow{B\pi_n} & BV(\Sigma_n\int FX) \\
\uparrow{\scriptstyle B\tau_n} & & \uparrow{\scriptstyle B\tau_n} & & \| \\
BV_{n+1}F_nX & \xrightarrow{BV_{n+1}\alpha_n} & BV_{n+1}(\Sigma_n\int FX) & \xrightarrow{B\pi_{n+1}} & BV(\Sigma_n\int FX) \\
\downarrow{\scriptstyle BV_{n+1}\nu_n} & & \downarrow{\scriptstyle BV_{n+1}\nu_n} & & \downarrow{\scriptstyle BV\nu_n} \\
BV_{n+1}F_{n+1}X & \xrightarrow{BV_{n+1}\alpha_{n+1}} & BV_{n+1}(\Sigma_{n+1}\int FX) & \xrightarrow{B\pi_{n+1}} & BV(\Sigma_{n+1}\int X)
\end{array}$$

The left horizontal arrows are homeomorphisms and the right horizontal arrows are homotopy equivalences by [11, A.2(ii) and A.4(ii)]. The diagram commutes by naturality and the facts that $\pi_n\tau_n = \pi_{n+1}$ and $\nu_n\alpha_n = \alpha_{n+1}\nu_n$. The required natural equivalence $PX \rightarrow \mathrm{Tel}\, BV(\Sigma_n\int FX)$ follows by passage to reduced double mapping cylinders and then to unions as in Definition 7.6.

§9. Some homotopical and homological properties of KX.

Again, let $(\mathcal{C}, \mathcal{G})$ be an A_∞ operad pair and let X be a grouplike $(\mathcal{C}, \mathcal{G})$-space. From the point of view of analysis of its invariants, the sophisticated functors V_n (and W) which entered into the construction of KX are of no significance. They simply replace a given structured space by a homotopy equivalent space with different structure. Thus, up to homotopy, only the classifying space functor, the telescope, and the plus construction are involved. These facts and Remarks 7.3 imply that the analysis of KX is considerably less refractory than the complicated theory necessary for its construction would suggest.

We begin with two elementary homotopy invariance properties, which will be seen later to be simultaneous generalizations of Waldhausen's assertions [22, 1.1 and 2.3] and [22, 1.3 and 2.4].

Recall that a map $f: X \to Y$ is said to be an n-equivalence if $\pi_i f$ is an isomorphism for $i < n$ and an epimorphism for $i = n$ for all choices of basepoint in X (and analogously for maps of pairs and for n-homology equivalences).

Proposition 9.1. If $f: X \to Y$ is a map of grouplike $(\mathcal{C}, \mathcal{G})$-spaces and an n-equivalence, then $Kf: KX \to KY$ is an (n+1)-equivalence.

Proof. By Remarks 7.3, $FM_\infty f: FM_\infty X \to FM_\infty Y$ is an n-equivalence. By Theorem 7.1, $Tf: TX \to TY$ is thus an (n+1)-equivalence. Therefore Tf and thus also Kf are (n+1)-homology equivalences. Since KX and KY are simple spaces, the conclusion follows by the Whitehead theorem.

We next want the relative version of this result, and we need some preliminaries in order to take account of the non-existence of an unstable relative Whitehead theorem and to handle some technical points ubiquitously ignored in the literature. Consider a homotopy commutative diagram of spaces

$$(*)\qquad \begin{array}{ccc} X & \xrightarrow{f} & Y \\ {\scriptstyle g}\downarrow & & \downarrow{\scriptstyle g'} \\ Z & \xrightarrow{f'} & W \end{array}$$

Definitions 9.2. The diagram (*) is said to be an (m, n)-equivalence if f is an m-equivalence and g is an n-equivalence. It is said to be q-homotopy Cartesion if there exists a map of triads

$$\overline{f} = (f; f', f): (Mg; Z, X) \to (Mg'; W, Y)$$

such that the map of pairs $\overline{f}: (Mg, X) \to (Mg', Y)$ is a q-equivalence, where Mg and Mg' denote the mapping cylinders of g and g'. If (*) commutes, with no homotopy required, we insist that this condition be satisfied with $\overline{f}(x,t) = (fx, t)$ on the cylinder, and it is then equivalent to require that the natural map $Fg \to Fg'$ of homotopy fibres be a (q+1)-equivalence for each choice of basepoint in X (by the standard verification that the two definitions of the relative homotopy groups of a map agree).

Remarks 9.3. In the general case, with based spaces and maps, the map of triads $\overline{f}$ induces a map

$$\tilde{f}: Fg = X \times_g PZ \to Y \times_{g'} PW = Fg'$$

via

$$\tilde{f}(x, \zeta) = (fx, \omega), \quad \text{where } \omega(t) = \begin{cases} f'\zeta(2t) & \text{if } 0 \le t \le \frac{1}{2} \\ r\overline{f}(x, 2t-1) & \text{if } \frac{1}{2} \le t \le 1 \end{cases}$$

with $r: Mg' \to W$ being the canonical retraction. If $\overline{f}$ is a q-equivalence of pairs then $\tilde{f}$ is a (q+1)-equivalence. A converse construction is not immediately obvious to me, and the definition has been given in the form we wish to use. Clearly a homotopy $h: f'g \simeq g'f$ induces a map of triads of the sort specified, via

$$\overline{f}(x,t) = \begin{cases} h(x, 2t) & \text{if } 0 \leq t \leq \frac{1}{2} \\ (fx, 2t-1) & \text{if } \frac{1}{2} \leq t \leq 1 \end{cases},$$

but whether or not $\overline{f}$ is a q-equivalence of pairs really does depend on the choice of homotopy.

Lemma 9.4. Assume that (*) is an (m,n)-equivalence, where $m \geq 0$ and $n \geq 1$. For $q \leq m+n$, (*) is q-homotopy Cartesian if and only if there exists a q-equivalence $\phi: M(g,f) \to W$ such that $\phi k \simeq f'$ and $\phi j \simeq g'$, where $M(g,f)$ is the double mapping cylinder of g and f and $k: Z \to M(g,f)$ and $j: Y \to M(g,f)$ are the natural inclusions. If (*) commutes, ϕ must be the natural map factoring through the quotient map to the pushout of f and g.

Proof. The last statement will be a consequence of the conventions in Definitions 9.2. By the homotopy excision theorem, the natural map $(Mg, X) \to (M(g,f), Y)$ is an (m+n)-equivalence. (The range is misstated in [22].) Clearly maps $\overline{f}: Mg \to Mg'$ as in Definitions 9.2 factor uniquely through maps

$$\psi = (\psi; f', 1): (M(g,f); Z, Y) \to (Mg'; W, Y).$$

If $r: Mg' \to W$ is the retraction and $r\psi = \phi$, then $\phi k = f'$ and $\phi j = g$. Conversely, given ϕ as in the statement, let ψ be the composite of ϕ and the inclusion $i: W \to Mg'$. Since $Y \amalg Z \to Mg'$ is a cofibration, ψ is homotopic to a map of triads ψ as displayed. By the five lemma, $\psi: (M(g,f), Y) \to (Mg', Y)$ is a q-equivalence if and only if $\psi: M(g,f) \to Mg'$ is a q-equivalence. The conclusion follows.

Proposition 9.5. If (*) is a (strictly) commutative diagram of grouplike $(\mathcal{C}, \mathcal{G})$-spaces which is a (q-1)-homotopy Cartesian (m-1, n-1)-equivalence with $m \geq 2$, $n \geq 2$, and $q \leq m+n$, then

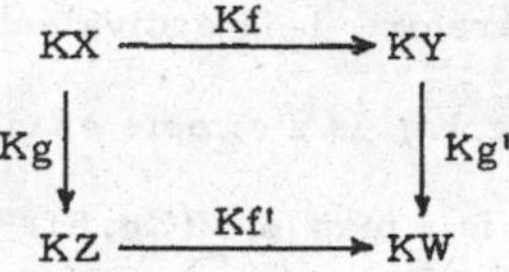

is a q-homotopy Cartesian (m, n)-equivalence.

Proof. By Remarks 7.3, application of FM_∞ to (*) gives a (q-1)-homotopy Cartesian (m-1, n-1)-equivalence. By Theorem 7.1 and a little standard argument with homotopy fibres, the (strictly) commutative diagram

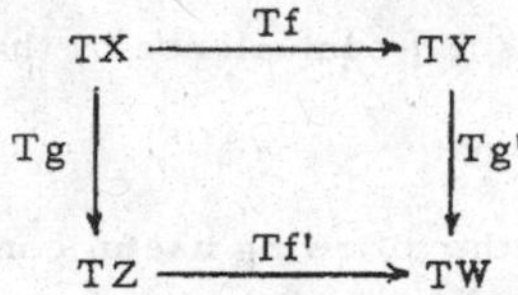

is a q-homotopy Cartesian (m, n)-equivalence. Consider the following diagram

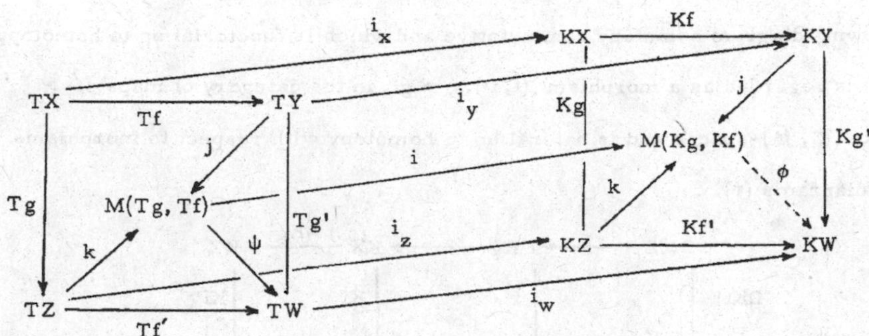

Breaking the cylinder of M(Tg, Tf) into three parts, mapping the middle third $TX \times [1/3, 2/3]$ to $KX \times [0,1]$ via i_x and expansion, and using homotopies $i_z \circ Tg \simeq Kg \circ i_x$ and $Kf \circ i_x \simeq i_y \circ Tf$ on $TX \times [0, 1/3]$ and $TX \times [2/3, 1]$, we obtain a map $i: M(Tg, Tf) \to M(Kg, Kf)$ such that $ij = ji_y$ and $ik = ki_z$ on the bases. By the van Kampen theorem and the fact that Tf, Tg, Kf, and Kg induce isomorphisms on π_1 since $m \geq 2$ and $n \geq 2$, $\pi_1 M(Tg, Tf) = \pi_1 TX$, $\pi_1 M(Kg, Kf) = H_1 TX$, and $\pi_1 i$ is Abelianization. By the Mayer-Vietoris sequence and the five lemma, i is a homology isomorphism. Therefore i is equivalent to the plus construction on $M(Tg, Tf)$. In particular, $M(Kg, Kf)$ is a simple space. By the universal property of the plus construction, there is a map $\phi: M(Kg, Kf) \to KW$, unique up to homotopy, such that $\phi i \simeq i_W \psi$, where $\psi: M(Tg, Tf) \to TW$ is the natural map. Since

$$\phi j i_Y = \phi i j \simeq i_W \psi j = i_W Tg' \simeq Kg' \circ i_Y ,$$

$\phi j \simeq Kg'$ by the universal property. Similarly $\phi k \simeq kf'$. Since ψ is a q-equivalence (by the lemma), it is a q-homology equivalence. Therefore ϕ is a q-homology equivalence and thus a q-equivalence by the Whitehead theorem. The conclusion follows from the lemma.

The proofs above have the following useful consequence.

Lemma 9.6. If (*) is a commutative diagram of grouplike $(\mathcal{C}, \mathcal{G})$-spaces, then there is a canonical homotopy class of maps $\tilde{f}: FKg \to FKg'$ which makes the following diagram homotopy commutative and which is functorial up to homotopy when (*) is regarded as a morphism $(f, f'): g \to g'$ in the category of maps of grouplike $(\mathcal{C}, \mathcal{G})$-spaces and is natural up to homotopy with respect to morphisms of such diagrams (*).

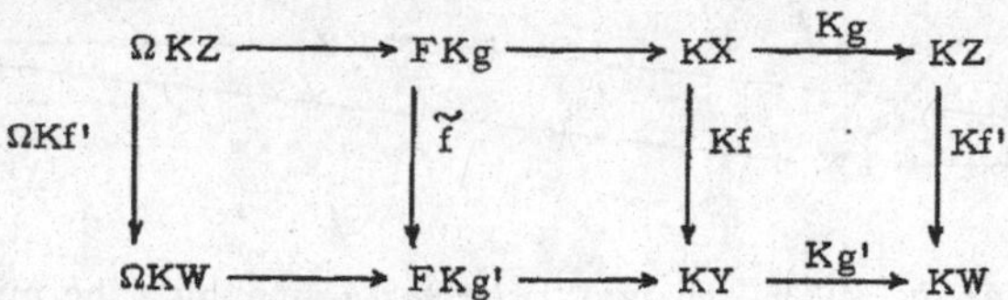

Here the unlabeled arrows are the natural maps of the displayed fibration sequences.

Proof. Of course, Barratt-Puppe sequence arguments give a map $\tilde{f}$, not uniquely determined up to homotopy. We ignore this. Construct $\phi: M(Kg, Kf) \to KW$ as in the previous proof, deform $i\phi$, $i: KW \to Mg'$ to a map of triads

$$\psi = (\psi; Kf', 1): (M(Kg, Kf); KZ, KY) \to (Mg'; KW, KY)$$

as in the proof of Lemma 9.4, let $\overline{f}$ be the composite of ψ and the natural map $MKg \to M(Kg, Kf)$, and construct $\tilde{f}$ as in Remarks 9.3. It is simple to check (by standard cofibration arguments for the passage from ϕ to ψ) that $\tilde{f}$ is a well-defined homotopy class which makes the displayed diagram homotopy commute. Its functoriality and naturality are then easily verified by the same sorts of homotopical arguments as those above.

Turning to homology, we first record the form taken in our context by the standard spectral sequence for the calculation of the homology of classifying spaces of topological monoids.

Since $\nu_n : FM_nX \to FM_{n+1}X$ only commutes up to homotopy with the multiplications, there is a slight ambiguity in giving $FM_\infty X$ an H-space structure. There is no ambiguity in its Pontryagin product, however, and the spectral sequences of the filtered spaces BV_nFM_nX pass to limits to give the following result. (See e.g. [9, 13.10].)

Proposition 9.7. Take homology with coefficients in a field k. There is then a natural spectral sequence $\{E^rX\}$ of differential coalgebras which converges from $E^2X = \mathrm{Tor}^{H_*FM_\infty X}(k,k)$ to H_*KX.

Field coefficients serve only to yield a conceptual description of E^2X. Since $i: TX \to KX$ induces an isomorphism on k_* for any connective homology theory k_* (by the Atiyah-Hirzebruch spectral sequence), we obtain a spectral sequence $\{E^rX\}$ converging to k_*KX with $E^1_{pq}X = k_q((FM_\infty X)^p)$ for any such k_*, d^1 being induced by application of k_* to the alternating sum of the standard bar construction face maps. (See [10, 11.14] for details.)

The proposition focuses attention on the problem of computing the Pontryagin algebras H_*FM_nX. It is clear from previous experience what procedures one should adopt: one should work in all of H_*M_nX and exploit the diagram of Proposition 5.2. Since $M_nX = X^{n^2}$ is an (additive) infinite loop space, all of the machinery of homology operations explained in [2, I] is available. Assuming that H_*M_nX is understood additively (as a Hopf algebra with product $*$, coproduct ψ, conjugation χ, augmentation ε, homology operations Q^s and Steenrod operations P^r_*, with $P^r_* = Sq^r_*$ at the prime 2), we can study its products by use of the following two results, the proofs of which are exactly the same as those in [2, p. 79-81].

Proposition 9.8. Take homology with coefficients in any field. Let $x, y, z \in H_*M_nX$, $n \geq 1$, and let $[0]$, $[1]$, and $[-1]$ be the classes of the zero matrix, the identity matrix I_n, and any matrix in the component additively inverse to that of I_n.

(i) $[0]x = (\varepsilon x)[0]$, $[1]x = x$, and $[-1]x = \chi x$.

(ii) $(x * y)z = \sum (-1)^{\deg y \deg z'} xz' * yz''$, where $\psi z = \sum z' \otimes z''$.

Proposition 9.9. Take homology with mod p coefficients, where p is any prime. Let $x, y \in H_*M_nX$, $n \geq 1$. Then

$$(Q^s x)y = \sum_i Q^{s+i}(xP^i_* y) \quad \text{and, if } p > 2,$$

$$(\beta Q^s x)y = \sum_i \beta Q^{s+i}(xP^i_* y) - \sum_i (-1)^{\deg x} Q^{s+i}(xP^i_* \beta y).$$

Remarks 9.10. Applied to the A_∞ ring spaces QX^+ of Examples 2.3, where X is an A_∞ space, the propositions above completely determine the Pontryagin algebra H_*FQX^+ in terms of the additive structure of H_*QX^+. Compare [2, II §4], where the analogous assertion for the E_∞ ring space QX^+ obtained from an E_∞ space X is explained in detail. To determine $H_*FM_nQX^+$ for $n > 1$, one would require additional formulas to explain the effect on homology of translations by more general matrices than in Proposition 9.8(i).

§10. Examples, natural maps, and formal properties of KX

Recall from section 1 that any A_∞ operad pair $(\mathcal{C}, \mathcal{G})$ admits an augmentation $(\mathcal{C}, \mathcal{G}) \rightarrow (\mathcal{N}, \mathcal{M})$. An $(\mathcal{N}, \mathcal{M})$-space is precisely a topological semi-ring and a grouplike $(\mathcal{N}, \mathcal{M})$-space is precisely a topological ring. Therefore a topological ring R is a grouplike $(\mathcal{C}, \mathcal{G})$-space by pullback. If R is discrete, then $FM_n R = GL(n, R)$, and the following example is immediate from the constructions of section 7.

Example 10.1. If R is a discrete ring, then KR is naturally equivalent to the plus construction on $BGL(\infty, R)$. Therefore $K_q R$, $q > 0$, is Quillen's q^{th} algebraic K-group of R [5, 18].

For general topological rings, our theory reduces to the topological version of Waldhuasen's [22, §1].

Example 10.2. For simplicial rings R_*, Waldhausen defined a certain functor KR_*. It would be immediate from the definitions that his KR_* is naturally equivalent to our $K|R_*|$, where $|R_*|$ is the geometric realization of R_*, were it not that he has chosen to throw in a discrete factor Z and we have not, so that his KR_* is our $K|R_*| \times Z$.

Remarks 10.3. There is one vital distinction to be made between KR for a topological ring R and KX for an arbitrary grouplike A_∞ ring space X. As is well-known in the discrete case [20; 21; 12] and will be proven in general at the end of this section, KR is an infinite loop space. In contrast, I see little reason to think that KX is an infinite loop space and have not yet been able to prove that it is even an H-space, although I believe this to be the case (compare Remarks 7.7).

It is immediate from the definitions that, for any $(\mathcal{C}, \mathcal{G})$-space X, the discretization map $d: X \to \pi_0 X$ is a map of $(\mathcal{C}, \mathcal{G})$-spaces. When X is grouplike, this has the following important consequence.

Proposition 10.4. For grouplike $(\mathcal{C}, \mathcal{G})$-spaces X, there is a natural augmentation $Kd: KX \to K\pi_0 X$.

This suggests the following reduced variant of KX.

Definition 10.5. Define $K'X$ to be the homotopy fibre of Kd. By Lemma 9.6, K' is then a functor of X such that the following diagram is homotopy commutative for any map $f: X \to Y$ of grouplike $(\mathcal{C}, \mathcal{G})$-spaces.

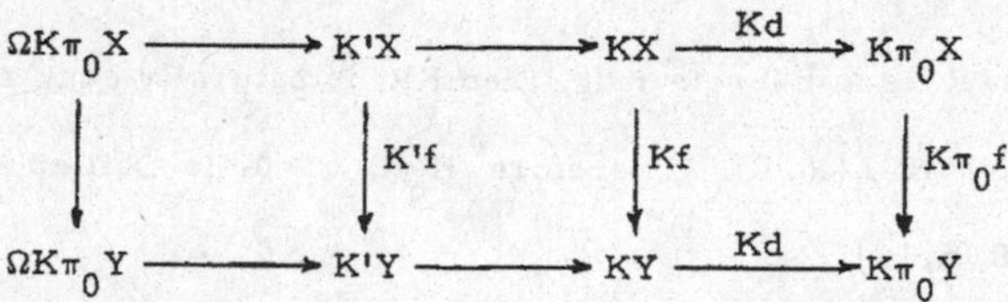

$$\begin{array}{ccccccc} \Omega K\pi_0 X & \longrightarrow & K'X & \longrightarrow & KX & \xrightarrow{Kd} & K\pi_0 X \\ \downarrow & & \downarrow K'f & & \downarrow Kf & & \downarrow K\pi_0 f \\ \Omega K\pi_0 Y & \longrightarrow & K'Y & \longrightarrow & KY & \xrightarrow{Kd} & K\pi_0 Y \end{array}$$

Of course, K'X is trivial if X is discrete.

We can mimic this construction after restriction to monomial matrices. The resulting functor may be described, up to equivalence, as follows.

Definition 10.6. Define $Q_0'(BFX^+)$, $BFX^+ = BFX \amalg \{0\}$, to be the homotopy fibre of $Q_0(Bd^+): Q_0(BFX^+) \to Q_0(BF\pi_0 X^+)$, where $F\pi_0 X$ is the group of units in the ring $\pi_0 X$. As in the previous definition, Lemma 9.6 (or rather its monomial matrix analog) implies that this is the object function of a functor of X. Since $SF\pi_0 X = \{1\}$, the corresponding functor obtained with FX replaced by SFX is equivalent to $Q_0 BSFX$.

At the risk of belaboring the obvious, we combine the definitions above with Theorem 8.3, Remarks 8.5, and further applications of Lemma 9.6 (and analogs thereof) in the following theorem.

Theorem 10.7. For grouplike $(\mathcal{C}, \mathcal{D})$-spaces X, the following is a homotopy commutative diagram and is natural up to homotopy.

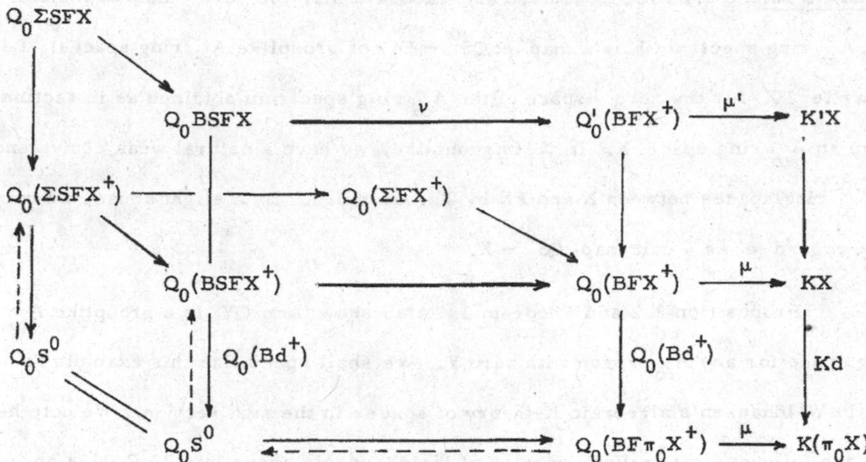

The dotted arrows denote the presence of evident sections, and the columns are fibrations; μ' and ν are obtained as in Lemma 9.6 and the remaining horizontal arrows are inclusions.

The diagram could be expanded further (by use of $Q_0(\Sigma F\pi_0X^+)$, etc.), but we desist. In view of its importance in the applications, we call special attention to the case $X = QS^0$, where FX and SFX are generally denoted F and SF (or G and SG) and where $\pi_0X = Z$ and thus $F\pi_0X = Z_2$. (Up to notation, the resulting diagram is an elaboration of one claimed by Waldhausen [22, above 2.3].) Here the bottom composite

$$Q_0S^0 \longrightarrow Q_0(BZ_2^+) \xrightarrow{\mu} KZ$$

is induced by the inclusions $\Sigma_n \to GL(n, Z)$ of permutation matrices and is thus the standard map studied by Quillen [19] (see also [12, VI §5 and VIII. 3.6]); the monomial matrix map μ has been studied in [12, VI. 5.9, VII. 4.6 and VIII. 3.6] and also in Loday [7]. As the following remarks make clear, this example is universal:

the diagram for QS^0 maps naturally to that for X for any grouplike A_∞ ring space X.

Remarks 10.8. By Proposition 2.2 and Theorem 3.1, the zeroth map of the unit of any A_∞ ring spectrum E is a map $e: QS^0 \to E_0$ of grouplike A_∞ ring spaces. Let us write ΓX for the zeroth space of the A_∞ ring spectrum obtained as in section 3 from an A_∞ ring space X. If X is grouplike, we have a natural weak equivalence of A_∞ ring spaces between X and ΓX by Theorem 3.3. By a slight abuse, we may thus regard e as a unit map $QS^0 \to X$.

Proposition 2.2 and Theorem 3.1 also show that QY is a grouplike A_∞ ring space for any A_∞ space with zero Y. We shall specialize this example to obtain Waldhausen's algebraic K-theory of spaces in the next section. We note here that the following generalized version of Waldhausen's assertion [22, 2.2] is an immediate consequence of Propositions 3.5 and 3.7 and Lemma 3.6 together with Theorem 7.1, Remarks 7.3, and Proposition 9.7.

Proposition 10.9. For any A_∞ space with zero Y, the Hurewicz map $h: QY \to \Gamma NY$ is a map of grouplike A_∞ ring spaces and a rational equivalence. Therefore $Kh: KQY \to K\Gamma NY$ is also a rational equivalence.

Remarks 10.10. NS^0 is precisely the additive monoid of non-negative integers, hence $\pi_0 \Gamma NS^0 = Z$ and $d: \Gamma NS^0 \to Z$ and therefore also $Kd: K\Gamma NS^0 \to KZ$ are equivalences. Thus we may view the Hurewicz map of KQS^0 as having target KZ.

The deepest source of examples is the theory of E_∞ ring spaces and E_∞ ring spectra. By neglect of structure, these are A_∞ ring spaces and A_∞ ring spectra. For example, the zeroth spaces of all of the various Thom spectra are A_∞ ring spaces [12, IV §2]. Of greater interest are the examples coming out of the chain of functors constructed in [12, VI and VII]:

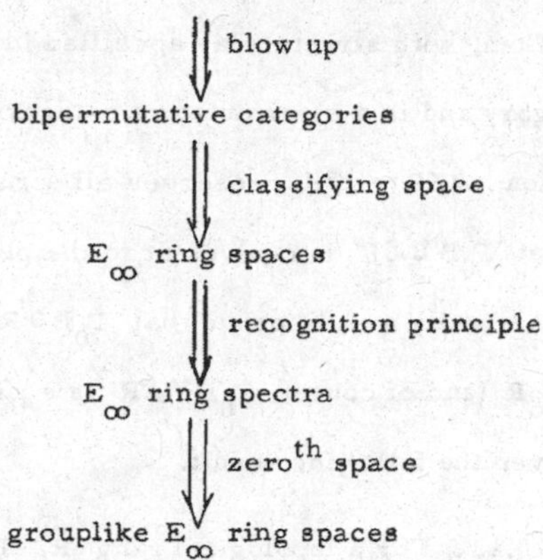

For instance, starting with the symmetric bimonoidal category $\mathscr{P}R$ of finitely generated projective modules over a commutative topological ring R, we arrive at an associated grouplike E_∞ ring space, denoted $\Gamma B\mathscr{P}R$, in which addition comes from the direct sum and multiplication comes from the tensor product. We interject the following note (compare [22, §1]).

Remarks 10.11. We are here faced with a conflict of definitions and notations. For a topological ring R (not necessarily commutative), I wrote K_qR for $\pi_q\Gamma B\mathscr{P}R$ in [10, VIII §1]. For discrete R, this agrees with the present K_qR for $q > 0$. For general R, it is quite different. I suggest writing K^tR for the plus construction on BGLR and writing $K_q^tR = \pi_qK^tR$, thinking of this as a topological K-theory (which it is when R is the topologized complex numbers for example). In practice, the functors K_*R and K_*^tR tend to be of interest for different choices of R.

In fact, the theory sketched above applies equally well to both generalizations of Quillen's theory from discrete rings to topological rings. To see this, let $\mathscr{GL}R$ denote the permutative category of finitely generated free R-modules, as described explicitly in [12, VI. 5. 2]. Define $\mathscr{F}R$ to be the category defined in precisely the same way, except that the space of morphisms $n \to n$ is the topological monoid FM_nR. Then, with structure as specified in the cited definition, $\mathcal{F}R$ is a permutative category and is a bipermutative category if R is commutative. Moreover, the inclusion $\mathcal{GL}R \to \mathcal{F}R$ preserves all structure in sight. The argument used to prove that $\Gamma_0 B\mathcal{GL}R$ is equivalent to the plus construction on BGLR in [12, VIII §1] applies equally well to prove that $\Gamma_0 B\mathcal{F}R$ is equivalent to the plus construction on $BFM_\infty R$ (and of course $\Gamma_0 B\mathcal{GL}R$ is equivalent to $\Gamma_0 B\mathcal{P}R$ [11, p. 85]). We have proven the following result.

Proposition 10.12. For a topological ring R, K^tR and KR are the zero components of infinite loop spaces $\Gamma B\mathcal{GL}R$ and $\Gamma B\mathcal{F}R$, and there is a natural infinite loop map $i: \Gamma B\mathcal{GL}R \to \Gamma B\mathcal{F}R$. If R is commutative, then $\Gamma B\mathcal{GL}R$ and $\Gamma B\mathcal{F}R$ are E_∞ ring spaces and i is an E_∞ ring map.

Here the additive infinite loop space structures associated to permutative categories are uniquely determined by the axioms in [14]; in particular, Segal's machine [20] and mine give equivalent spectra.

At the moment, nothing is known about the resulting "second order" algebraic K-theories $K_*\Gamma B\mathcal{GL}R$ and $K_*\Gamma B\mathcal{F}R$ of commutative topological rings. They do not appear to be very closely related to the "first order" theories K_*R or K_*^tR. Since $\pi_0\Gamma B\mathcal{GL}R = \pi_0\Gamma B\mathcal{F}R = Z$, Proposition 10.4 and naturality give maps

$$K\Gamma B\mathcal{GL}R \to K\Gamma B\mathcal{F}R \to KZ \to K^tR \to KR.$$

If we had followed Waldhausen and crossed everything with Z, we could consistently write

$$KK^tR = K\Gamma B\mathcal{GL}R \quad \text{and} \quad KKR = K\Gamma B\mathcal{F}R,$$

hence our view of these as second order theories.

§11. The algebraic K-theory of spaces

Finally, we specialize our theory to make rigorous the algebraic K-theory of spaces suggested by Waldhausen [22]. In this section, X will denote a non-degenerately based CW-space with basepoint 1. There are at least three ways that ΩX can be interpreted as an A_∞-space.

(1) The ordinary loop space of X is a $\mathcal{C}_1$-space, where $\mathcal{C}_1$ is the little 1-cubes A_∞ operad of [10, §4].

(2) The Moore loop space of X is a topological monoid, or $\mathcal{M}$-space.

(3) $|GSX|$, the geometric realization of the Kan loop group of the total singular complex of X is also an $\mathcal{M}$-space.

We lean towards the first choice, but the theory works equally well with any choice. As in Example 2.3, we construct from ΩX an A_∞ space with zero $\Omega X^+ = \Omega X \amalg \{0\}$ and thus a grouplike A_∞ ring space $Q(\Omega X^+)$. For any based space Y,

$$\pi_0 QY = \operatorname{colim} \pi_n \Omega^n \Sigma^n Y = \operatorname{colim} \tilde{H}_n \Sigma^n Y = \tilde{H}_0 Y,$$

this isomorphism being realized by $h\colon QY \to \Gamma NY$ on passage to components. It follows easily that, as a ring, $\pi_0 Q(\Omega X^+) = \pi_0 \Gamma N(\Omega X^+)$ is the integral group ring $Z[\pi_1 X]$. In reading the following definitions, it will be useful to keep in mind the following commutative diagram of A_∞ ring spaces (see Propositions 3.7, 10.4, and 10.9 and Remarks 10.10).

$$\begin{array}{ccccc}
QS^0 & \xrightarrow{\ \eta\ } & Q(\Omega X^+) & \xrightarrow{\ \varepsilon\ } & QS^0 \\
\downarrow h & & \downarrow h & & \downarrow h \\
\Gamma NS^0 & \xrightarrow{\ \eta\ } & \Gamma N(\Omega X^+) & \xrightarrow{\ \varepsilon\ } & \Gamma NS^0 \\
d \downarrow \simeq & & \downarrow d & & \simeq \downarrow d \\
Z & \xrightarrow{\ \eta\ } & Z[\pi_1 X] & \xrightarrow{\ \varepsilon\ } & Z
\end{array}$$

For uniformity of notation, we write η for maps induced by inclusions of basepoints of spaces or trivial subgroups of groups and write ε for the corresponding projections, so that $\varepsilon\eta = 1$. The vertical composites in the diagram are again discretization maps d. We shall continue to write η, ε, d, and h upon application of the functor K.

Definitions 11.1. Define AX, the connected algebraic K-space of X, to be $KQ(\Omega X^+)$. Further, define the following reduced variants.

$$\tilde{A}X = \text{fibre}\ (\varepsilon : AX \to A\{*\}), \quad A\{*\} = KQS^0$$

$$A'X = \text{fibre}\ (d : AX \to KZ[\pi_1 X]) = K'Q(\Omega X^+)$$

$$\tilde{A}'X = \text{fibre}\ (\varepsilon' : A'X \to A'\{*\}), \quad A'\{*\} = K'QS^0$$

For $q > 0$, define $A_q X$, the qth algebraic K-group of X, to be $\pi_q AX$, and introduce similar notation for the reduced variants.

We have an analogous algebraic K-theory of X with coefficients in Z.

Definition 11.2. Define $A(X; Z) = K\Gamma N(\Omega X^+)$ and define the reduced variants

$$\tilde{A}(X; Z) = \text{fibre}\ (\varepsilon : A(X; Z) \to A(*; Z)), \quad A(*; Z) = KZ$$

$$A'(X; Z) = \text{fibre}\ (\varepsilon' : A(X; Z) \to KZ[\pi_1 X]) = K'\Gamma N(\Omega X^+).$$

Here $A'(*; Z) \simeq \{*\}$, hence we set $\tilde{A}'(X; Z) = A'(X; Z)$. Define $A_q(X; Z) = \pi_q A(X; Z)$ and similarly for the reduced variants.

For a (discrete) group π and commutative ring R, write $\tilde{K}R[\pi] = \text{fibre}\ (\varepsilon : KR[\pi] \to KR)$.

By Lemma 9.6, these are all well-defined functors of their variables such that the various canonical maps in sight are natural. We summarize the relationships between these functors in the following result.

Theorem 11.3. The rows and columns are fibration sequences in the following natural homotopy commutative diagram.

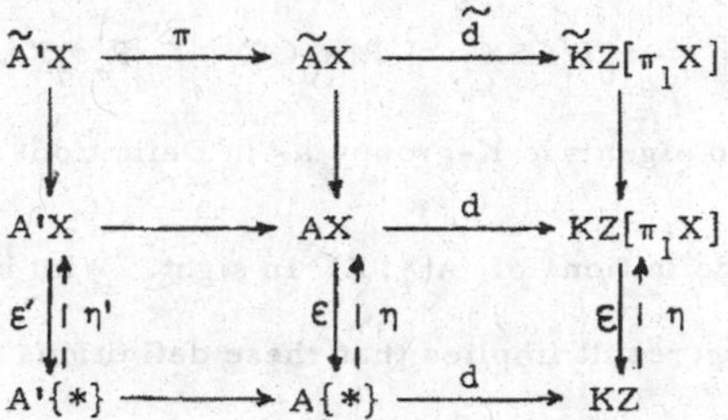

The dotted arrows denote the presence of sections. The Hurewicz map sends this diagram naturally, via a rational equivalence, to the corresponding natural 3×3 diagram of fibration sequences with AX and its variants replaced by $A(X; Z)$ and its variants.

Proof. In view of our earlier results, only the construction of π and the verification that it is equivalent to the fibre of $\tilde{d}$ are needed. This would be obvious enough if we knew that AX were an H-space. In the absence of this, a technical argument with Barratt-Puppe sequences, which we defer to the appendix, is required.

Of course, Theorem 10.7 applies naturally to all A_∞ ring spaces in sight and produces a maze of commutative diagrams. In particular, we have the maps

$$Q_0 X_1^+ \simeq Q_0(B\Omega X^+) \xrightarrow{Q_0(B\eta^+)} Q_0(BFQ(\Omega X^+)^+) \xrightarrow{\mu} AX ,$$

where X_1 is the component of the basepoint of X, $X_1 \simeq B\Omega X$ by [9,14.3 or 15.4], $\eta: \Omega X \to FQ(\Omega X^+)$ is the natural A_∞ map of Example 2.3, and μ is the monomial matrix map of section 8. This agrees with assertions of Waldhausen [22, §2].

It is desirable to have algebraic K-theories of spaces with coefficients in arbitrary commutative rings R. For many rings, this can be obtained topologically. (For Z_n, compare Dold and Thom [3] to the material at the end of section 3.) However it is most natural to follow Waldhausen [22,§1] and do this simplicially.

Definitions 11.4. Define $A(X;R) = K|R[GSX]|$, where the group ring of a simplicial group is formed degreewise. Clearly

$$\pi_0|R[GSX]| = \pi_0 R[GSX] = R[\pi_0 GSX] = R[\pi_1 X].$$

Define reduced variants and algebraic K-groups as in Definitions 11.2.

We now have two definitions of $A(X;Z)$ in sight. With interpretation (3) above for ΩX, the following result implies that these definitions agree up to natural equivalence.

Proposition 11.5. For simplicial groups G, there is a natural weak equivalence of grouplike A_∞ ring spaces between $\Gamma N|G|^+$ and $|Z[G]|$.

Proof. $N|G|^+$ is the free topological Abelian monoid generated by $|G|$ (the disjoint basepoint being identified to zero). The inclusion $G \to Z[G]$ induces $|G| \to |Z[G]|$ and thus, by freeness, $N|G|^+ \to |Z[G]|$. Moreover, this map clearly extends over the free Abelian group $\tilde{N}|G|^+$ generated by $|G|$, and it is easy to verify from the fact that realization commutes with products that $\tilde{N}|G|^+ \to |Z[G]|$ is actually a homeomorphism. The natural map

$$\lambda : N|G|^+ \to \tilde{N}|G|^+ \cong |Z[G]|$$

is clearly a map of semi-rings, hence of $(\mathcal{C}, \mathcal{G})$-spaces for any A_∞ operad pair $(\mathcal{C}, \mathcal{G})$. By Corollary 3.8 (recall that $\Gamma = B_0$), $\Gamma\lambda$ is a weak equivalence. By Theorem 3.3 and the fact that $|Z[G]|$ is grouplike, $\Gamma|Z[G]|$ is naturally weakly equivalent as an A_∞ ring space to $|Z[G]|$.

In view of the unit map $Z \to R$, all of our diagrams remain present with Z replaced by R, and the resulting diagrams are natural in R as well as X.

The simplicial approach has the advantage of giving us infinite loop space structures on our algebraic K-spaces, by Proposition 10.12 applied to $K|R[G]|$ for a simplicial group G. When G is the trivial simplicial group, $|R[G]| = R$. We thus have $\varepsilon : K|R[G]| \to KR$ and $\eta : KR \to K|R[G]|$. Write $\tilde{K}|R[G]| = \text{fibre } \varepsilon$

and give it the induced infinite loop structure. The following result is immediate by a glance at homotopy groups.

Proposition 11.6. For simplicial groups G and commutative rings R, the composite

$$\widetilde{K}|R[G]| \times KR \xrightarrow{p \times \eta} K|R[G]| \times K|R[G]| \xrightarrow{\phi} K|R[G]|,$$

where p is the canonical map and ϕ is the product, is a natural equivalence of infinite loop spaces.

The following is a special case.

Corollary 11.7. A(X; R) is naturally equivalent as an infinite loop space to $\widetilde{A}(X;R) \times KR$.

All of these algebraic K-theories on spaces admit stabilizations to generalized homology theories.

Theorem 11.8. There are natural homomorphisms $\widetilde{A}_n X \to \widetilde{A}_{n+1} \Sigma X$ such that if $\widetilde{A}^s_q X = \operatorname{colim} \widetilde{A}_{q+r} \Sigma^r X$, then $\widetilde{A}^s_*$ is a reduced homology theory. There are analogous theories $\widetilde{A}^s_*(X;R)$ defined and natural on R, and there is a Hurewicz homomorphism $\widetilde{h}^s_* : \widetilde{A}^s_* X \to \widetilde{A}^s_*(X;Z)$ which is a rational equivalence.

We need only construct the theories. The rest will follow from evident naturality arguments. Of course, $\widetilde{A}_* X$ maps naturally to $\widetilde{A}^s_* X$, and similarly and compatibly for $\widetilde{A}_*(X;R)$. The following consequence is immediate (see Corollary A.3).

Corollary 11.9. For unbased spaces X, the definitions $A^s_* X = \widetilde{A}^s_*(X^+)$ and $A^s_*(X;R) = \widetilde{A}^s_*(X^+;R)$ specify unreduced homology theories.

Note here that $\Omega(X^+)$ is a point, hence $\widetilde{A}_*(X^+) = \widetilde{A}_*(S^0)$. The corollary was asserted by Waldhausen [22, 1.4 and 2.8] (but with a rather misleading sketch proof). We shall prove a general result, Theorem A.2, on the stabilization of functors to homology theories. Theorem 11.8 will be an immediate special case in view of the following two results.

Proposition 11.10. The functors AX and A(X; R) preserve n-equivalences and q-homotopy Cartesian (m, n)-equivalences with $m \geq 2$, $n \geq 2$, and $q \leq m+n$ provided that, for the latter, the domain square is strictly commutative.

Proof. We lose a dimension upon appliction of Ω and gain it back upon passage to K and quotation of Propositions 9.1 and 9.5. We need only verify that the intervening functors Q, ΓN, or R[?] have the appropriate preservation properties, where R[C] for a simplicial set C is the free simplicial R-module generated by C. Since $\pi_* \Gamma N(X^+) = H_* X$ and, by [8, §22], $\pi_* R[C] = H_*(C; R)$, the conclusions are obvious in these cases. Since $\pi_* QX^+ = \pi_*^s(X^+)$ is the unreduced homology theory associated to stable homotopy, the conclusion here follows by use of the Atiyah-Hirzebruch spectral sequence.

Lemma 11.11. The functors AX and A(X; R) from based CW-spaces to the homotopy category of based CW-spaces are homotopy preserving.

Proof. This is not obvious. Recall that a functor T from based spaces to based spaces is said to be continuous if the function $T: F(X, Y) \to F(TX, TY)$ on function spaces is continuous. Since a homotopy between maps $X \to Y$ is a map $I \to F(X, Y)$, continuous functors are homotopy preserving. The various monads, bar constructions, and telescopes which entered into the construction of the functor T on $(\mathcal{C}, \mathcal{G})$-spaces are all continuous. (In particular, this uses the fact that geometric realization of simplicial spaces is a continuous functor [12, p. 21].) We pass from T to K by first converting to CW-complexes by applying geometric realization on the total singular complex, this composite being homotopy preserving although not continuous, and then applying the plus construction (which is a homotopy functor by definition). The functors Q, ΓN, and also Ω when interpreted as in (1) or (2) are continuous, and the functors R[GSX] are homotopy preserving by standard facts on the relationship between simplicial and topological homotopy theory (e.g. [8, §16 and §26]).

§12. Notes on Waldhausen's work

Since this work started with an attempt to understand Waldhausen's, a rundown of those things in [22, §1 and 2] not considered above may not be taken amiss.

There are two calculational results in [22, §1] concerning simplicial (or topological) rings. Our theory adds nothing new to the foundations here except for the infinite loop space structure on KR of Proposition 10.12 and the concomitant splittings of Proposition 11.6. Since the H-space level of these additions provides some clarification of Waldhausen's arguments, I shall run through the details (modulo the relevant algebra; these details are included at Rothenberg's request).

Proposition 12.1 ([22, 1.2]). Let $f: R \to R'$ be an (n-1)-equivalence of topological rings, where $n \geq 2$. Let F and F' be the homotopy fibres in the following diagram and choose $F \to F'$ which makes the top square homotopy commute.

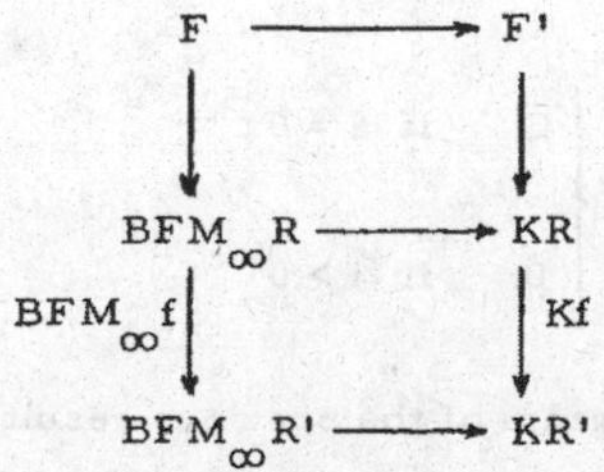

For any Abelian group A, the diagram induces an isomorphism

$$H_0(GL(\pi_0 R'); M_\infty(\pi_{n-1}) \otimes A) \cong H_0(BFM_\infty R'; H_n(F; A)) \to H_n(F'; A),$$

where π_{n-1} is the $(n-1)^{st}$ homotopy group of the fibre of f.

Proof. Waldhausen asserts further that the left side is clearly isomorphic, via the trace, to the Hochschild homology group $H_0(\pi_0 R'; \pi_{n-1} \otimes A)$; I haven't checked the algebra. By the Hurewicz theorem, Remarks 7.3, and universal coefficients, $H_n(F;A) \cong M_\infty(\pi_{n-1}) \otimes A$. The first isomorphism follows from the definition of local coefficients, the fact that $\pi_1 BFM_\infty R' \cong GL(\pi_0 R')$, and a check that the action on $H_n(F;A)$ agrees with the natural action on $M_\infty(\pi_{n-1}) \otimes A$. Let $\{E^r\} \to \{'E^r\}$ be the map of Serre spectral sequences (with coefficients in A) induced by the diagram.. The essential topological fact is that, since Kf is a map of connected H-spaces which induces an epimorphism on fundamental groups, it has trivial local coefficients. Our assertion is thus that $E^2_{0n} \to {}'E^2_{0n}$ is an isomorphism. Since

$E^2_{*0} = H_*(BGLR';A) \to H_*(KR';A) = {}'E^2_{*0}$ and $H_*(BGLR;A) \to H_*(KR;A)$

are isomorphisms and $E^r_{*q} = {}'E^r_{*q} = 0$ for $0 < q < n$, $E^2_{n0} = E^\infty_{n,0}$ and ${}'E^2_{n0} = {}'E^\infty_{n0}$, hence the five lemma gives that $E^\infty_{0n} \to {}'E^\infty_{0n}$ is an isomorphism. A diagram chase from the edge homomorphism gives that $E^\infty_{*0} \to {}'E^\infty_{*0}$ is an isomorphism, and another five lemma argument (involving the transgression $d^{n+1}: E^2_{n+1,0} \to E^2_{0n}$) gives the conclusion.

Proposition 12.2 ([22, 1.5])

$$A^s_i(\{*\};Z) \otimes Q \cong \begin{cases} Q & \text{if } i = 0 \\ 0 & \text{if } i > 0 \end{cases}$$

Proof. Specialize the diagram of the previous result to

$$\begin{array}{ccccc}
F & \longrightarrow & F' & = & \widetilde{A}(S^n; Z) \quad , \\
\downarrow & & \downarrow & & \downarrow \\
BFM_\infty R_n & \longrightarrow & KR_n & = & A(S^n; Z) \\
{\scriptstyle BFM_\infty d}\downarrow & & \downarrow & & \downarrow \\
BGLZ & \longrightarrow & KZ & = & A(\{*\}; Z)
\end{array}$$

where $R_n = |Z[GS\Sigma^n S^0]|$ and $d: R_n \to Z$ is the discretization. F' is (n-1)-connected. We claim that $H_n F' = Q$ and $H_q F' = 0$ for $n < q \leq 2n-2$ (Q coefficients understood). By the Whitehead theorem applied to a π_n-isomorphism $F_0' \to K(Q, n)$, where F_0' is the rationalization of F', it will follow that $\pi_n F' \otimes Q = Q$ and $\pi_q F' \otimes Q = 0$ for $n < q \leq 2n-2$, hence the conclusion. By Remarks 7.3 and [8, §22], for $q \geq 2$ and $n \geq 1$ we have

$$\pi_q F = \pi_q BFM_\infty R_n = \pi_{q-1} FM_\infty R_n = \pi_{q-1} SFM_\infty R_n$$

$$\cong \pi_{q-1} M_\infty R_n = M_\infty \pi_{q-1} R_n = M_\infty H_{q-1}(\Omega S^n; Z),$$

which is zero for $q \leq 2n-2$ and $q \neq n$. By the Whitehead theorem applied to a π_n-isomorphism $F \to K(M_\infty Z, n)$, $H_n(F; Z) = M_\infty Z$ and $H_q(F; Z) = 0$ otherwise, $0 < q \leq 2n-2$, hence similarly with Z replaced by Q. The key algebraic fact, due to Farrell and Hsiang [4] and based on work of Borel [1], is that $H_*(BGLZ; M_\infty Q)$ is isomorphic to $H_*(BGLZ; Q)$; the definition of the isomorphism, via the trace, is irrelevant to the argument here. Consider the rational homology Serre spectral sequences. Certainly E^2 is finite-dimensional in each degree $\leq 2n-2$ (say by Borel's calculations of $H_*(BGLZ)$ [1]) and, in this range,

$$E^2 = E^2_{*0} \oplus E^2_{*n} \quad (E^2_{*0} = H_* BGLZ, E^2_{*n} \cong H_* BGLZ).$$

By the previous result, $Q \cong E^2_{0n} \to {'E}^2_{0n}$ is an isomorphism. By Corollary 11.7, $A(S^n; Z) \simeq A(S^n; Z) \times KZ$ and therefore

$${'E}^2 = {'E}^\infty = H_* KZ \otimes H_* F' \quad ({'E}^2_{*0} = H_* KZ, {'E}^2_{*n} \cong H_* KZ).$$

Since $\{E^r\}$ and $\{'E^r\}$ converge to isomorphic homologies, $H_qF' = 0$ for $n < q \leq 2n-2$ by a trivial comparison of dimensions.

The following is an immediate consequence, by Theorem 11.8.

Corollary 12.3

$$A_i^s\{*\} \otimes Q = \begin{cases} Q & \text{if } i = 0 \\ 0 & \text{if } i > 0 \end{cases} .$$

The results claimed about Postnikov systems in [22, §2] seem much more problematical (and are fortunately much less essential to the overall program).

Remarks 12.4. The nth term $R_*^{(n)}$ of the natural Postnikov system of a simplicial set R_* is $R_*/(\overset{n}{\sim})$, where $x \overset{n}{\sim} y$ for q-simplices x and y if all of their iterated faces of dimension $\leq n$ are equal [8, §8]. Visibly each $R_*^{(n)}$ is a simplicial ring if R_* is so, and the natural maps $R_*^{(n)} \to R_*^{(m)}$ for $n > m$ are maps of simplicial rings. As Waldhausen states [22, §1], there results a spectral sequence the E^2-term of which is given by the homotopy groups of the fibres of the maps $K|R_*^{(n)}| \to K|R_*^{(n-1)}|$ and which converges to $K_*|R_*|$. He asserts further [22, 2.5 and sequel] that the same conclusions hold with R_* replaced by an arbitrary ring up to homotopy, that is, in our terminology, by an arbitrary grouplike A_∞ ring space X. If true, anything like this would be enormously difficult to prove. Certainly, the coskeleta $X^{(n)}$ of X could at best be strong homotopy A_∞ ring spaces of some sort (with more homotopies in sight than in Definition 5.1; see the discussion following that notion). He also asserts [22, 2.6 and sequel] that the coskeleta of QS^0 give rise to a spectral sequence the E^2-term of which is given by the homotopy groups of certain fibres and which converges to A_*X for any space X. Here he thinks of QS^0 as the "coefficient ring" of AX, in analogy with the role of R in Definitions 11.4. Since this is at best only a metaphor, rigor seems still further away. The infinite loop space splitting $Q(\Omega X^+) \simeq QS^0 \times Q\Omega X$ does not seem relevant. Even if they do exist, there seems to be little reason to believe that such spectral sequences would help much with explicit calculations.

Of course, it is conceivable that there is a simplicial analog of our theory for which this difficulty disappears, but I am skeptical (and certain that other technical difficulties would appear in any such approach).

It is time to discuss the main issue. Waldhausen proposed our AX as a nice description of what he wanted, if it were to exist, but he gave an alternative definition in terms of which the proofs were all to proceed. We write WX for Waldhausen's functor (or rather its connected version). If GSX is the simplicial group of (11.3), then WX is the plus construction on the classifying space of the colimit over n and k of certain categories $(h\mathscr{S}GSX)^n_k$ with objects simplicial GSX-sets suitably related to the wedge of X and k copies of S^n. In the absence of any indications of proof, I for one find it hard to see how analogs for WX of some of the results above for AX are to be made rigorous from this definition. The technical details, for example of the rational equivalence required for Corollary 12.3, must surely be considerable. It would seem preferable to compare AX and WX. Waldhausen asserts (without proof, [22, 2.1]) that the loop of the classifying space of the colimit over n of the categories $(h\mathscr{S}GSX)^n_k$ is equivalent to $FM_kQ(|GSX|^+)$. While this certainly seems plausible, his further claim that the equivalence is one of H-spaces seems much more difficult, and this in turn is nowhere near strong enough to prove the following assertion.

<u>Conjecture 12.5</u>. AX and WX are naturally equivalent.

Except that the definitions of AX and WX seem farther apart, one might view this as analogous to the equivalence between his two definitions that was the pivotal result in Quillen's development of algebraic K-theory [5]. The point is that it is AX which is most naturally connected with Quillen's algebraic K-theory, but it is WX and its various equivalents in [22] which Waldhausen's arguments relate to the Whitehead groups for stable PL concordance.

Appendix. Stabilizations of functors to homology theories

We first give the technical lemma needed to complete the proof of Theorem 11.3 and then give a very general theorem (presumably part of the folklore) on the stabilization of homotopy functors to generalized homology theories.

We work in the category $\mathcal{V}$ of nondegenerately based spaces of the homotopy type of a CW-complex and in its homotopy category $h\mathcal{V}$. The proofs below use well-known facts about fibration sequences but, annoyingly, I know of no published source which contains everything we need; such details will appear in [15, I §1].

Lemma A.1. Consider the following diagram in $\mathcal{V}$, in which i and p are written generically for the canonical maps of fibration sequences, the solid arrow parts of the diagram homotopy commute, the bottom squares with solid vertical arrows erased also homotopy commute, and the dotted arrows ζ, η, and θ are homotopy sections ($\varepsilon\theta \simeq 1$, etc.).

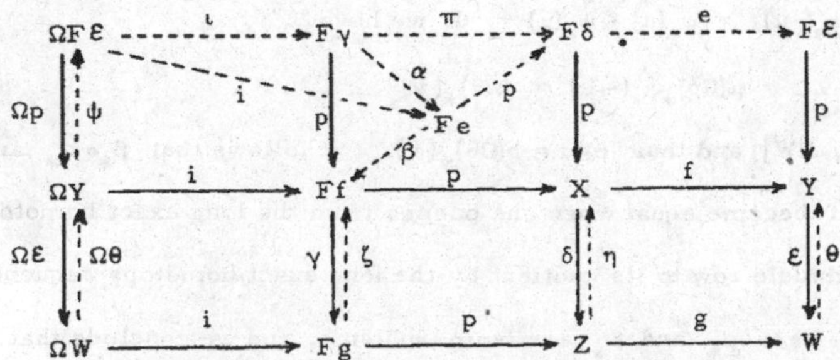

There exist maps e, π, and ι, unique up to homotopy, such that the top three squares homotopy commute and there exists an equivalence $\alpha: F\gamma \to Fe$ such that $p\alpha \simeq \pi$.

Proof. The homotopy commutativity of the lower three squares implies the existence of e, π and ι. Here e is unique since two such maps differ by the action of a map $F\delta \to \Omega W$ and the action of $[F\delta, \Omega W]$ on $[F\delta, F\varepsilon]$ is trivial since

$i: \Omega W \to F\mathcal{E}$ is null homotopic (because of the section). Similarly π and ι are unique. Since $pe\pi$ is null homotopic by the diagram, pe is null homotopic (by the exact sequence of the right column). Thus there exists $\alpha': F\gamma \to Fe$ such that $p\alpha' \simeq \pi$. The top row clearly induces a long exact sequence of homotopy groups mapping onto a direct summand of that of the middle row, and the desired conclusion that α' is an equivalence would be immediate from the five lemma if $\alpha'\iota$ were homotopic to i. However, there seems to be no reason to suppose that α' can be so chosen. Choose a map $\beta: Fe \to Ff$ such that $\beta i \simeq i\Omega p$ and $pp \simeq p\beta$. Again, α' would be an equivalence if $\beta\alpha'$ were homotopic to $p: F\gamma \to Ff$. Here we have more room for maneuver. Since ΩY is an H-space, the sum of Ωp and $\Omega\theta$ is an equivalence $\Omega F\mathcal{E} \times \Omega W \to \Omega Y$. Let $\psi: \Omega Y \to \Omega F\mathcal{E}$ be an inverse projection. Certainly $\beta \circ \alpha' = p\mu$ for some $\mu: F\gamma \to \Omega Y$, where $p\mu$ is given by the right action of $[F\gamma, \Omega Y]$ on $[F\gamma, Ff]$ coming from $Ff \times \Omega Y \to Ff$. Let $\alpha = \alpha'\psi_*(-\mu)$. Then

$$\beta \circ \alpha = \beta \circ (\alpha'\psi_*(-\mu)) = (\beta \circ \alpha')(\Omega p)_*\psi_*(-\mu) = p(\mu(\Omega p)_*\psi_*(-\mu)) .$$

Since $\psi_*(\mu(\Omega p)_*\psi_*(-\mu)) = \psi_*(\mu) - \psi_*(\mu) = 0$, we have

$$\mu(\Omega p)_*\psi_*(-\mu) = (\Omega\theta)_*(\nu)$$

for some $\nu \in [F\gamma, \Omega W]$ and thus $\beta \circ \alpha = p(\Omega\theta)_*(\nu)$. It follows that $\beta_* \circ \alpha_*$ and $p_*: \pi_* F\gamma \to \pi_* Ff$ become equal when one passes from the long exact homotopy sequence of the middle row to its quotient by the long exact homotopy sequence of the bottom row. Here p_* and β_* are isomorphisms, and we conclude that α_* is an isomorphism and thus that α is an equivalence.

The naturality of π in Theorem 11.3 follows from the argument used to prove the uniqueness of e in the lemma.

Turning to the desired construction of homology theories, we define a stability sequence $\{a_n\}$ to be a strictly increasing sequence of positive integers a_n such that $a_n - n$ tends to infinity.

Theorem A.2. Let $k: h\mathcal{V} \to h\mathcal{V}$ be a functor with the following properties.

(1) Application of k to an n-equivalence yields a b_n-equivalence, where $\{2b_n\}$ is a stability sequence.

(2) Application of k to a strictly commutative 2n-homotopy Cartesian (n, n)-equivalence yields an a_n-homotopy Cartesian square, where $\{a_n\}$ is a stability sequence.

Let $\varepsilon: kX \to k\{*\}$ be induced by $X \to \{*\}$ and let $\tilde{k}X$ be the fibre of ε. Then there exist natural maps $\sigma: \tilde{k}X \to \Omega\tilde{k}\Sigma X$ such that if $\tilde{k}^s X$ is defined to be the telescope of the spaces $\Omega^n \tilde{k}\Sigma^n X$ with respect to the maps $\Omega^n\sigma$, $\sigma: \tilde{k}\Sigma^n X \to \Omega\tilde{k}\Sigma^{n+1}X$, and if $\tilde{k}^s_q X$ is defined to be $\pi_q \tilde{k}^s X$, then $\tilde{k}^s_*$ is a reduced homology theory which satisfies the wedge axiom.

The following is a standard consequence.

Corollary A.3. On unbased spaces X, define $k^s_* X = \tilde{k}^s_*(X^+)$. On unbased pairs (X, A), define $k^s_*(X, A) = \tilde{k}^s_*((X \cup CA)^+)$ where CA is the (unreduced) cone on A. Then k^s_* is a generalized homology theory in the classical sense.

Returning to based spaces, we first discuss the statement of the theorem. It will turn out that property (1) is only needed for the wedge axiom, and then only for maps $X \to \{*\}$, hence may be omitted in obtaining a homology theory on finite complexes. Property (2) will also only be needed for a few simple types of diagrams, to be displayed in the proof. Since ε is only given as a homotopy class of maps, we must choose a representative before constructing $\tilde{k}X$. The first part of the proof of Lemma A.1 gives the following result.

Lemma A.4. For $f: X \to Y$, there is a unique homotopy class $\tilde{k}f: \tilde{k}X \to \tilde{k}Y$ such that the following diagram commutes in $h\mathcal{V}$.

$$\begin{array}{ccc} \tilde{k}X & \xrightarrow{\tilde{k}f} & \tilde{k}Y \\ \downarrow & & \downarrow \\ kX & \xrightarrow{kf} & kY . \end{array}$$

It follows that $\tilde{k}$ is a well-defined functor $h\mathcal{V} \to h\mathcal{V}$ such that $\tilde{k} \to k$ is natural. We also need the following analog.

Lemma A.5. Let $\eta: k\{*\} \to kX$ be induced by $\{*\} \to X$ and let $\hat{k}X$ be the fibre of η. For $f: X \to Y$, there is a unique homotopy class $\hat{k}f: \hat{k}X \to \hat{k}Y$ such that the following diagram commutes in $h\mathcal{V}$.

$$\begin{array}{ccc} \Omega kX & \xrightarrow{\Omega kf} & \Omega kY \\ \downarrow & & \downarrow \\ \hat{k}X & \xrightarrow{\hat{k}f} & \hat{k}Y . \end{array}$$

Proof. The map $\Omega kX \to \Omega k\{*\} \times \hat{k}X$ with first coordinate $\Omega\varepsilon$ and second coordinate the canonical map is an equivalence.

It follows that $\hat{k}$ is a functor and $\Omega k \to \hat{k}$ is natural.

Lemma A.6. The composite $\Omega\tilde{k}X \to \Omega kX \to \hat{k}X$ is a natural equivalence.

Proof. The map $\Omega k\{*\} \times \Omega\tilde{k}X \to \Omega kX$ given by the sum of $\Omega\eta$ and the canonical map is also an equivalence.

These observations suffice for the construction of σ.

Lemma A.7. There is a natural map $\sigma: \tilde{k}X \to \Omega\tilde{k}\Sigma X$ such that σ is an $(a_{n+1} - 1)$-equivalence if X is n-connected.

Proof. We define σ to be the top composite in the diagram

$$\begin{array}{ccccc} \tilde{k}X & \dashrightarrow & \hat{k}\Sigma X & \xrightarrow{\cong} & \Omega\tilde{k}\Sigma X \\ \downarrow & & \downarrow & & \\ kX & \xrightarrow{\varepsilon} & k\{*\} & & \\ \varepsilon\downarrow & & \downarrow\eta & & \\ k\{*\} & \xrightarrow{\eta} & k\Sigma X & & \end{array}$$

Here we have the tautological strict equality $\eta\varepsilon = \eta\varepsilon$, and the dotted arrow is canonical; its naturality up to homotopy is easily checked by direct inspection. If X is n-connected, then the commutative square

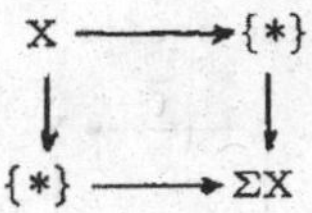

is a (2n+2)-homotopy Cartesian (n+1, n+1)-equivalence by the homotopy excision theorem (compare Definition 9.2). By a mild interpretation, property (2) implies the result.

The spaces $\tilde{k}^s X$ of Theorem A.2 are now defined. Since Ω commutes with telescopes, there is an evident homotopy equivalence

$$\tau: \tilde{k}^s X = \operatorname*{Tel}_{n\geq 0} \Omega^n \tilde{k}\Sigma^n X \rightarrow \operatorname*{Tel}_{n\geq 0} \Omega^{n+1}\tilde{k}\Sigma^{n+1} X = \Omega\tilde{k}^s \Sigma X .$$

While $\tilde{k}^s$ need not be a functor and τ need not be natural, since $\lim^1$ terms might well be present, they induce functors $\tilde{k}^s_q X$ and natural isomorphisms $\tau_q : \tilde{k}^s_q X \rightarrow \tilde{k}^s_{q+1}\Sigma X$ on passage to homotopy groups. Alternatively, with $\tilde{k}_q X = \pi_q \tilde{k} X$, we could equally well define

$$\tilde{k}^s_q X = \operatorname{colim} \tilde{k}_{q+n}\Sigma^n X$$

and not bother with the telescopes, the isomorphisms τ_q then being evident.

For reduced homology theories, excision reduces to the suspension axiom just verified on trivial formal grounds, without use of properties (1) and (2). The things to be proven are exactness and the wedge axiom. The following lemma verifies the appropriate exactness axiom.

Lemma A.8. If $\iota: A \rightarrow X$ is a cofibration, then the sequence

$$k^s_q A \xrightarrow{k^s_q \iota} k^s_q X \xrightarrow{k^s_q \pi} k^s_q(X/A)$$

is exact for all q, where π is the quotient map.

Proof. Since the functors Σ^n commute with cofibration sequences, a glance at the relevant colimit systems shows that it suffices to prove

$$\tilde{k}_q A \xrightarrow{\tilde{k}_q \iota} \tilde{k}_q X \xrightarrow{\tilde{k}_q \pi} \tilde{k}_q(X/A)$$

to be exact in a suitable range when A, X, and X/A are n-connected. By the homotopy excision theorem again, the square

$$\begin{array}{ccc} A & \xrightarrow{\iota} & X \\ \downarrow & & \downarrow \pi \\ \{*\} & \longrightarrow & X/A \end{array}$$

is a $2n$-homotopy Cartesian (n, n)-equivalence. Consider the following diagram

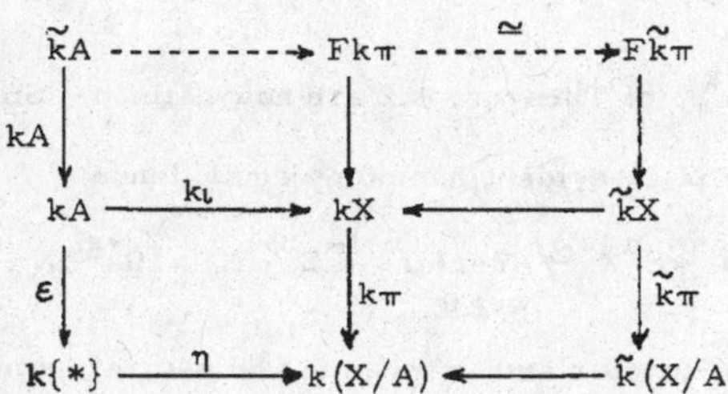

Verdier's axiom for fibration sequences applied to the triangle

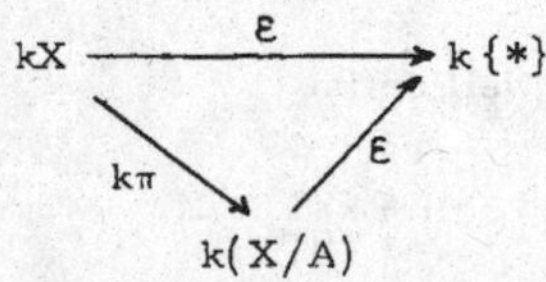

gives an equivalence $Fk\pi \to F\tilde{k}\pi$ such that the upper right square homotopy commutes. Property (2) and Remarks 9.3 give an $(a_n + 1)$-equivalence $\tilde{k}A \to Fk\pi$ such that the upper left square homotopy commutes. By Lemma A.4, the composite $\tilde{k}A \to \tilde{k}X$ in the diagram is $\tilde{k}\iota$. The conclusion follows from the long exact sequence of homotopy groups of the right column.

It remains only to verify the wedge axiom.

<u>Lemma A.9.</u> For any set of spaces $\{X_i\}$, the natural map

$$\bigoplus_i \tilde{k}^s_q X_i \to \tilde{k}^s_q (\bigvee_i X_i)$$

is an isomorphism for all q.

<u>Proof</u>. Since the functors Σ^n commute with wedges, a glance at the relevant colimit systems shows that it suffices to prove

$$\bigoplus_i \tilde{k}_q(X_i) \to \tilde{k}_q(\bigvee_i X_i)$$

to be an isomorphism in a suitable range when each X_i is n-connected. If X is n-connected, then $X \to \{*\}$ is an $(n+1)$-equivalence, hence $\varepsilon: kX \to k\{*\}$ is a b_{n+1}-equivalence by property (1), hence $\tilde{k}X$ is $(b_{n+1}-1)$-connected. Therefore the inclusion of $\bigvee_i \tilde{k}X_i$ in the weak direct product of the kX_i (all but finitely many coordinates at the basepoint) is a $(2b_{n+1}-1)$-equivalence, and the conclusion follows.

Bibliography

1. A. Borel. Stable real cohomology of arithmetic groups. Ann. Sci. Ecole Normale Sup. 4^{e} serie t. 7 (1974), 235-272.

2. F. Cohen, T. Lada, and J. P. May. The Homology of Iterated Loop Spaces. Springer Lecture Notes in Mathematics, Vol. 533, 1976.

3. A. Dold and R. Thom. Quasifaserungen und unendliche symmetrische produkte. Annals of Math. 67 (1958), 239-281.

4. F. T. Farrell and W. C. Hsiang. Proc. Amer. Math. Soc. Summer Institute. Stanford, 1976. To appear.

5. D. Grayson. Higher algebraic K-theory: II (after D. Quillen). Springer Lecture Notes in Mathematics. Vol. 551, 216-240, 1976.

6. T. Lada. An operad action on infinite loop space multiplication. Canadian J. Math. To appear.

7. J. L. Loday. Les matrices monomiales et le groupe de Whitehead Wh_2. Springer Lecture Notes in Mathematics, Vol. 551, 155-163, 1976.

8. J. P. May. Simplicial Objects in Algebraic Topology. D. van Nostrand, 1967.

9. J. P. May. Classifying Spaces and Fibrations. Memoirs Amer. Math. Soc. 155, 1975.

10. J. P. May. The Geometry of Iterated Loop Spaces. Springer Lecture Notes in Mathematics, Vol. 271, 1972.

11. J. P. May. E_∞ spaces, group completions, and permutative categories. London Math. Soc. Lecture Note Series 11, 61-94, 1974.

12. J. P. May (with contributions by N. Ray, F. Quinn, and J. Tornehave). E_∞ Ring Spaces and E_∞ Ring Spectra. Springer Lecture Notes in Mathematics, Vol. 577, 1977.

13. J. P. May. Infinite loop space theory. Bull. Amer. Math. Soc. 83(1977), 456-494.

14. J. P. May. The spectra associated to permutative categories. To appear.

15. J. P. May. The Homotopical Foundations of Algebraic Topology. Academic Press. In preparation.

16. J.P. May and R.Thomason. The uniqueness of infinite loop space machines. To appear.

17. R.J. Milgram. The bar construction and Abelian H-spaces. Illinois J. Math. 11(1957), 242-250.

18. D. Quillen. Higher algebraic K-theory I. Springer Lecture Notes in Mathematics, Vol. 341, 85-147, 1973.

19. D. Quillen. Letter from Quillen to Milnor on $\mathrm{Im}(\pi_i 0 \xrightarrow{J} \pi_i^s \longrightarrow K_i Z)$. Springer Lecture Notes in Mathematics, Vol. 551, 182-188, 1976.

20. G. Segal. Categories and cohomology theories. Topology 13(1974), 293-312.

21. J.B. Wagoner. Delooping classifying spaces in algebraic K-theory. Topology 11(1972), 349-370.

22. F. Waldhausen. Algebraic K-theory of topological spaces, I. Proc. Amer. Math. Soc. Summer Institute. Stanford, 1976. To appear.

COHOMOLOGY OPERATIONS AND VECTOR FIELDS

by

M.L. Michelsohn

University of California, Berkeley

§1. *Introduction*

A problem of longstanding interest has been to compute for a given manifold M the largest number r such that there exist r everywhere linearly independent vector fields on M; this number is known as the span of M. The problem has been resolved for spheres, though even here its history is long. It was known by Poincaré in the nineteenth century that $\operatorname{span} S^{2n} = 0$ and $\operatorname{span} S^{2n+1} \geqslant 1$. Then in 1923 Hurwitz [9] and Radon [16] found by explicit construction a lower bound for $\operatorname{span} S^n$. To each integer n they associate an integer $\phi(n)$ in the following way. Suppose n is of the form:

$$(1) \quad n = 2^m(2t + 1)$$

and that the exponent m is expressed as

$$(2) \quad m = 4a + b \quad \text{with} \quad 0 \leqslant b \leqslant 3 .$$

Then $\phi(n)$ is defined by

$$(3) \quad \phi(n) = 8a + 2^b .$$

They then prove that

$$(4) \quad \operatorname{span} S^{n-1} \geqslant \phi(n) - 1 .$$

(We shall fix the notation of (1), (2) and (3) for the remainder of this paper.)

In 1951 Steenrod and J.H.C. Whitehead [17] proved $\operatorname{span} S^{n-1} < 2^m$ thereby showing that the lower bound $\phi(n) - 1$ is sharp for exponents $m \leqslant 3$; in 1961 Toda [18] proved equality for $m \leqslant 10$; and finally in 1962 Adams [1] proved equality for all n.

Now let M be any smooth, closed, connected manifold of dimension $n - 1$. Poincaré [15] proved that if M is a surface, the Euler-Poincaré characteristic is the obstruction to the existence of one vector field and Hopf [7] generalized Poincaré' theorem to arbitrary dimensions. Frank and Thomas [6] generalized the Steenrod-Whitehead theorem proving that if:

(i) $H_i(M; Z_2) = 0$ for all $i < 2^m$ where m is the exponent defined in (1) above,

(ii) the Stiefel-Whitney class $w_{2^m}(M) = 0$

and (iii) $\hat{\chi}(M) = (\sum_{i=0}^{\frac{n}{2}-1} \dim H_i(M; Z_2)) \bmod 2 = 0$,

then (iv) span $M^{n-1} < 2^m$.

They do this by showing that a certain decomposition of Sq^n (simply an Adam relation) yields an unstable secondary operation which gives an obstruction to the existence of 2^m vector fields.

In this paper we only assume that the tangent bundle of M , when restricted to the $\phi(n)$-skeleton of M , is trivial. We indicate how to find a decomposition of Sq^n in terms of higher order cohomology operations. This decomposition yields an unstable operation which is the last obstruction for the existence of only $\phi(n)$ linearly independent vector fields. We make the standing assumptions in this paper that cohomology, when not otherwise stated, is with Z_2 coefficients and that the integer k , whenever it appears, is less than $n/2$.

In particular, we prove the following

Theorem 1. *Let* $n = 2^m(2t + 1)$, $m \geqslant 4$, $n \neq 16$. *There exists a decomposition*

$$Sq^n = \sum a_{i,n} \Pi_{i,n} + b_{i,n} \xi_{i,n}$$

which holds on classes u *for which the right side is defined. The operations occurring on the right side have the following properties:*

(i) *The* $a_{i,n}$ *and* $b_{i,n}$ *are primary cohomology operations.*

(ii) *The* $\pi_{i,n}$ *and* $\xi_{i,n}$ *are cohomology operations of order* $m - 1$ *if* $m > 4$ *and secondary operations if* $m = 4$.

(iii) *The* $\pi_{i,n}$ *are of degree* $\geqslant n - \phi(n)$ *and come from the Adams spectral sequence of* $\Sigma^n P^{n-1}_{n-\phi(n)-1}$. *They are zero on* $(\phi(n) + 1)$*-fold suspension classes.*

(iv) *The* $\xi_{i,n}$ *are of degree* $\leqslant \phi(n)$ *and come from the Adams spectral sequence of a sphere.*

Let ψ_{n-1} be the unstable operation yielded by the above decomposition. Let U_M be the Thom class of the manifold M and let ε be the trivial line bundle over M .

Theorem 2. *Let* M^{n-1} *be a smooth, closed, connected* $(n-1)$*-dimensional manifold,* n *as above, such that its tangent bundle,* $\tau(M)$, *is trivial over the* $\phi(n)$*-skeleton of* M *and* $\tau(M) + \varepsilon$ *has* $\phi(n)$ *sections. Then* span $M^{n-1} \geqslant \phi(n)$ *if and only if* $\psi_{n-1}(U_M) = 0$.

More details may be found in [12] and [13].

Sections 2 through 5 constitute an outline of the proof of theorem 1. In section 2 we recall the meaning of higher order cohomology operations and construct a partial Adams resolution of the sphere. In section 3 we discuss some fibrations and their properties which will be necessary in the construction of the decomposition. In section 4 we construct a large diagram of spaces using the partial resolution of section 2 and a resolution of stunted projective spaces. We then show that the results of section 3 imply that this basic construction has certain properties which in section 5 are shown to imply the theorem.

In section 6 we outline the proof of theorem 2. The main point is to show that the lifting problem corresponding to the existence of linearly independent vector fields is equivalent to a certain lifting problem for Thom complexes.

§2. *A partial Adams resolution of the sphere* S^n

We recall the construction of secondary cohomology operations. A cohomology class of X, $u \in H^n(X; Z)$ may be thought of as a map $X \xrightarrow{u} K(Z, n)$. A class $\nu \in H^{n+i}(K(Z, n); Z_2)$ may likewise be thought of as a map $K(Z, n) \to K(Z_2, n + i)$ and in fact as a primary cohomology operation on classes such as u. Now suppose we have a relation $\sum_{i=1}^{k} a_i b_i = 0$ where a_i and b_i are primary operations. Then we may construct the fibration

$$\begin{array}{ccc} F = \prod_{i=1}^{k} K(Z_2, n + m_i - 1) \xrightarrow{i} Y & & \\ \downarrow & & \\ K(Z, n) & \xrightarrow{b_1, \ldots, b_k} & \prod_{i-1}^{k} K(Z_2, n + m_i) \end{array}$$

by pulling the loop-path space fibration over $B = \prod_{i=1}^{k} K(Z_2, n + m_i)$ back over $K(Z, n)$. Then a map $\phi: Y \to \Pi\, K(Z_2, n + m_i + n_i - 1)$ such that $\phi \circ i$ is a map corresponding to the stable primary operations $a_1, \ldots, a_k$ is the secondary operation corresponding to the relation $\sum_{i=1}^{k} a_i b_i = 0$. Thus a secondary operation is a cohomology class of Y. Note that the secondary operation is defined on a class such as u if and only if there is a lift of u to Y which holds if and only if the b_i's are all zero on u. The construction of a tertiary operation based on a relation of secondary operations $\sum c_i d_i = 0$, c_i primary and the d_i secondary is done in a similar way pulling a loop path space fibration back over Y.

Consider the Postnikov system

$$\begin{array}{ccc} F_1 = \prod_{i=2}^{k-1} K(Z_2, n + i - 1) \to Y_{1,n} & & \\ \downarrow & & \\ K(Z, n) & \xrightarrow{Sq^2, \ldots, Sq^{k-1}} & \prod_{i=2}^{k-1} K(Z_2, n + i) . \end{array}$$

If we have a cell decomposition of $K(Z, n)$ which is S^n with various cells of dimension greater than n attached then $Y_{1,n}$ is a space which has a cell decomposition which looks like that of $K(Z, n)$ but without the attaching maps which are homotopy classes which are detected by $Sq^2, \ldots, Sq^{k-1}$ (as η detects Sq^2 and ν detects Sq^4). Cells which were attached by such maps are now just attached to the base point. Although in the construction of $Y_{1,n}$ we have killed $Sq^2, \ldots, Sq^{k-1}$ in $K(Z, n)$, $Y_{1,n}$ will not, of course, be without cohomology in dimensions $n + 1$ through $n + k - 1$. In fact, attaching maps in $K(Z, n)$ which are detected by secondary operations will be detected by operations which are cohomology classes of $Y_{1,n}$. We proceed then to kill the attaching maps which detect secondary operations and so forth, constructing a Postnikov tower

$$\ldots \to Y_{n,s} \to Y_{n,s-1} \to \ldots \to Y_{n,1} \to Y_{n,0} = K(Z, n)$$

which is a partial Adams resolution of S^n .

§3. *The fibrations*

Just as $K(Z, n)$ is the universal space for primary cohomology operations, $\Sigma^k K(Z, n - k)$ is the universal space for primary operations on k-fold suspension classes. We are led to consider the fibration

$$F \to \Sigma^k K(Z, n - k) \to K(Z, n) .$$

It will be to our benefit to consider more generally the fibration

$$F_X \to \Sigma^k \Omega^k X \xrightarrow{e_i} X$$

where e_i is the evaluation map. By looping this fibration k times we get the more familiar fibration

$$\Omega^k F_x \to \Omega^k \Sigma^k \Omega^k X \to \Omega^k X$$

whose fibre $\Omega^k F_x$ is known [3], [10], [14] to be the quadratic construction on $\Omega^k X$ $= Q_{k-1}(\Omega^k X) = S^{k-1} \ltimes_{Z_2} (\Omega^k X \wedge \Omega^k X)$ where the Z_2 actions are the antipodal map on S^{k-1} and the flip map on the smash product. The vertical bar in $\ltimes$ means collapse $S^{k-1} \times$ (base pt.) to the base point. It follows from considering homology cell decompositions that $H^*(Q_{k-1}Y)$ consists of three types of elements: if $u \in H^j(Y)$ then u generates a copy of $H^*(\Sigma^j P_j^{j+k-1})$ and if u, v are distinct $u \in H^i(Y)$, $v \in H^j(Y)$ then there also exist $(u, v)_0 \in H^{i+j}(Q_k(Y))$ and $(u, v)_{k-1} \in H^{i+j+k-1}(Q_{k-1}Y)$.

Suppose we have two such fibrations

$$F_{x_i} \to \Sigma^k \Omega^k X_i \xrightarrow{e_i} X_i \qquad i = 1, 2 .$$

By looking at the homology decomposition of $\Omega^k X$ [2], [4], [5] and that of $Q_{k-1}(\Omega_k X)$ we can prove the following lemma:

<u>Lemma</u>. *Suppose the n-skeleta* $(X_1)^n = (X_2)^n = S^n$ *and let* $f: X_1 \to X_2$ *be a map which induces an isomorphism on* $H^n(X_2)$ *and the 0-map on* $H^j(X_2)$ *for* $n < j < n + q$ *for some* $q < n - k$. *Let* $\tilde{f}: F_{x_1} \to F_{x_2}$ *be the induced map on the fibres. Then*

(1) $B \subseteq H^*(F_{x_i})$ *where*

$$B \cong H^*(\Sigma^k P_{n-k}^{n-1})$$

(2) $\tilde{f}^* | H^*(F_{x_2}) - B$ *is the zero map through dimension* $2n - k + q$

(3) $\tilde{f} | B$ *is an isomorphism.*

§4. *The resolution construction*

Our plan is to construct a resolution

$$\to E_s \to E_{s-1} \to \dots \to E_1 \to E_0 = K(Z, n)$$

which, when the homotopy functor is applied, yields a spectral sequence which splits into a direct sum of two spectral sequences one of which will be the Adams spectral sequence of $\Sigma^n P^{n-1}_{n-k}$, the other the Adams spectral sequence of S^n . E_s will be, in fact, a diagonal element $E_{s,s}$ in a construction of spaces $E_{t,s}$, $t \leqslant s + 1$. We will begin with $E_{0,s} = Y_{n,s}$ all s . We have the fibration

$$F_{0,0} \to \Sigma^k \Omega^k E_{0,0} \to E_{0,0}$$

which is just

$$F \to \Sigma^k K(Z, n - k) \to K(Z, n)$$

and $H^*(\Sigma^n P^{n-1}_{n-k}) \subseteq H^*(F)$. Let $\dots \to P_s \to \dots \to P_1 \to P_0 = \Sigma^n P^{n-1}_{n-k}$ be an Adams resolution of $\Sigma^n P^{n-1}_{n-k}$. Let $B_s = H^*(P_s)$.

We shall construct the following diagram of spaces

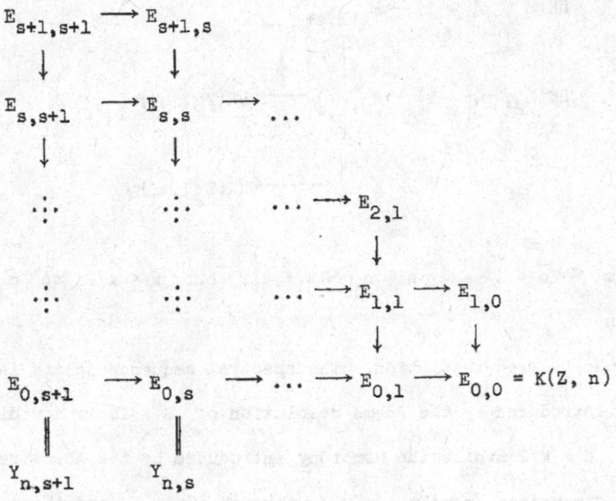

such that the fibrations $\Sigma^k\Omega^k Y_{n,s} \to Y_{n,s}$ factorize into $\Sigma^k\Omega^k Y_{n,s} \to E_{t,s} \to Y_{n,s}$ and the map $\Sigma^k\Omega^k Y_{n,s} \to E_{t,s}$ is a fibration with fibre denoted by $F_{t,s}$. Furthermore, these fibres are to have the following properties

(1) $B_s \subseteq H^*(F_{s,s})$ and

(2) $B_s \subseteq H^*(F_{s,s+1})$ as an A-module where A is the mod 2 Steenrod algebra

The lemma in the previous section allows us to conclude property (2) from property (1). Property (2) in one row allows us to conclude property (1) in the next. To construct the $E_{s,t}$ consider $v \in B_{s-1} \subset H^j(F_{s-1,s-1})$ which transgresses to some $\tilde{v} \in H^{j+1}(E_{s-1,s-1})$. Then construct $E_{s,s-1}$ using the $\tilde{v}$'s as k-invariants. Likewise if $u \in B_{s-1} \in H^j\ (F_{s-1,s})$ transgresses to $\tilde{u} \in H^{j+1}\ (E_{s-1,s})$ construct $E_{s,s}$ using the $\tilde{u}$'s as k-invariants. Construct the rest of the row by pulling back.

Note that this construction yields the space $E_{s+1,s}$. We know nothing about its cohomology. Its only role is to let us conclude that $E_{s+1,s+1} \to E_{s,s}$ is a one-stage Postnikov system, that is a fibration by a generalized Eilenberg-MacLane space:

$$\begin{array}{ccccc}
\Pi K(Z_2, n_j - 1) \to E_{s+1,s+1} & & \\
\downarrow & & \\
\Pi K(Z_2, m_j - 1) \to E_{s+1,s} & \longrightarrow & \Pi K(Z_2, n_j) \\
\downarrow & & \downarrow \\
E_{s,s} & \longrightarrow & \Pi K(Z_2, m_j)
\end{array}$$

We have $2n - k \leq m_j \leq 2n - 1$ and $n \leq n_j \leq n + k/2$ but $k < n/2$ so $m_j - 1 > n_i$ for any m_j and n_i.

Thus we have the desired resolution. The spectral sequence splits into two because the homotopy introduced by the Adams resolution of S^n is in low dimensions namely n through $n + k/2$ while the homotopy introduced by the Adams resolution of $\Sigma^n P^{n-1}_{n-k}$ is in high dimensions, namely $2n - k$ through $2n - 1$ and these do not overlap because of the stability assumption $k < n/2$.

§5. *The decomposition*

In the sequence

$$F \to \Sigma^k K(Z, n-k) \to K(Z, n)$$

we have $H^*(\Sigma^n P^n_{n-k}) \subset H^*(F)$ and the non-zero element $\alpha_{n+j} \in H^{n+j}(\Sigma^n P^n_{n-k})$ transgresses to $Sq^{j+1} \iota_n$ where ι_n is the fundamental class of $K(Z, n)$. Now while $1_{2^m(2t+1)-1}$ is a permanent cycle in the Adams spectral sequence for $P_{2^m(2t+1)-\phi(n)}$ it is not a permanent cycle in the Adams spectral sequence for $P_{2^m(2t+1)-\phi(n)-1}$ and $d_r 1_{2^m(2t+1)-1} \neq 0$ in the spectral sequence for $P_{2^m(2t+1)-\phi(n)-1}$ where $r = m - 1$

if $m > 4$ and $r = 2$ if $m = 4$ [1], [11]. Thus we let k be $\phi(n) + 1$ in the preceeding theory.

The image of d_r on 1_{n-1}, had it not been in the image of d_r, would have been a cohomology operation of order $r + 1$ based on a relation

$$\sum \alpha_i \beta_i = 0$$

of operations β_i of one line below in the Adams spectral sequence, that is, of operations of order r. Thus instead we have a decomposition of Sq^n in terms of these operations

$$Sq^n = \sum \alpha_i \beta_i .$$

When Sq^n is zero we have the unstable operation, ψ_{n-1}, given by the relation

$$0 = \sum \alpha_i \beta_i .$$

The operations α_i are primary cohomology operations. The operations α_i are of two kinds:

(1) Those coming from the Adams spectral sequence of $\Sigma^n P^{n-1}_{n-(n)-1}$. These are of degree $\geqslant n - \phi(n)$. They are zero on $(\phi(n)+1)$-fold suspension classes. We will denote these $\Pi_{i,n}$.

(2) Those coming from the Adams spectral sequence of a sphere. These are of degree $\leqslant \phi(n)$. We will denote these $\xi_{i,n}$.

Thus we have Theorem 1.

§6. *Vector Fields*

Let M^{n-1} be a smooth, closed connected manifold of dimension $n - 1$. Denote the tangent bundle of M^{n-1} by τM and the k-skeleton of M by $M^{(k)}$. Suppose $\tau M|M^{(\phi(n))}$ is trivial, $\tau(m) + \varepsilon$ has $\phi(u)$ sections, and $n = 2^{4a+b}$, $a \geqslant 1$, $n \neq 16$.

Let τM be classified by $M^n \xrightarrow{\hat{f}} BSO_{n-1}$ and the stable tangent bundle be classified by $\tilde{f}: M \to BSO_n$. We let the $k - 1$ connective cover of BSO_n be denoted by $BSO_n[k, \ldots, \infty]$ see [8], p. 155. Since $\tau M|M^{(\phi(n))}$ is trivial $\tilde{f}$ lifts to $f: M \to BSO_n[\phi(n) + 1, \ldots, \infty]$. Now τM has $\phi(n)$-cross sections, that is, M has $\phi(n)$ linearly independent vector fields if and only if f lifts to some $\bar{f}: M^{n-1} \to BSO_{n-\phi(n)-1}[\phi(n) + 1, \ldots, \infty]$. The fibre of $BO_{n-\phi(n)-1}[\phi(n)+1, \ldots, \infty] \to BO_n[\phi(n) + 1, \ldots, \infty]$ is the Stiefel manifold $V_{n,\phi(n)+1}$ of $(\phi(n) + 1)$-frames in Euclidean n-space. But the n-skeleton of $V_{n,\phi(n)+1}$, which is all that is relevant to this lifting problem, is the truncated real projective space $P^{n-1}/P^{n-\phi(n)-2}$, that is P^{n-1} with $P^{n-\phi(n)-2}$ collapsed to the base point, which we will denote by $P^{n-1}_{n-\phi(n)-1}$. Since $BSO_{n-\phi(n)-1}[\phi(n) + 1, \ldots, \infty] \to BSO_n[\phi(n) + 1, \ldots, \infty]$ is a fibration it is also a cofibration in the stable range and $\Sigma P^{n-1}_{n-\phi(n)-1}$ is the cofibre in that range. We insured that we are in that range by assuming $\phi(n) + 1 < n/2$. We summarize the situation in the following diagram

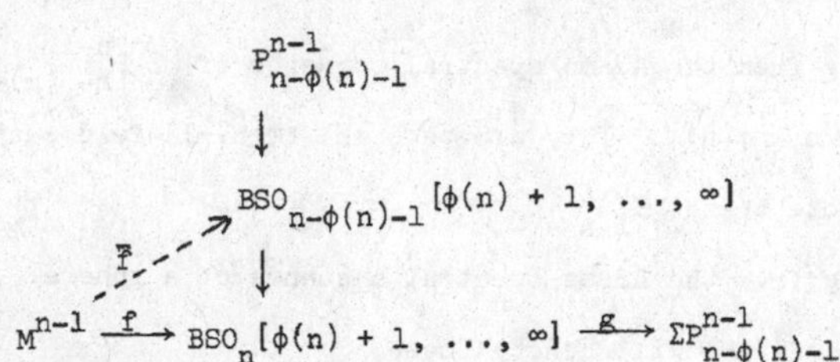

Diag. 6.1

The existence of $\phi(n)$ vector fields on M^{n-1} is equivalent to the existence of a lifting $\bar{f}$ of f which is equivalent to the existence of a null homotopy of $g \circ f$.

Diagram 6.1 induces the following diagram of Thom complexes

$$\begin{array}{ccc} & & \Sigma^{\phi(n)+1} MSO_{n-\phi(n)-1}[\phi(n)+1, \ldots, \infty] \\ & \nearrow & \\ TM^{n-1} & \xrightarrow{Tf} & MSO_n[\phi(n)+1, \ldots, \infty] \end{array}$$

and the existence of $\bar{f}$ implies the existence of a lifting $\tilde{f}$ of Tf. However, in this case the converse is also true.

One may prove this via the following considerations.

Let E be the universal bundle over BSO_n and E' be the universal bundle over $BSO_{n-\phi(n)-1}$ and consider the following diagram where the vertical sequences are fibrations and the third column consists of the appropriate Thom complexes.

$$\begin{array}{ccccc} S^{\phi(n)} * E'' & \longrightarrow & V_{n,\phi(n)+1} & \longrightarrow & T = S^n \vee \Sigma^n V \\ \downarrow & & \downarrow & & \\ S^{\phi(n)} * E' & \longrightarrow & BSO_{n-\phi(n)-1} & \longrightarrow & \Sigma^{\phi(n)+1} MSO_{n-\phi(n)-1} \\ \downarrow & & \downarrow & & \\ E & \longrightarrow & BSO_n & \longrightarrow & MSO_n \end{array}$$

Since $\tau M | M^{(\phi(n))}$ is trivial we may consider $BSO_{n-\phi(n)-1}[\phi(n)+1, \ldots, \infty]$ and $BSO_n[\phi(n)+1, \ldots, \infty]$, which we will for simplicity denote $\overline{BSO}_{n-\phi(n)-1}$ and $\overline{BSO}_n$, and their corresponding Thom complexes which we will denote $\Sigma^{\phi(n)+1}\,\overline{MSO}_{n-\phi(n)-1}$ and $\overline{MSO}_n$. Since there is always a lifting of the Thom class we go one step to the right in the diagram and pinch out S^n. Then it can be shown, using relative homology, that T/S^n which is equivalent to $\Sigma^n V$ is equivalent, through the relevant range of dimensions, to the fibre of $\Sigma^{\phi(n)+1}(\overline{MSO}_{n-\phi(n)-1}/S^{n-\phi(n)-1}) \to \overline{MSO}_n/S^n$.

The construction may be continued by letting ${}_1X$ be the firt term after X in an Adams resolution of X :

$$\begin{array}{ccccccc}
E'' & & {}_1V & \longrightarrow & {}_1T & \longrightarrow & {}_1T/S^n \equiv {}_1F_1 \equiv \Sigma^n_1 V \\
\downarrow & & \downarrow & & & & \downarrow \\
S^{\phi(n)} * E'' & \longrightarrow & V & \longrightarrow & T \quad F & & T/S^n \equiv F_1 \equiv \Sigma^n V \\
\downarrow & & \downarrow & & \downarrow & & \downarrow \\
S^{\phi(n)} * E' & \longrightarrow & \overline{BSO}_{n-\phi(n)-1} & \longrightarrow & \Sigma^{\phi(n)+1}\,\overline{MSO}_{n-\phi(n)-1} & \longrightarrow & \Sigma^{\phi(n)-1}(MSO_{n-\phi(n)-1}/S^{n-\phi(n)-1}) \\
\downarrow & & \downarrow & & \downarrow & & \downarrow \\
E & \longrightarrow & \overline{BSO}_n & \longrightarrow & \overline{MSO}_n & \longrightarrow & \overline{MSO}_n/S^n
\end{array}$$

Thus if there is an obstruction at some level to the lifting in the BSO problem there will be an obstruction to the lifting in the MSO problem.

Thus the fibration relevant to the lifting problem is

$$\Sigma^n V_{n,\phi(n)+1} \to \Sigma^{\phi(n)+1}\, MSO_{n-\phi(n)-1} \to MSO_n \, .$$

However, the dimension of TM^{n-1} lets us consider instead the fibration induced by restricting the above to (2n-1)-skeleta which we write

$$\Sigma^n P^{n-1}_{n-\phi(n)-1} \to \Sigma^{\phi(n)+1}\, \widehat{MSO}_{n-\phi(n)-1} \to \widehat{MSO}_n$$

In this fibration the connectivity of the fibre is $2n - \phi(n) - 2$ and the connectivity of the base is $n - 1$ so there is a Serre exact sequence through dimension $3n - \phi(n) - 2$. We have the standing stability assumption that $\phi(n) < n/2$ so $3n - \phi(n) - 2 > 2n - 1$ and we are in the stable range:

$$\ldots \to H^i(\Sigma^n P^{n-1}_{n-\phi(n)-1}; Z) \longrightarrow H^{i+1}(MSO_n; Z)$$

$$\to H^{i+1}(MSO_{n-\phi(n)-1}; Z) \longrightarrow H^{i+1}(\Sigma^n P^{n-1}_{n-\phi(n)-1}; Z)$$

$$\to \ldots$$

is exact for $i < 2n - 1$. Moreover every $u \in H^*(\Sigma^n P^{n-1}_{n-\phi(n)-1})$ transgresses to some $\tilde{v} \in H^{*+1}(\widehat{MSO}_n)$. Using the isomorphism $H^i(X;\pi) \cong [X, K(\pi, i]$ the v's can be taken to be k-invariants in the following construction:

$$\begin{array}{c} D_{1,0} \\ \downarrow \\ MSO_n \xrightarrow{\tilde{v}} K(H_*(\Sigma^{n+1}P^{n-1}_{n-\phi(n)}; Z) \end{array}$$

where $K(\Sigma^{n+1}P^{n-1}_{n-\phi(n)}; Z)$ means a product of $K(m)$'s one for each class $\alpha \in H_m(\Sigma^{n+1}P^{n-1}_{n-\phi(m)})$. This construction may now be completed precisely in the manner employed in the construction of ψ_{n-1} of sections 2-5 yielding

$$\begin{array}{ccccccc} & & D_{2,2} & & D_{2,1} & & \\ & & \downarrow & & \downarrow & & \\ & & D_{1,2} & \to & D_{1,1} & \to & D_{1,0} \\ & & \downarrow & & \downarrow & & \\ X_{n,S} \to \ldots \to & & X_{n,2} & \to & X_{n,3} & \to & MSO_n = X_{n,0} \end{array}$$

Note that for S large enough $X_{n,S}$ is $MSO_n[\phi(n) + 1, \ldots, \infty]$ and $X_{n-\phi(n)-1,s}$ is

$$\Sigma^{\phi(n)+1} MSO_{n-\phi(n)-1}[\phi(n) + 1, \ldots, \infty] .$$

The k-invariants here are precisely the same as those in the construction of ψ_{n-1} . Therefore we can lift $Tf: TM^{n-1} \to MSO_n$ to $\tilde{f}: TM^{n-1} \to \Sigma^{\phi(n)+1}MSO_{n-\phi(n)-1}$ precisely if ψ_{n-1} is zero on U , the Thom class of M , since the assumed existence of $\phi(n)$ cross-sections in $\tau M + \varepsilon$ implies that the lower obstructions vanish.

REFERENCES

[1] Adams, J.F. "Vector fields on spheres," *Ann. Math.*, *75*(1962).

[2] Araki, S. and T. Kudo. "Topology of H_n-spaces and H-squaring operations," *Mem. Fac. Sci. Kyūsyū Univ. Ser. A*, *10*(1956), 85-120.

[3] Barratt, M.G. "The quadratic construction" (unpublished), Conference on Algebraic Topology at the University of Illinois, Chicago Circle, 1968.

[4] Browder, W. "Homology operations and loop spaces," *Ill. J. of Math.*, *4*(1961), 347-357.

[5] Dyer, E. and R. Lashoff. "Homology of iterated loop spaces," *Amer. J. of Math.*, *84*(1962), 35-88.

[6] Frank, D. and E. Thomas. "A generalization of the Steenrod-Whitehead vector field theorem," *Top.* *7*(1968), 311-316.

[7] Hopf, H. "Vectorfelder in n-dimensionaler Mannigfaltigkeiten," *Math. Annln.* *96*(1927), 225-250.

[8] Hu, S.T. *Homotopy Theory*, Academic Press, New York (1959).

[9] Hurwitz, A. "Über die Komposition der quadratischen Formen," *Math. Ann. Bol.*, *88*(1923), 1-25. (Math. Werke II, 641-666; Seiten angaben in unserem Text beziehen sich auf den Abdruck in den Math. Werken.)

[10] Kahn, D.S. "Cup-i products and the Adams spectral sequence," *Top.* *9*(1970), 1-9.

[11] Mahowald, M.E. "On the metastable homotopy of S^n," *Mem. Amer. Math. Soc.* No. 72 (1967).

[12] Michelsohn, M.L. "Decompositions of Steenrod Squares," preprint (1976), University of California, Berkeley.

[13] ____________. "Vector fields on manifolds," preprint (1977), University of California, Berkeley.

[14] Milgram, R.J. "Unstable homotopy from the stable point of view," *Lecture Notes in Mathematics*, No. 368, Springer-Verlag (1974).

[15] Poincaré, H. "Analysis Situs," *Paris J. Ecole Polytechn.*, *ser. 2*, *1*(1895), 1-121.

[16] Radon, J. "Lineare Scharen orthogonaler Matrizen," *Abh. Sem. Hamburg I* (1923), 1-14.

[17] Steenrod, N.E. and J.H.C. Whitehead. "Vector fields on the n-sphere," *Proc. Nat'l. Acad. Sci. U.S.A.*, *37*(1951), 58-63.

[18] Toda, H. "Vector fields on spheres," *Bull. Amer. Math. Soc.* *67*(1961), 408-412.

On G and the Stable Adams Conjecture

by

Haynes R. Miller and Stewart B. Priddy

The purpose of this note is to record the results of our study of the spectrum of G, the space of stable homotopy equivalences of spheres. Because of the J homomorphism and the fibration of infinite loop spaces

(1) $$O \xrightarrow{J} G \to G/O$$

one is reduced to studying G/O. We compute a summand of the cohomology of the spectrum of G/O. We also establish a fibration of infinite loop spaces

$$BU \to X \to IBO$$

where X = G/O with a possibly different infinite loop space structure and IBO is the fiber of the unit map $QS^0 \to BO \times \mathbb{Z}$. Finally we formulate a stable version of the real Adams Conjecture the truth of which is shown to imply that X is G/O with the standard infinite loop space structure. Thus a proof of our conjecture will determine G in terms of more elementary infinite loop spaces.

We would like to thank Mark Mahowald for his interest and for numerous helpful conversations during the course of this research. We also thank the NSF for its support under MCS76-07051. The first author wishes to thank Northwestern University for its hospitality during his very enjoyable year's visit.

§1. Preliminaries

Let $QX = \lim \Omega^n\Sigma^n X$. Then QX is an infinite loop space; i.e., the zero space of the Ω-spectrum $\{Q\Sigma^n X\}$. For $X = S^0$, QS^0 has components $Q_k S^0$, $k \in \mathbb{Z}$, determined by the degree of self maps of spheres.

Stable spherical fibration theory is classified by BG where $G = Q_{\pm 1}S^0$; for oriented theory one uses BSG where $SG = Q_1 S^0$. Both G and SG are infinite loop spaces under composition. On the other hand, reduced stable cohomotopy theory is classified by $Q_0 S^0$, itself an infinite loop space under loop sum. Since SG and $Q_0 S^0$ are equivalent as spaces one would like to understand the relationship between these two basic (and apparently very different) infinite loop structures.

We remind the reader that in the case of oriented real (or complex) K-theory the zero and one components $BSO_{\oplus}$ and $BSO_{\otimes}$ are actually equivalent as infinite loop spaces when localized at any prime [AP]. Certainly nothing so simple is true for SG and $Q_0 S^0$ because their Pontryagin algebras differ.

To give all of this a focus the reader may wish to keep in mind the old problem of computing the homology of the spectrum sg associated to SG

$$H_* sg = \lim H_{*+n} B^n SG$$

Throughout this note we shall use (co-)homology with coefficients in $\mathbb{Z}/2$. All spaces will be localized at 2. The case of odd primes is fundamentally different as we shall indicate in §2.

We recall that any infinite loop space has Dyer-Lashof homology operations; in the case of QS^0

$$Q^k\colon H_* Q_n S^0 \to H_{*+k} Q_{2n} S^0$$

Let $[n] \in H_0 Q_n S^0 = \mathbb{Z}/2$ denote the generator. Then Browder [B] computed

$$H_*Q_0S^0 = \mathbb{Z}/2[Q^I[1]*[-2^{\ell(I)}]]$$

where $*$ denotes the Pontryagin product (under loop sum) and $I = (i_1,\ldots,i_\ell)$ runs over those sequences of positive integers with $i_j \le 2i_{j+1}$, $i_1 > i_2 + \cdots + i_\ell$ and $\ell(I) \ge 1$. Such sequences are called allowable. The length of I, $\ell(I)$, is defined to be ℓ.

Later, Milgram [Mg] described H_*SG in terms of the Dyer-Lashof operations for QS^0 as

$$H_*SG = E[Q^k[1]*[-1]] \otimes \mathbb{Z}/2[Q^kQ^k[1] * [-3]] \otimes \mathbb{Z}/2[Q^I[1]*[1-2^{\ell(I)}]]$$

where $k \ge 1$ and I runs over the same sequences as above except $\ell(I) > 1$. The exterior classes $Q^k[1]*[-1]$ are easily shown to come from SO under the J-homomorphism. Further, fibration (1) and the Eilenberg-Moore spectral sequence show that

$$H_*SG = H_*SO \otimes H_*G/O$$

with

$$H_*SO = E[Q^k[1]*[-1]]$$

$$H_*G/O = \mathbb{Z}/2[Q^kQ^k[1]*[-3]] \otimes \mathbb{Z}/2[Q^I[1]*[1-2^{\ell(I)}]]$$

Thus one may naively explain the difference between the Pontryagin algebras $H_*Q_0S^0$ and H_*SG by saying that the exterior classes $Q^k[1]*[-1]$ force the existence of new generators $Q^kQ^k[1]*[-3]$ to compensate for the fact that the ranks must be equal. It is important to note that these elements are <u>decomposable</u> in Q_0S^0, i.e.

$$Q^kQ^k[1]*[-4] = (Q^k[1]*[-2])^{*2}.$$

The rest of H_*SG looks like $H_*Q_0S^0$ (superficially at least).

By using the Dyer-Lashof operations of SG (derived from the composition product) a stronger statement is possible. We denote these operations by

$$\tilde{Q}^k: H_*SG \to H_{*+k}SG$$

On SO, Kochman [K] has determined these operations while on G/O one has Madsen's formula [Md]: let $x_I = Q^I[1]*[1-2^{\ell(I)}]$ then

$$\tilde{Q}^k x_I = x_{(k,I)} + \sum_{\ell(J)<\ell(k,I)} x_J + \circ\text{-decomposables} \tag{2}$$

where $\circ$ denotes the Pontryagin product. Thus modulo lower length terms and decomposables the $\tilde{Q}$ operations correspond precisely to the Q operations. This strongly suggests some geometric relation between SG and Q_0S^0 as infinite loop spaces. We shall return to this in §3.

§2. A summand of H_*g/O

In studying G/O it is natural to consider the Adams Conjecture

$$\begin{array}{ccccc} G/O & \to & BSO & \xrightarrow{BJ} & BSG \\ & \nwarrow^{\alpha} & \uparrow{\scriptstyle \psi^3-1} & & \\ & & BO & & \end{array} \tag{3}$$

According to Quillen and Sullivan [Q,S] $BJ\circ(\psi^3-1) \simeq 0$ and so one has the indicated lift α. However, Madsen [Md] using (2) has shown that no choice of α is an H-map and so α is of little use in studying G/O as an infinite loop space. It appears that the most one can say is that the infinite loop map

$$QBO(2) \xrightarrow{\bar{\alpha}} G/O$$

(induced by α restricted to BO(2)) splits up to homotopy [P2]. The deviation of α from additivity has recently been analyzed by Tonehave; it involves the Bott map $BO \xrightarrow{\eta} SO$.

In the complex case, there is no such obstruction to additivity and Friedlander and Seymour [FS] have recently solved the Stable Complex Adams Conjecture; i.e.

$$\begin{array}{ccccc} SG/U & \to & BU & \xrightarrow{BJ} & BSG \\ & \nwarrow^{\beta} & \uparrow{\scriptstyle \psi^3-1} & & \\ & & BU & & \end{array}$$

with $BJ\cdot(\psi^3-1) \simeq 0$ as infinite loop maps. (They prove the analogous assertion also at an odd prime. It follows that at an odd prime the analogue of α in (3) can be taken to be an infinite loop map). We define f to be the resulting infinite loop map

$$f\colon BU \xrightarrow{\beta} SG/U \xrightarrow{\rho} G/O$$

where ρ is the natural map. Recalling that $H_*BU = \mathbb{Z}/2[a_k]$, $\dim a_k = 2k$, we have

<u>Proposition</u>. $f_*(a_k) = x_{kk}$ in QH_*G/O, the module of $\circ$-indecomposables ($x_{kk} = Q^kQ^k[1]*[-3]$).

<u>Proof</u>: Consider the homotopy commutative diagram

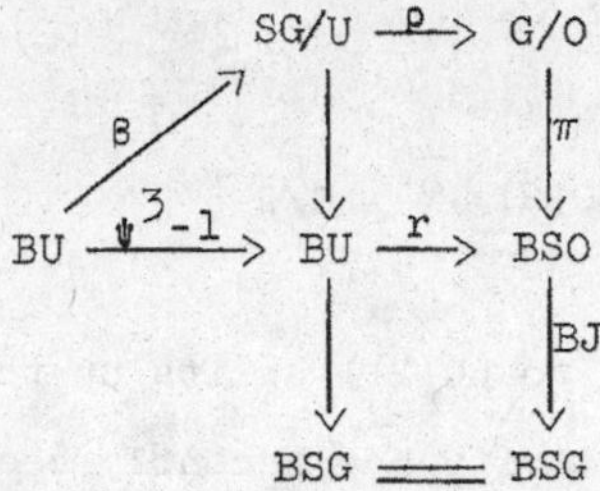

where the vertical maps form fiber sequences and r is realification.

Let $a_2' \in H_4(BU; \mathbb{Z})$ be a class which reduces (mod 2) to a_2. Then in integral homology, $(\psi^3-1)_*(a_2') = 8a_2'$ [A2]. Let $b_2 \in H_4(BSO; \mathbb{Z})/\text{Torsion}$ be a generator. Then $r_*(a_2') = n \cdot b_2$ where n is odd [C]. Let $c_2 \in H_4(G/O; \mathbb{Z})/\text{Torsion}$ be a generator. Then using a solution of the real Adams Conjecture [Q,S]

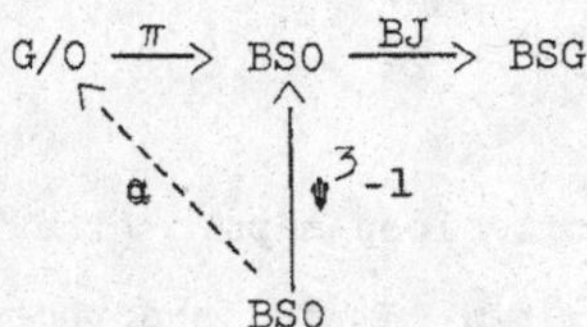

one can deduce that $\pi_*(c_2) = 8k \cdot b_2$, k odd. Hence $f_*(a_2') = \ell \cdot c_2$, ℓ odd. Thus, reducing mod 2 and using a standard Bockstein argument we find $f_*(a_2) = x_{22}$ in QH_*G/O (in mod 2 homology). Also $f_*(a_1) = f_*(Sq_*^2 a_2) = Sq_*^2 x_{22} = x_{11}$.

Using this fact and examining the diagonal map it is easy to see that in QH_*G/O

$$f_*(a_k) = x_{k,k} + \Sigma\, x_I$$

for some (possibly empty) set of allowable sequences I with $\ell(I) > 2$ and $|I| = \Sigma i_j$ even. We wish to show that $\Sigma\, x_I = 0$.

$$0 = Sq_*^1 f_*(a_k) = Sq_*^1(\Sigma\, x_I) = \Sigma(i_1-1)x_{I-\Delta_1}$$

where $I = (i_1,\ldots)$ and $\Delta_1 = (1,0,\ldots,0)$. If i_1 is even then $x_{I-\Delta_1}$ is allowable, hence I appears only if i_1 is odd. However, let

$$m = \max\{\ell(I) \mid x_I \text{ is a summand of } f_*(a_k)\}$$

and suppose $(i_1,i_2,\ldots,i_m)$ occurs. Then

$$0 = f_*(0) = f_*(Q^{2i_1-1}a_k) = \tilde{Q}^{2i_1-1}f_*(a_k)$$

$$= \sum_{\ell(I)=m} \tilde{Q}^{2i_1-1}x_I + \sum_{\ell(I)<m} \tilde{Q}^{2i_1-1}x_I = \Sigma\, x_{(2i_1-1,I)} +$$

terms of length $\leq m$ (using (2)). Since each of the terms $x_{(2i_1-1,I)}$ is allowable this completes the proof.

This proposition has immediate implications for the (co-)homology of the Ω-spectrum g/O with zero space G/O. Let bu denote the Ω-spectrum with zero space BU; i.e. connective reduced complex K-theory. Adams [A1] has computed

$$H^*\,bu = \Sigma^2 A/A(Sq^1,Sq^3)$$

where A denotes the mod 2 Steenrod algebra as usual. Now f induces a map of spectra

$$f\colon bu \to g/O$$

and we have

<u>Corollary</u>. $H_*\,bu$ is a $\mathbf{Z}/2$-summand of $H_*\,g/O$.

<u>Proof</u>: Equivalently we show f^* is surjective. Since H^*bu is monogenic

over A we need only show that the generator in dimension 2 is in the image. But this is the Hurewicz dimension so the result follows from the Proposition.

Remark: H^*bu is not a summand over A because the 2 and 3 dimensional classes of H^*g/O are connected by Sq^1.

§3. Two Conjectures

In this section we study the cokernel of the infinite loop map $f: BU \to G/O$ of §2.

Let ko denote the Ω-spectrum representing connective unreduced real K-theory; i.e. the zero space of ko is $BO \times \mathbb{Z}$. The unit map $S \to ko$ gives rise to an infinite loop map

$$u: QS^0 \to BO \times \mathbb{Z}$$

Recalling that $H_*BO = \mathbb{Z}/2[\bar{e}_k]$ and $u_*(Q^k[1]*[-2]) = \bar{e}_k\,[P2]$, we have an exact sequence of $\mathbb{Z}/2$-modules

$$0 \to QH_*BU \xrightarrow{f_*} QH_*G/O \xrightarrow{\gamma} QH_*Q_0S^0 \xrightarrow{u_*} QH_*BO \to 0$$

where $Q(\cdot)$ is the algebra indecomposables functor. The map γ is defined on basis elements by $x_I \to Q^I[1]*[-2^{\ell(I)}]$. Since f_* and u_* are induced by infinite loop maps, both preserve Dyer-Lashof operations. By (2), γ preserves Dyer-Lashof operations up to a length filtration. Of course, a priori, γ is just an algebraic map, but the first author has a spectral sequence for computing the homology of a spectrum and the E^2 term depends on the homology indecomposables of the zero space as an unstable module over the Dyer-Lashof algebra. Thus it seems plausible to make the following conjecture. Let IBO denote the fibre of $u: QS^0 \to BO \times \mathbb{Z}$.

Conjecture A. There exists a fibration

$$BU \xrightarrow{f} G/O \to IBO$$

of infinite loop spaces.

One consequence of this conjecture is a complete calculation of H^*g/O. From the cofibration sequence

$$\cdots \to \Sigma^{-1}(ko/S) \to S \to ko \to ko/S \to \cdots$$

and Stong's calculation [St], $H^*ko = A/A(Sq^1,Sq^2)$ we have

$$H^*\Sigma^{-1}(ko/S) = \Sigma^{-1}I(A/A(Sq^1,Sq^2))$$

Hence from the Corollary of §2 we have

Corollary of Conjecture A

$$H^*g/O = \Sigma^2(A/A(Sq^1,Sq^3)) \oplus \Sigma^{-1}I(A/A(Sq^1,Sq^2))$$

with the 2 and 3 dimensional generators connected by Sq^1.

We now construct a candidate for a solution to Conjecture A. Since BSpin is connected, the composite

$$QS^0 \xrightarrow{u} BO \times Z \xrightarrow{\psi^3 - 1} BSpin$$

is null homotopic as an infinite loop map. Hence there is an induced map of infinite loop space fibrations

(4)

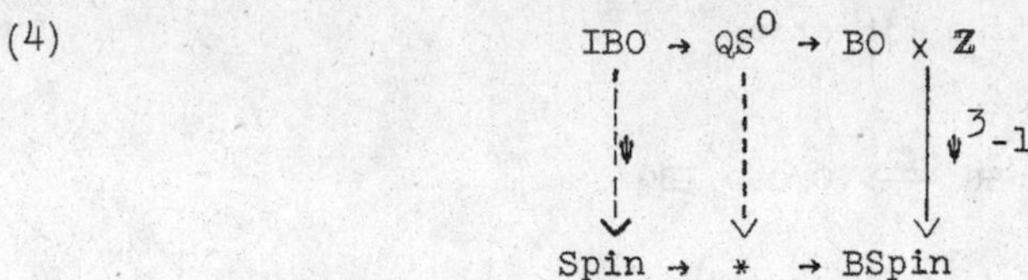

Let X be the fiber of the composite IBO $\xrightarrow{\psi}$ Spin $\xrightarrow{c}$ SU where c is complexification. Then from the Bott sequence BSO $\xrightarrow{\eta}$ Spin $\xrightarrow{c}$ SU we have an induced map of infinite loop space fibrations

(5)

$$\begin{array}{ccccccc} BU & \longrightarrow & X & \longrightarrow & IBO & \longrightarrow & SU \\ \| & & \downarrow{\scriptstyle \epsilon} & & \downarrow{\scriptstyle \psi} & & \| \\ BU & \xrightarrow{r} & BSO & \xrightarrow{\eta} & Spin & \xrightarrow{c} & SU \end{array}$$

where r is realification.

We have been unable to prove that

(6) $$BU \to X \to IBO$$

is a solution to Conjecture A. However in §4 we show that as a space X is equivalent to G/O and so X provides some delooping of G/O (possibly non standard). We also show that both X and G/O provide infinite loop space factorizations of r. Thus (6) seems a very good candidate for Conjecture A.

First we show how Conjecture A relates to the Stable Adams Conjecture. Consider the diagram

$$(7)\qquad \begin{array}{ccccc} BU \to & X & \overset{\varphi}{\dashrightarrow} & G/O & \\ {\scriptstyle r}\searrow & \downarrow{\scriptstyle \epsilon} & & \downarrow{\scriptstyle \pi} & \\ & BSO & \xrightarrow{\psi^3-1} & BSO & \xrightarrow{BJ} BSG \end{array}$$

where r is realification and π is inclusion of the fiber of BJ. By the Adams Conjecture $BJ\cdot(\psi^3-1) \simeq 0$ as maps of spaces but not H-spaces. By the Stable Adams Conjecture $BJ\cdot(\psi^3-1)\cdot r \simeq 0$ as infinite loop space maps (see §2). We propose the intermediate conjecture.

<u>Conjecture B</u>. $BJ\cdot(\psi^3-1)\cdot\epsilon \simeq 0$ as infinite loop space maps.

This immediately implies Conjecture A

<u>Lemma</u>. Any infinite loop map φ completing diagram (7) is an equivalence (at 2).

<u>Proof</u>: In §4 we show $X \simeq G/O$ as a space, hence it is enough to prove φ_* is surjective in mod-2 homology. Over the Dyer-Lashof algebra, QH_*G/O is generated by the coalgebra C with basis $\{x_{a,b}\colon a \le 2b,\ a \ge b \ge 0\}$ (see [Md]). Thus it suffices to show $x_{a,b} \in \operatorname{Im} \varphi_*$ modulo terms of higher length. As an algebra $C^* = \mathbb{Z}/2[x,y]$ where x and y are dual to x_{11} and x_{21} respectively. Since $X \simeq G/O$ as a space, H^*X is a polynomial algebra and thus it suffices to show $x_{11}, x_{21} \in \operatorname{Im} \varphi_*$. By the argument of the Proposition of §2, $x_{11} \in \operatorname{Im} \varphi_*$. The relation $Sq^1 x = y$ implies $x_{21} \in \operatorname{Im} \varphi_*$. This completes the proof.

§4. <u>Properties of</u> X.

<u>Proposition</u>. $X \simeq G/O$ as spaces.

Let $\operatorname{ImJ} \times \mathbb{Z}$ denote the fiber of $\psi^3-1\colon BO \times \mathbb{Z} \to BSpin$. Then from diagram (4) and the 3 x 3 Lemma for infinite loop spaces (or spectra) we have the following homotopy commutative diagram of infinite loop spaces and maps.

(8)

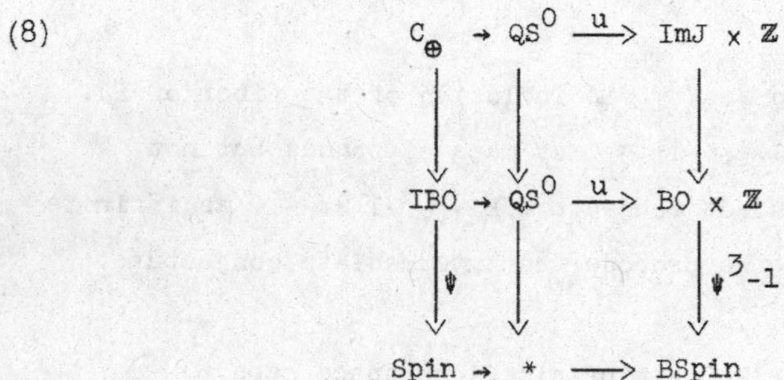

where the vertical and horizontal sequences are fibrations and where the common fiber, $C_{\oplus}$, is called the (additive) coker J. A (multiplicative) coker J, $C_{\otimes}$, is defined as the fiber of the unit map $u: QS^0 \to \mathrm{ImJ} \times \mathbb{Z}$ restricted to the 1-components. As spaces $C_{\oplus} \simeq C_{\otimes}$.

Proof of Proposition: Combining diagrams (5) and (8) we have

(9)

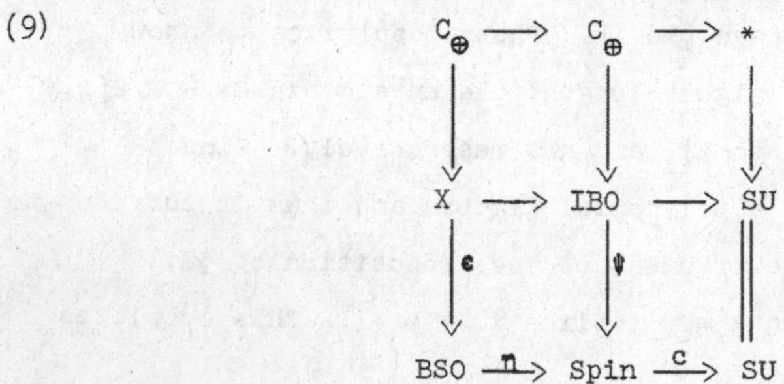

Let $IBO_{\otimes}$ be the fiber of $u: QS^0 \to BO \times \mathbb{Z}$ restricted to the 1-components. May [My] has shown $IBO_{\otimes} \simeq C_{\otimes} \times \mathrm{Spin}$ as infinite loop spaces. Since $IBO_{\otimes} \simeq IBO$ as spaces and since $KO^*(C_{\otimes}) = 0$ [Sn] we have a splitting $IBO \xrightarrow{\psi} \mathrm{Spin}$ (as spaces) and thus from (9) a splitting $X \underset{\leftarrow}{\xrightarrow{\varepsilon}} BSO$;

i.e. $X \simeq C_{\oplus} \times BSO$ as spaces. However $G/O \simeq C_{\otimes} \times BSO$ [MST] and so this completes the proof.

From diagram (5) we see that X factors realification. Next we show that G/O shares this property. The Atiyah-Bott-Shapiro orientation of Spin bundles defines a KO-characteristic class $e\colon G/O \to BSO_{\otimes}$ which is an infinite loop map [MST]. The Adams cannibalistic class $\rho_3\colon BSO \to BSO_{\otimes}$ is an infinite loop equivalence [MST].

Proposition. G/O factors realification; i.e.

$$\begin{array}{ccc} BU & \xrightarrow{f} & G/O \\ & \searrow^{r} & \downarrow{\scriptstyle \rho_3^{-1}\circ e} \\ & & BSO \end{array}$$

is homotopy commutative as infinite loop maps.

First we establish

Lemma.

$$\begin{array}{ccc} G/O & \xrightarrow{e} & BSO_{\otimes} \\ \downarrow{\scriptstyle \pi} & & \downarrow{\scriptstyle \rho_3^{-1}} \\ BSO & \xleftarrow{\psi^3-1} & BSO \end{array}$$

is homotopy commutative as infinite loop maps.

Proof: The diagram

$$\begin{array}{ccc} G/O & \xrightarrow{e} & BSO_{\otimes} \\ \downarrow{\scriptstyle \pi} & & \downarrow{\scriptstyle \psi^3/1} \\ BSO & \xrightarrow{\rho_3} & BSO_{\otimes} \end{array}$$

is homotopy commutative on the space level [MST]. Let $d = [(\psi^3/1) \cdot e]/(\rho_3 \cdot \pi): G/O \to BSO_\otimes$. Since the set of homotopy classes of infinite loop maps $C_\otimes \to BSO_\otimes$ is trivial [MST] there is an infinite loop map δ factoring d

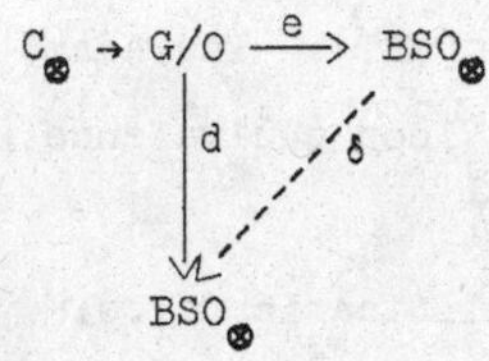

An infinite loop self map of BSO is determined by its restriction to the zero space [MST]. ($BSO_\otimes$ is equivalent to BSO [AP].) Similarly a self map of BSO is determined by its induced morphism in rational homology [My]. Since $d \simeq 0$ and e is a rational equivalence it follows that $\delta \simeq 0$ as an infinite loop map.

The lemma now follows from the homotopy commutativity of

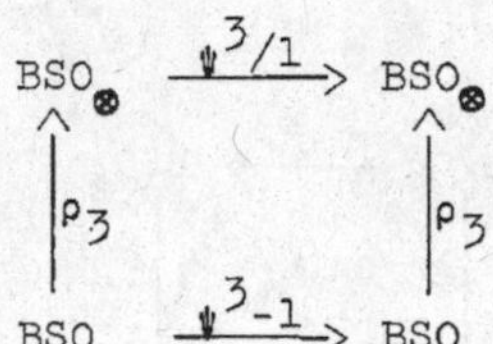

as infinite loop maps [My]. This completes the proof.

<u>Proof of Proposition:</u> Consider the diagram

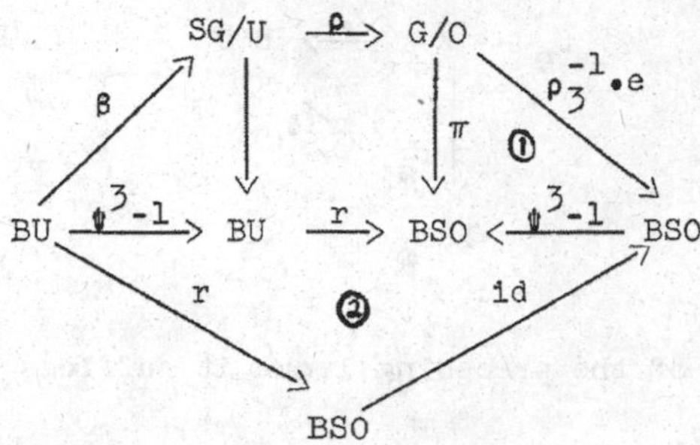

By definition the unlabeled squares commute up to homotopy as infinite loop maps. Similarly for square 1 by the preceding lemma and for square 2 by Adams [A2]. This completes the proof.

Finally we mention another

Corollary of Conjecture B: $C_{\oplus} \simeq C_{\otimes}$ as infinite loop spaces.

Proof: By the Lemma of §3, φ is an equivalence. Since ρ_3 is also an equivalence, it suffices to show that φ fits into a map of infinite loop space fibrations

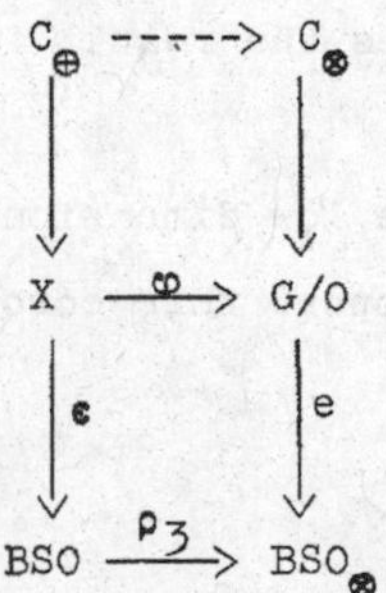

Let $d = e \cdot \varphi / \rho_3 \cdot \varepsilon \colon X \to BSO_{\otimes}$. Since the set of homotopy classes of infinite loop space maps $C_{\oplus} \to BSO_{\otimes}$ is trivial [MST] there is an infinite loop map δ factoring d

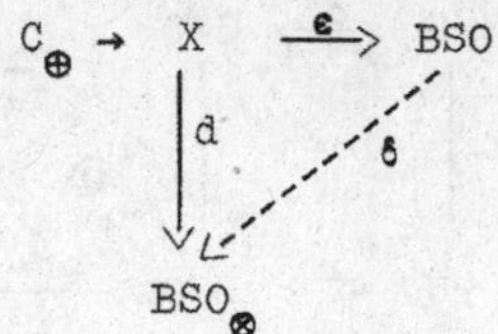

Now as in the proof of the preceding lemma it suffices to show $\delta_* = 0$ in rational homology.

Consider the diagram

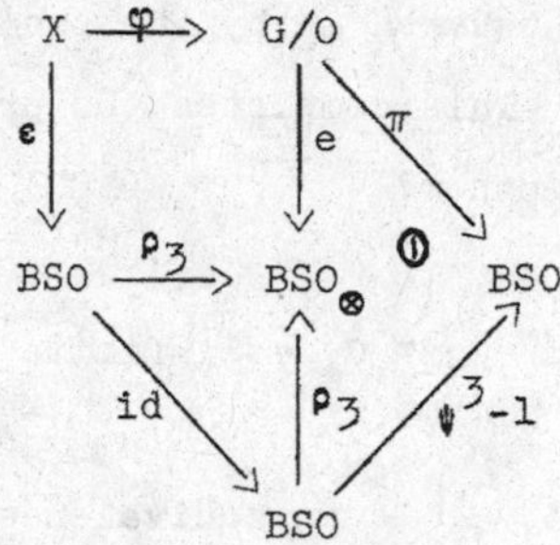

Square 1 commutes up to homotopy by the preceding lemma. The outer diagram commutes up to homotopy by definition of φ. Since ρ_3 and ψ^3-1 are rational equivalences the result follows.

Remark: P. May has made some low dimensional calculations with homology operations which support this corollary.

References

[A1] J. F. Adams, On Chern characters and the structure of the unitary group, Proc. Camb. Phil. Soc. 57 (1961), 189-199.

[A2] ___________, Vector fields on spheres, Ann. of Math. 75 (1962), 603-632.

[AP] J. F. Adams and S. B. Priddy, Uniqueness of BSO, Math. Proc. Camb. Phil. Soc. 80 (1976), 475-509.

[B] W. Browder, Homology operations and loop spaces, Ill. J. Math. 4 (1960), 347-357.

[C] H. Cartan, Démonstration homologique des théorèmes de periodicite de Bott II, Sem. H. Cartan (1959/60).

[FS] E. Friedlander and R. Seymour, Two proofs of the stable Adams Conjecture, Bull. Amer. Math. Soc. (to appear).

[K] S. O. Kochman, Homology of the classical groups over the Dyer-Lashof algebra, Trans. Amer. Math. Soc. 185 (1973), 83-136.

[Md] I. Madsen, On the action of the Dyer-Lashof algebra in $H_*(G)$, Pacific J. Math. 69 (1975), 235-275.

[MST] I. Madsen, V. Snaith, and J. Tornehave, Infinite loop maps in geometric topology, Math. Proc. Camb. Phil. Soc. 81 (1977), 399-429.

[My] J. P. May (with contributions by F. Quinn, N. Ray, and J. Tornehave), E_∞ ring spaces and E_∞ ring spectra, Lecture Notes in Math. vol. 577, Springer-Verlag, Berlin and New York, 1977.

[Mg] R. J. Milgram, The mod 2 spherical characteristic classes, Ann. of Math. 92 (1970), 238-261.

[P1] S. B. Priddy, Dyer-Lashof operations for the classifying spaces of certain matrix groups, Quart. J. Math. 26 (1975), 179-193.

[P2] ___________, Homotopy splittings involving G and G/O, (to appear).

[Q] D. G. Quillen, The Adams conjecture, Topology 10 (1971), 67-80.

[Sn] V. P. Snaith, Dyer-Lashof operations in K-theory, Lecture Notes in Math. vol. 496, Springer-Verlag, Berlin and New York, 1975.

[St] R. E. Stong, Notes on Cobordism Theory, Princeton University Press, Princeton, N. J., 1968.

[S] D. Sullivan, Genetics of homotopy theory and the Adams conjecture, Ann. of Math. (2) 100 (1974), 1-79.

Harvard University
Northwestern University

COMPLETIONS OF COMPLEX COBORDISM

Jack Morava

A <u>genus</u> $\chi : U \to k$ of complex-oriented manifolds [5, 10] is a ring-homomorphism from the ungraded complex bordism ring $U = \oplus U_*(pt)$ to a ring k.

We assume k is a field, and write U_χ for the (Hausdorff) completion of U in the topology $\{\underline{m}_\chi^i \mid i \geq 0\}$, where the maximal ideal $\underline{m}_\chi$ is the kernel of χ.

0.1 <u>Proposition</u>. The χ-adic completion $U_\chi(-) = U_\chi \otimes_U (\oplus U_*(-))$ of the complex bordism functor is an ungraded homology theory on CW-complexes.

<u>Proof</u>: If U were noetherian, U_χ would be flat [2, §10.14]. However, U is coherent, and if X is a finite complex, then $U(X) = \oplus U_*(X)$ is a finitely-presented U-module, and on the category of such modules, $U_\chi \otimes -$ is exact. [Every short exact sequence of such modules can be pulled back from a short exact sequence of modules over a noetherian subring [14].] Now both $U_\chi \otimes_U -$ and $U(-)$ commute with direct limits, so $U_\chi(-)$ is exact on arbitrary CW-complexes. ||

0.2 <u>Proposition</u>. $U_\chi(-)$ is a comodule over the (Hausdorff) completion $U_\chi U$ of the two-sided U-algebra $UU = \oplus U_* U$ in the bilateral χ-adic topology $\{\underline{m}_\chi^i UU \underline{m}_\chi^j \mid i,j \geq 0\}$.

<u>Proof</u>: The bialgebra structure of UU can be summarized by a diagram

$$U \underset{e}{\overset{\eta_L}{\underset{\eta_R}{\rightrightarrows}}\!\!\longleftarrow} UU$$

[1, §11] and the diagram

$$\begin{array}{ccc} U & \underset{e}{\overset{\eta_L}{\underset{\eta_R}{\rightrightarrows}}\!\!\longleftarrow} & UU \\ \downarrow & & \downarrow \\ U_\chi & \underset{\hat{e}}{\overset{\hat{\eta}_L}{\underset{\hat{\eta}_R}{\rightrightarrows}}\!\!\longleftarrow} & U_\chi U \end{array}$$

commutes. The completions of the structure maps of the Hopf algebroid [9, 10] (U,UU) define a Hopf algebroid $(U_\chi, U_\chi U)$, while the natural composition

$$\psi_X : U_\chi \otimes_U U(X) \to U_\chi \otimes_U UU \otimes_U U \otimes_U U(X) \to U_\chi \otimes_U UU \otimes_U U_\chi \otimes_U U(X) \longrightarrow$$
$$\to U_\chi U \otimes_{U_\chi} U_\chi \otimes_U U(X)$$

makes of $U_\chi(-)$ a $(U_\chi, U_\chi U)$-comodule. ||

Section one of the following paper is general nonsense about groupoid schemes and their linear representations, which is used in section 2 to derive consequences of the theory of deformations of formal group laws. There we describe the category of comodules over the U_χ-cooperations, where χ is a genus of finite height [4, I §3, Prop. 5]. In particular, we construct a natural splitting of $(U_\chi, U_\chi U)$. A previously known structure theorem for cobordism comodules is an easy corollary.

§1 A groupoid is a category in which every morphism is an isomorphism. If for example a group G acts on a set S, then the category S/G with S as objects, G×S as morphisms, and structure maps ('source', 'target', 'identity')

$$\text{mor } S/G = G\times S \underset{e}{\overset{\eta_L}{\underset{\eta_R}{\rightrightarrows}}} S = \text{obj } S/G$$ given by $\eta_L(g,s) = s$,

$\eta_R(g,s) = gs$, $e(s) = (1,s)$ is a groupoid.

A homomorphism of groupoids is a functor of categories. Two groupoids are equivalent if they are equivalent as categories; this is not the same as isomorphy.

1.1 Definition. A groupoidscheme $\underline{\underline{G}} : B \underset{e}{\overset{\eta_L}{\underset{\eta_R}{\rightrightarrows}}} C$ is a diagram of commutative rings and homomorphisms such that for any commutative ring k, the diagram

$$\text{mor } \underline{\underline{G}}(k) = \text{Hom}_{\text{Rings}}(C,k) \underset{e}{\overset{\eta_L}{\underset{\eta_R}{\rightrightarrows}}} \text{Hom}_{\text{Rings}}(B,k) = \underline{\underline{\text{obj}}}\ G(k)$$

is a groupoid, i.e. a groupoidscheme is a representable functor from commutative rings to groupoids. The composition of morphisms in $\underline{\underline{G}}(k)$ is represented by a ring-homomorphism $C \to C\otimes_B C$, making (B,C) into a Hopf algebroid. See for example [7].

1.2 Example. If the affine groupscheme G (with Hopf algebra A_G of functions) acts on the affine scheme S (with algebra A_S of functions) then the groupoidscheme $\underline{\underline{S/G}}$ is represented by the diagram

$$B_{S/G} = A_S \underset{e}{\overset{\eta_L}{\underset{\eta_R}{\rightrightarrows}}} A_G \otimes A_S = C_{S/G}.$$

If G acts trivially on S, then $C_{S/G}$ is a Hopf $B_{S/G}$-algebra in the usual sense.

A groupoidscheme equivalent to a groupoidscheme of the form S/G will be said to split.

1.3 Definition. A linear representation V of the groupoidscheme $\underline{\underline{G}} : B \rightrightarrows C$ is a left B-module V together with an associative, unitary "cooperation" homomorphism $\psi_V : V \to C \otimes_B V$ of left B-modules, [12, II §3.0.3]. We write $\underline{\underline{G}}$-Mod for the category of such representations, with the obvious morphisms; if G is as in 1.2, and S is the spectrum of a field k, then a linear representation of $\underline{\underline{S/G}}$ is a linear representation of G over k [3, II §2, no. 2].

1.4 Definition. If C is a flat left (or, equivalently, right) B-module, then $\underline{\underline{G}}$ is a flat groupoidscheme. If $\underline{\underline{G}}$ is flat, then $\underline{\underline{G}}$-Mod is an abelian category [12, II 2.0.6] where injectives are direct summands of $-\otimes_B$ (some B-module) [9, §2.2]. (Any split groupoidscheme is flat.)

1.5 Construction. Suppose $\phi : \underline{\underline{G}} \to \underline{\underline{G}}'$ is a homomorphism of groupoidschemes, represented by ringhomomorphisms $\phi_B : B' \to B$, $\phi_C : C' \to C$. Suppose V is a linear representation of $\underline{\underline{G}}$. Then $\phi^*V = B\otimes_{B'}V$ is a linear representation of V via the cooperation

$$\psi_{\phi^*V} : B\otimes_{B'}V \to B\otimes_{B'}C'\otimes_{B'}V \to B\otimes_B C\otimes_{B'}V \cong C\otimes_B B\otimes_{B'}V,$$

and $\phi^* : \underline{\underline{G}}'\text{-}\underline{\text{Mod}} \to \underline{\underline{G}}\text{-}\underline{\text{Mod}}$ is a covariant functor.

1.6 Proposition: Suppose $\phi : \underline{\underline{G}} \to \underline{\underline{G}}'$, $\gamma : \underline{\underline{G}}' \to \underline{\underline{G}}$ are homomorphisms between flat groupoidschemes, such that the composite functors $\phi\,\gamma$, $\gamma\,\phi$ are naturally equivalent to the appropriate identities, i.e. ϕ and γ are inverse natural equivalences of groupoid-valued functors. Then ϕ^* and γ^* are inverse equivalences of categories.

Proof: Evidently $(\phi\,\gamma)^* = \gamma^*\,\phi^*$, so it suffices to assume that $\underline{\underline{G}}' = \underline{\underline{G}}$, and that γ is the identity. Call the postulated natural transformation $\theta : 1_{\underline{\underline{G}}} \to \phi$, and write $\theta : C \to B$ for the representing ring-homomorphism, satisfying $\theta\eta_L = 1_B$, $\theta\eta_R = \phi_B$. Then the composition

$$\theta_V^* : V \xrightarrow{\psi_V} C\otimes_B V \xrightarrow{\theta\phi\otimes 1_V} B^{\phi}\otimes_B V = \phi^*V$$

is a homomorphism of left B-modules, where B^{ϕ} is the B-bimodule on symbols $(b)\phi$, with product $b'((b)\phi)b'' = ((b'b)\phi_B(b''))\phi$, $b, b', b'' \in A$; for $c \mapsto (\theta(c))\phi$ is a B-bimodule homomorphism from C to B^{ϕ}. Suppose first that V = C, given its natural cooperation $C \to C\otimes_B C$; then θ_C^* is a ring-homomorphism. The set-theoretic map $\theta^*(k)$ sends an element $(x,g) \in \mathrm{Hom}_{\mathrm{Rings}}(B^{\phi}\otimes_B C,k)$ (interpreted as a pair $x \in \mathrm{obj}\ \underline{\underline{G}}(k)$, $g \in \mathrm{mor}\ \underline{\underline{G}}(k)$, such that $\phi(x) = \mathrm{source}\ (g)$)

to the composition $(x \xmapsto{\theta(x)} \phi(x) \xrightarrow{g} y) \in \mathrm{Hom}_{\mathrm{Rings}}(C,k)$. Now $\theta^*_C(k)$ has an inverse $[(x \xrightarrow{g} y) \mapsto (\phi(x) \xrightarrow{g\cdot\theta(x)^{-1}} y)]$ and is thus a bijection, so θ^*_C is an isomorphism; consequently θ^*_I is an isomorphism for any injective $\underline{\underline{G}}$-module I. Any linear representation V of $\underline{\underline{G}}$ possesses an injective resolution in $\underline{\underline{G}}$-$\underline{\underline{\mathrm{Mod}}}$, yielding a commutative diagram

$$\begin{array}{ccccccccc} 0 & \longrightarrow & V & \longrightarrow & I_0 & \longrightarrow & I_1 & \longrightarrow & \cdots \\ & & \downarrow{\scriptstyle \theta^*_V} & & \wr\downarrow{\scriptstyle \theta^*_{I_0}} & & \wr\downarrow{\scriptstyle \theta^*_{I_1}} & & \\ 0 & \longrightarrow & \phi^*V & \longrightarrow & \phi^*I_0 & \longrightarrow & \phi^*I_1 & \longrightarrow & \cdots \end{array}$$

and the assertion follows from the lemma. ‖

We need a formal analogue of 1.6 below:

Let $\underline{\mathrm{Art}}_k$ denote the category of Artin local rings, whose residue fields contain k, and whose morphisms are homomorphisms of local rings [13].

The completion $U \to U_\chi$ can be characterized universally: if $A \in \underline{\mathrm{Art}}_k$, then any lifting $\chi' : U \to A$ of χ factors through a unique continuous ring homomorphism $\chi'' : U_\chi \to A$ (with A discrete) such that the diagram

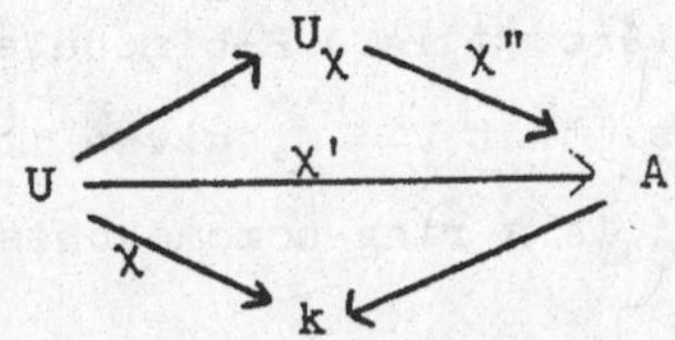

commutes; in other words, $\mathrm{Hom}^{\mathrm{cont}}_{\mathrm{Rings}}(U_\chi, A) = \underline{\mathrm{lifts}}_\chi(A)$.

1.6' Proposition. As above, for pro-representable functors $\hat{\underline{\underline{G}}} : \underline{\mathrm{Art}}_k \to$ (Groupoids), whose morphism algebra is flat over its object algebra.

Proof: A pro-representable functor is representable by an inverse limit of objects in $\underline{\mathrm{Art}}_k$. The proof is as above, replacing only the set-theoretic map $\theta^*_C(-)$ by its restriction to $\underline{\mathrm{Art}}_k$. ||

1.7 Proposition. The Hopf algebroid $(U_\chi, U_\chi U)$ pro-represents the functor $\underline{\mathrm{lifts}}_\chi : \underline{\mathrm{Art}}_k \to$ (Groupoids).

Proof: If χ is identified with a formal group law $F(X,Y) \in k[[X,Y]]$, then an object of $\underline{\mathrm{lifts}}_\chi(A)$ may be interpreted as a formal group law $G(X,Y) \in A[[X,Y]]$ such that $\bar{G} = F$, where $A \ni x \mapsto \bar{x} \in k$ denotes reduction modulo the maximal ideal of A. The functor Hom(UU,-) represents the morphisms of the category of one-parameter formal grouplaws [7]; thus we may interpret a continuous homomorphism $\zeta : U_\chi U \to A$ as a pair G_1, G_2 of formal grouplaws over A (induced by $\hat{\eta}_L(\zeta), \hat{\eta}_R(\zeta)$ respectively, and which thus both lift F) together with a morphism $f : G_1 \to G_2$ of grouplaws over A, i.e. $f = \sum_{i \geq 0} f_i T^{i+1} \in A[[T]]$ with $\bar{f}_0 \neq 0$, such that

$$f(G_1(X,Y)) = G_2(fX, fY).$$

We denote this groupoid $\underline{\text{lifts}}_X(A)$.

§2.

Suppose that the characteristic p of k is not zero. By a theorem of Cohen, we identify $\underline{\text{Art}}_k$ with a category of W(k)-algebras, where W(k) is the ring of Witt vectors over k.

2.1 Proposition. If G_1, G_2 are formal group laws over $A \in \underline{\text{Art}}_k$, such that $\bar{G}_1$ is of finite height, then the homomorphism $f \mapsto \bar{f} : \text{Hom}_{\text{Grouplaws}/A}(G_1, G_2) \to \text{Hom}_{\text{Grouplaws}/k}(\bar{G}_1, \bar{G}_2)$ is injective.

Proof: We assume temporarily that $[p]_{\bar{G}_1}(T) = T^q$ where $q = p^n$, n = height of $\bar{G}_1$. We show that if $\bar{f} = 0$, i.e. $f \in \underline{m}_A[[T]]$, then $f = 0$; it suffices to show that $f \in \underline{m}_A^r[[T]]$ implies $f \in \underline{m}_A^{r+1}[[T]]$, since A is Artin, with maximal ideal $\underline{m}_A$. But if $f \in \underline{m}_A^r[[T]]$ then $fG_1(X,Y) = G_2(fX, fY) = fX + fY$ modulo m_A^{r+1}, since $p \in \underline{m}_A$. But then $f(T^q) = 0 \bmod \underline{m}_A^{r+1}$, i.e. $f \in \underline{m}_A^{r+1}[[T]]$.

In general there exists an invertible power series $g \in \bar{k}[[T]]$ over a separable closure of k, such that $F^g(X,Y) = gF(g^{-1}X, g^{-1}Y)$ satisfies the [p]-condition above [4, III §2, Lemma 3]. Let $\bar{A} = A \otimes_{W(k)} W(\bar{k}) \in \underline{\text{Art}}_{\bar{k}}$; choose $h \in \bar{A}[[T]]$ lifting g, and argue as before, using $h\, f\, h^{-1}$ in place of f. ‖

2.2 Definition. Two lifts G_1, G_2 of F to A are *-isomorphic if there exists an isomorphism $f : G_1 \to G_2$ of formal grouplaws such that $\bar{f}(T) = T$. If F is of finite height, then a $*$-isomorphism $f : G_1 \to G_2$ is necessarily unique [if f_0, f_1 are two such, then $f_0^{-1}f_1 \in \mathrm{Hom}_{\mathrm{Grouplaws}/A}(G_1,G_2)$ maps to $1_F \in \mathrm{Hom}_{\mathrm{Grouplaws}/k}(F,F)$, so $f_0 = f_1$ by the above].

2.3 Construction. The $*$-isomorphism classes of lifts of F to A form a groupoid $\underline{\mathrm{lifts}}_\chi^*(A)$.

If G_1^*, G_2^* denote $*$-isomorphism classes, and G_1, G_2 representative elements, then by 2.1 the image of $f \mapsto \bar{f}$: (isomorphisms of G_1 with G_2) $\to$ (Endomorphisms of F) is independent of the choice of representatives; we denote it $\mathrm{mor}_{\mathrm{lifts}}(G_1^*, G_2^*)$. ‖

2.4 Proposition. If F is of finite height, then the functors $\underline{\mathrm{lifts}}_\chi$ and $\underline{\mathrm{lifts}}_\chi^*$ are naturally equivalent.

Proof: Examine the category $\underline{\mathrm{lifts}}_\chi'$ whose objects are those of $\underline{\mathrm{lifts}}_\chi$ and whose morphisms are those of $\underline{\mathrm{lifts}}_\chi^*$; by 2.1, $\underline{\mathrm{lifts}}_\chi$ is equivalent to $\underline{\mathrm{lifts}}_\chi^*$ by 2.2 and the axiom of choice. ‖

2.5 Construction. Let $\mathrm{Aut}\,\chi(A)$ denote the group of automorphisms of F, with coefficients in the field $A/\underline{m}_A = k_A \supset k$, i.e. invertible power series $f \in k_A[[T]]$ such that $fF(\bar{f}^{-1}X, \bar{f}^{-1}Y) = F(X,Y)$. If G is a lift of F to A, and $g \in A[[T]]$ is a lift of $f \in \mathrm{Aut}\,\chi(A)$, then g is invertible, and $G^g(X,Y) = G(g^{-1}X, g^{-1}Y)$ is a lift of F to A.

If $h \in A[[T]]$ is another lift to A, then the composition $g\bar{h}' : G^h \to G^g$ is a *-isomorphism of lifts, and the equivalence class depends only on the class G* and the series f. Consequently $(f,G^*) \mapsto G^{*f} (= G^{g^*})$ defines an action of the covariant group-valued functor Aut χ on the set-valued functor lifts_χ^* of objects of $\underline{\text{lifts}}_\chi^*$; in fact $\text{mor}_{\text{lifts}}(G_1^*, G_2^*) = \{f \in \text{Aut}\ \chi \mid G_1^{*f} = G_2^*\}$. The functors $\underline{\text{lifts}}_\chi^*$ (and hence $\underline{\text{lifts}}_\chi$) are equivalent to the split groupoid-valued functor $\text{lifts}_\chi^*/\text{Aut}\ \chi$.

2.7 Theorem. The functors lifts_χ^* and Aut χ are pro- representable, if χ is of finite height.

2.8 Corollary. The category of $(U_\chi, U_\chi U)$-comodules is equivalent to the category of $(E_\chi, E_\chi \otimes_{W(k)} H_\chi)$-comodules, where $E_\chi, H_\chi \in$ (W(k)-Algs) are naturally identified by isomorphisms

$$\text{Hom}_{\text{Rings}}^{\text{cont}}(E_\chi, A) \cong \text{lifts}_\chi^*(A),$$

$$\text{Hom}_{\text{Rings}}^{\text{cont}}(H_\chi, A) \cong \text{Aut}_\chi(A).$$

Proof: There exists a formal group law $\tilde{F}$ over the formal power series algebra $E_\chi = W(k)[[t_1, \ldots, t_{n-1}]]$, n = height of F, with the following "*-universal" property: for any lift G of F over A, there exists a unique W(k)-algebra homomorphism $\theta_G : E_\chi \to A$ and *-isomorphism $a_G : \theta_G \tilde{F} \to G$ [8, 15], i.e. there is a natural bijection $\text{Hom}_{\text{Rings}}^{\text{cont}}(E_\chi, A) \cong \text{lifts}_\chi^*(A)$. To see that Aut χ is representable, recall that if the height of χ is finite, the group S_χ of automorphisms of F defined over a separable closure $\bar{k}$ of k is a compact totally disconnected topological group, with a continuous action of $Gal(\bar{k}/k)$: if $f = \Sigma f_i T^{i+1} \in \bar{k}[[T]] \in S_\chi$, then $f^\sigma = \Sigma f_i^\sigma T^{i+1} \in S_\chi$.

The functor

$$\underline{Art}_k \ni A \mapsto Aut\ \chi(A) = Aut_{grouplaws/k_A}(F \otimes_k k_A)$$

$$= Gal(\bar{k}/k_A)\text{-invariants of } S_\chi$$

$$= Top_{Gal(\bar{k}/k)}(Spec\ A, S_\chi) \in (Groups)$$

is pro-represented by the Hopf W(k)-algebra H_χ of continuous functions $f : S_\chi \to W(\bar{k})$ such that $f(\delta^\sigma) = (f(\delta))^\sigma$ for $\sigma \in Gal(\bar{k}/k)$, $\delta \in S_\chi$ [3, II §1 no. 2.12, §5, no. 1.7]. ||

2.9 Let C denote the category of graded $(U_*(pt), U_*U)$-comodules; let $I_{p,n} = (p, v_1, \ldots, v_{n-1})$ denote a canonical invariant prime ideal of $U_*(pt)$, and let $C(I_{p,n})$ denote the full subcategory of $M \in C$ such that $I_{p,n}M = 0$ and v_n-multiplication is an automorphism of M in C. The universal Lubin-Tate lift $\tilde{F}$ of F is classified by a ring of homomorphism $U \to U_\chi \to E_\chi$; if $F(X,Y) = X+Y$ modulo terms of degree p^n, then v_i maps to $c_i t_i$ modulo $(t_1, \ldots, t_{i-1})$, where $\overline{c_i} \neq 0 \in k$, [8, Prop. 1.1, part 2]; in particular $\underline{m}_\chi \supset I_{p,n}$, and if $k = F_p$, we may regard $C(I_{p,n})$ as a full subcategory of $(U_\chi, U_\chi U)$-comodules.

Consequently $M \in C(I_{p,n})$ if and only if $(p,t_1,\ldots,t_{n-1})\phi^*M = 0$, where $(p,t_1,\ldots,t_{n-1})$ is the maximal ideal of E_χ, i.e. if and only if ϕ^*M is a linear representation of Aut χ over the residue field F_p.

Corollary. The category $C(I_{p,n})$ is equivalent to the category of linear representations of Aut χ over F_p, if χ is as above.

2.10 Remark. When a Lie group G acts smoothly on a manifold X, a slice through x in X is an open disc D of maximal dimension containing x and transverse to its orbit; i.e. T_xD and $T_x(Gx)$ are orthogonal subspaces of T_xX. The stabilizer of x in G acts naturally on the germ of a slice through x. The Lubin-Tate theorem asserts the existence of a formal slice for the action of the group of coordinate-changes on the moduli space for formal groups at a point of finite height. ‖

[After this paper was finished, Peter Landweber proved that the completion of U at a genus of finite height is indeed flat, and Haynes Miller showed that the hypothesis of flatness in 1.6 is superfluous.]

SUNY, Stony Brook 11790, 24 September 1976

References

1. J. F. Adams, Stable Homotopy and Generalized Homology, U. of Chicago Press.

2. M. F. Atiyah, I. MacDonald, Commutative Algebra, Addison-Wesley.

3. M. Demazure, P. Gabriel, Groupes Algebriques, North-Holland.

4. A. Frohlich, Formal Groups, Springer Lecture Notes #74.

5. F. Hirzebruch, Topological Methods in Algebraic Geometry, Springer.

6. P. S. Landweber, BP_*BP and Typical Formal Groups, Osaka J. Math. 12 (1975) 357-363.

7. ________, Associated prime ideals and Hopf algebras, J. Pure and Appl. Algebra 3 (1973) 43-58.

8. J. Lubin, J. Tate, Formal moduli for one-parameter formal Lie groups, Bull. Soc. Math., France, 94 (1966) 49-60.

9. H. Miller, Some algebraic aspects of the Adams-Novikov spectral sequence, dissertation, Princeton University, 1974.

10. H. Miller, W. S. Wilson, On Novikov's Ext^1 modulo an invariant prime ideal, Topology, 15 (1976) 131-141.

11. D. G. Quillen, Elementary proofs of some results of cobordism theory, Adv. Math. 7 (1971) 29-56.

12. N. Saavedra Rivano, Categories Tannakiennes, Springer Lecture Notes 265.

13. M. Schlessinger, Functions of Artin Rings, Trans. A.M.S. 130 (1968) 208-222.

14. L. Smith, On the finite generation of $\Omega_*^U(X)$, J. Math. Mech. 18 (1969) 1017-1024.

15. H. Umemura, Nagoya Math. J. 42 (1971) 1-7.

THE E_2-TERM OF NOVIKOV'S SPECTRAL SEQUENCE

Jack Morava

Introduction.

S. P. Novikov [6] proved the following

THEOREM. There is a spectral sequence with

$$E_2^{s,t} = \mathrm{Ext}_C^s(U_*(pt), U_*(S^t)) \text{ converging to}$$

$$E_\infty^{s,t} = \mathrm{Gr}^s \prod_{t-s}(pt).$$

Here $\prod_*$ denotes the stable homotopy functor, U_* is the complex bordism functor, and C is the category of modules over the ring of stable cobordism operations.

This spectral sequence seems to give a good hold on $\prod_*(pt)$; for example, the first line $E_2^{1,*}$ is the image of the J-homomorphism (at odd primes); see [6:§10]. For direct computations, see [13].

In this paper we show (1.4) that the E_2-term of Novikòv's spectral sequence can be computed from certain groups $\mathcal{E}^i(n)$, which estimate the $2(p^n-1)$-periodic phenomena in $E_2^{i,*}$.

The main result (1.7) describes the $\mathcal{E}^*(n)$ qualitatively. Their structure is related to questions of arithmetic; see Venkov's "noncommutative Dirichlet unit theorem" [12].

[According to Larry Smith [9], a finite complex X is a $V(n)$-space if $U_*(X) = U_*(pt)/I_{p,n}$; see §1 for notation. The group $\mathcal{E}^*(n)$ would be the E_2-term of the Novikov spectral sequence abutting to $\prod_*(V(n-1))$, i.e.,

$$\mathcal{E}^i(n) = \mathrm{Ext}_C^i(U_*(pt); U_*(pt)/I_{p,n-1}).$$

Toda [10; see also 11] conjectures that a V(n)-space exists iff $0 \leq n \leq \frac{1}{2}(p-1)$, but for the (purely algebraic) study of the E_2-term, the actual existence of a V(n)-space is irrelevant.]

By convention, $\mathcal{E}^*(0) = E_2^{*,*}$; $\mathcal{E}^*(1)$ is the E_2-term of the Novikov spectral sequence for the homotopy of a mod p Moore space. There is a short exact sequence

$$0 \to E_2^{i,*} \otimes \mathbb{F}_p \to \mathcal{E}^i(1) \to \mathrm{Tor}(E_2^{i-1,*}, \mathbb{F}_p) \to 0 .$$

If $n \geq 1$, $\mathcal{E}^i(n)$ is a module over the polynomial ring $A_n = \mathbb{F}_p[\phi_n]$, where ϕ_n is a "periodicity operator" of degree $(0, 2(p^n-1))$. We show

(1.5): there is a short exact sequence

$$0 \to \mathcal{E}^i(n) \otimes_{A_n} \mathbb{F}_p \to \mathcal{E}^i(n+1) \to \mathrm{Tor}_{A_n}(\mathcal{E}^{i-1}(n), \mathbb{F}_p) \to 0;$$

(2.4, 2.8): there is a natural isomorphism

$$\mathcal{E}^i(n)[\phi_n^{-1}] \cong H^i(\mathbb{S}(D_n); \mathbb{F}_p) \otimes_{\mathbb{F}_p} \mathbb{F}_p[\phi_n, \phi_n^{-1}] .$$

Here D_n is a certain p-adic division algebra [2.1], $\mathbb{S}(D_n)$ its (profinite) group of strict units. In [5, §8] the cohomology of such unit groups was examined; if, e.g., p-1 does not divide n, $\mathcal{E}^*(n)[\phi_n^{-1}]$ is Poincaré duality algebra of formal dimension n^2. [In general see 1.7 or 2.9.]

<u>Example</u>: If n = 1, then $D_n = \mathbb{Q}_p$, $\mathbb{S}(D_n) \cong \hat{\mathbb{Z}}_p$, and $H^*(\hat{\mathbb{Z}}_p; \mathbb{F}_p) = E(e)$ is an exterior algebra on one generator. We have $\mathcal{E}(1)^*[\phi_1^{-1}] = E(e) \otimes \mathbb{F}_p[\phi_1, \phi_1^{-1}]$.

If p is <u>odd</u>, and V(0) denotes a mod p Moore space, there exists [cf. 9] a stable map $\alpha : S^{2(p-1)}V(0) \to V(0)$ which induces the endomorphism ϕ_1 of the Novikov spectral sequence for V(0). On the E_∞ term, this map corresponds to multiplication (in the ring structure of $\prod_*(V(0)) = \prod_*(pt;\mathbb{F}_p)$) with a class $\alpha \in \prod_{2(p-1)}(pt;\mathbb{F}_p)$. It can be shown that $e\cdot\phi_1$ corresponds to a class $\delta \in \prod_{2p-3}(pt;\mathbb{F}_p)$ of mod p Hopf invariant 1.

Corollary: The ring $\prod_*(pt;\mathbb{F}_p)$ contains $E(\delta) \otimes P(\alpha)$ (the product of an exterior algebra on δ and a polynomial algebra on α). The conjecture that $E(\delta) \otimes P(\alpha) \subset \prod^*(pt;\mathbb{F}_p)$ is an isomorphism modulo α-torsion. has since been proved by Michael Barratt and Haynes Miller [who observes that this is an odd-primary analogue of theorem 7 in Mahowald's 1970 AMS Bulletin announcement [p.1311].]

N.B. J. Cohen [3: 5.7, example] shows that the E_∞ term of the (graded) localization of a spectral sequence can properly contain the localization of the E_∞ term.

The next example ($(\mathcal{E}^*(2))$) requires the study of the p-adic quaternion algebra; see §3.

Remarks on the Proof:

In §1 we discuss homological algebra in the category C and in certain related categories C(n), $\mathcal{C}$(n). The main result (1.9) is a theorem on change of rings and localization. In §2 we discuss division algebras, and prove the results above; our main technical tool is a study [5] of the category $\mathcal{C}$(n) suggested by the theory of formal groups. The finiteness theorem (2.9.1) used to prove 1.7 comes from [5; §8]; it is not very precise if n is divisible by p-1, but there is hope of improving it [5: conjecture 8.7]. Thus our results improve as $p \to \infty$, and are weak when $p = 2$.

Acknowledgements:

This paper grew out of many discussions with W. Steve Wilson and Haynes Miller. Besides thanking them for their interest, I must thank them for their patience and ideas. I wish also to thank Professor Milnor for his help, and the Institute for Advanced Study for its aid during the writing of this paper.

1.1. Rappels [5,7].

We write $X \mapsto U_*(X)$ for the homotopy functor which assigns to the finite complex X, the group $U_*(X)$ of bordism classes of complex-oriented manifolds mapped to X, graded by (real) dimension.

The group $U = U_*(pt)$ is in fact a graded ring, and $U_*(X)$ is a $U_*(pt)$-module, which can be shown to be finitely-presented. According to Thom, $U_*(pt) \otimes \mathbb{Q}$ is the graded polynomial $\mathbb{Q}$-algebra generated by the projective spaces $\mathbb{C}P(n)$ of dimension 2n. According to Milnor, Novikov or Quillen, $U_*(pt)$ is a polynomial ring over the integers, with a generator in each even dimension.

We write $\mathbb{Z}[\underline{t}] = \mathbb{Z}[t_0^{-1}, t_0, t_1, \ldots]$ for the polynomial ring over $\mathbb{Z}$ on generators t_i of dimension -2i, and abbreviate

$$U_*(X)[\underline{t}] = U_*(X) \otimes_{\mathbb{Z}} \mathbb{Z}[\underline{t}] = U_*(X)[t_0^{-1}, t_0, t_1, \ldots] .$$

There is a natural homomorphism

$$s_{\underline{t}} : U_*(X) \to U_*(X)[\underline{t}] \text{ , } s_{\underline{t}}(X) = \Sigma s^{\alpha} X \cdot t^{\alpha}$$

called the [giant] Landweber-Novikov operation [here $t^{\alpha} = t_0^{\alpha_0} \ldots t_r^{\alpha_r}$; α is a multi-index, with $\alpha_0 \in \mathbb{Z}$]. If X = pt, the relation

$$\sum_{n \geq 1} s_{\underline{t}}(\mathbb{C}P(n-1)) \frac{T^n}{n} = \sum_{n \geq 1} \mathbb{C}P(n-1) \frac{t(T)^n}{n}$$

in $U_*(pt)[\underline{t}][[T]]$ defines $s_{\underline{t}}$ on $U_*(pt) = U$ [where $t(T) = \sum_{j \geq 0} t_j T^{j+1}$ is the "generating function" of $\underline{t}$]. In general, we have relations:

i) If $u \in U$, $x \in U_*(M)$, then $s_{\underline{t}}(u \cdot x) = s_{\underline{t}}(u) \cdot s_{\underline{t}}(x)$ with $s_{\underline{t}}(u)$ as above.

ii) $s_{\underline{t}}(s_{\underline{t'}}x) = s_{\underline{t''}}x$, where $\underline{t''}$ is defined by the generating function

$$t''(T) = \sum_{i\geq 0} t''_i T^{i+1} = t(t'(T)) \in \mathbb{Z}[\underline{t}][\underline{t'}][[T]] .$$

We write C for the category of U-modules M_* endowed with $s_{\underline{t}} : M_* \to M_*[\underline{t}]$ satisfying i), ii) above.

The category C is equivalent to the category of (evenly) graded modules over $U_*(pt) \hat{\otimes} S^*$, where S^* is the ring of Landweber-Novikov operations [with $\mathbb{Z}$-basis s^{α}]; hence C is an abelian category.

1.2. Definition: If $M, N \in C$, we write $\mathrm{Ext}^i_C(M, N)$ for the usual group of extensions. [If M, N belong to the (thick) subcategory C_{fp} of finitely-presented U-modules in C, these Ext's can be computed in C_{fp}.] We write $\mathrm{Ext}^{i,2k}_C(M, N) = \mathrm{Ext}^i_C(M, S^{2k}N)$, where $S^{2k}N$ has (for $x \in N$, $S^{2k}(x) \in S^{2k}N$)

$$\text{U-module structure } u \cdot S^{2k}(x) = S^{2k}(u \cdot x)$$

$$s_{\underline{t}}\text{-structure defined by } s_{\underline{t}}[S^{2k}(x)] = S^{2k}[s_{\underline{t}}(x)] \cdot t_0^k .$$

1.3. Proposition: A finitely-generated prime ideal $I \subset U$ is invariant under the operations $s_{\underline{t}}$ (i.e., $s_{\underline{t}}(I) \subset I[\underline{t}]$) if and only if I is of the form $I = I_{p,n} = (p, v_1, \ldots, v_n)$ for some prime p and integer $n \geq 0$.

Proof: See [5:§3]. Here $v_n \in U_{2(p^n-1)}(pt)$ is a Milnor generator, which can be taken [by a theorem of Floyd] to be the p-dric hypersurface $z_0^p + \ldots + z_q^p = 0$ $(q = p^n)$ in $\mathbb{C}P(q)$. //

1.4. Proposition: For any prime p and integer $n \geq 1$, there is a long exact sequence

$$\cdots \to \mathrm{Ext}_C^{i,k-2(p^n-1)}(U;U/I_{p,n-1}) \xrightarrow{\phi_n} \mathrm{Ext}_C^{i,k}(U;U/I_{p,n-1}) \xrightarrow{\mu}$$

$$\to \mathrm{Ext}_C^{i,k}(U;U/I_{p,n}) \xrightarrow{\delta_n} \mathrm{Ext}_C^{i+1,k}(U;U/I_{p,n-1}) \xrightarrow{\phi_n}$$

$$\to \mathrm{Ext}_C^{i+1,k+2(p^n-1)}(U;U/I_{p,n-1}) \to \cdots .$$

Proof: From 1.3 it is clear that $s_{\underline{t}}(v_n) = v_n t_0^{p^n-1} \bmod I_{p,n-1}[\underline{t}]$; consequently, the short exact sequence of U-modules

$$0 \to S^{-2(p^n-1)} U/I_{p,n-1} \xrightarrow{\phi_n} U/I_{p,n-1} \to U/I_{p,n} \to 0$$

defined by $\phi_n(x) = v_n \cdot x$ is exact in C. Standard properties of the functor $\mathrm{Ext}_C^*(U;-)$ yield the result. //

1.5. Definition: In the following we will write $\mathcal{E}^i(n) = \mathrm{Ext}_C^{i,*}(U;U/I_{p,n-1}) \in$ (A_n-modules), where $A_n = \mathbb{F}_p[\phi_n]$ is the polynomial ring generated by ϕ_n. The exact sequence above can be reformulated as

$$0 \to \mathcal{E}^i(n) \otimes_{A_n} \mathbb{F}_p \to \mathcal{E}^i(n+1) \to \mathrm{Tor}_{A_n}(\mathcal{E}^{i-1}(n), \mathbb{F}_p) \to 0$$

where $\phi_n \mapsto 0 : A_n \to \mathbb{F}_p$ defines the A_n-module structure on $\mathbb{F}_p$.

1.6. Remark: We regard $\mathcal{E}^*(n+1)$ as a measure of the ϕ_n-torsion of $\mathcal{E}^*(n)$. Our main result estimates $\mathcal{E}^*(n)$ modulo its ϕ_n-torsion. Thus 1.5 and 1.7 together give a rough picture of the size of $\mathrm{Ext}_C^{*,*}(U;U)$.

1.7. Main Theorem: The localization $\mathcal{E}^*(n)[\phi_n^{-1}] = \mathcal{E}^*(n) \otimes_{A_n} A_n[\phi_n^{-1}]$ is

i) a Poincaré duality algebra of formal dimension n^2 over $A_n[\phi_n^{-1}]$, provided p-1 does not divide n; while

ii) if $n = m(p-1)$, there exists $\tau \in \mathcal{E}^*(n)[\phi_n^{-1}]$ which generates a polynomial subalgebra $A_n[\phi_n^{-1}][\tau]$ of $\mathcal{E}^*(n)[\phi_n^{-1}]$, such that $\mathcal{E}^*(n)[\phi_n^{-1}]$ is a finite $A_n[\phi_n^{-1}][\tau]$-module.

Proof Outline: By a sequence of change of rings lemmas [1, XVI, §5] we reduce the computation of $\mathrm{Ext}^*_C(U; U/I_{p,n-1})[\phi_n^{-1}]$ to that of $\mathrm{Ext}^*_{\mathcal{C}(n)}(U(n), U(n))$, where $U(n)$ lies in a category $\mathcal{C}(n)$ of cobordism comodules "of height n". In §2.9 we complete the proof using a structure theorem for $\mathcal{C}(n)$ proved elsewhere [5].

1.8. Definition: A module $M \in C$ has height $\geq n$ (at p) if multiplication by $v_i : M \to M$ is zero for $i = 0, \ldots, n-1$; we write $C(n)$ for the full subcategory of modules of height $\geq n$; evidently $C(n)$ is equivalent to the category of modules over $(U/I_{p,n-1}) \hat{\otimes} S^*$. A module $M \in C$ has height (exactly) n if $M \in C(n)$ and, moreover, multiplication by $v_n : M \xrightarrow{\sim} M$ is an isomorphism. We write $\mathcal{C}(n) \subset C(n) \subset C$ for the full subcategory of such modules, which is evidently equivalent to the category of modules over $U(n) \hat{\otimes} S^*$, where we abbreviate

$$U(n) = (U/I_{p,n-1})[v_n^{-1}] .$$

In the remainder of §1 we prove

1.9. Proposition: For any $M \in C(n)$ finitely-presented over U, there is a natural isomorphism

$$\mathrm{Ext}^*_{\mathcal{C}(n)}(U(n), M[v_n^{-1}]) \xrightarrow{\sim} \mathrm{Ext}^*_{\mathcal{C}}(U;M)[\phi_n^{-1}].$$

Proof: We break the proof into three lemmas:

1.9.1. Lemma: If $M \in C(n)$, the composition

$$\mathrm{Ext}^*_{C(n)}(U/I_{p,n-1};M) \to \mathrm{Ext}^*_C(U/I_{p,n-1};M) \to \mathrm{Ext}^*_C(U;M)$$

is an isomorphism.

Proof: An extension in $C(n)$ is an extension in C; this defines the first homomorphism of the composition, and the second is induced by $U \to U/I_{p,n-1}$. According to [1, XVI, §5, case 3] the natural isomorphism $[N \in C,\ M \in C(n)]$

$$\mathrm{Hom}_{C(n)}(U/I_{p,n-1} \otimes_U N;M) \xrightarrow{\sim} \mathrm{Hom}_C(N;M)$$

yields a composite-functor spectral sequence

$$\mathrm{Ext}^*_{C(n)}(\mathrm{Tor}^U_*(U/I_{p,n-1};N);M) \Rightarrow \mathrm{Ext}^*_C(N;M).$$

The lemma is the special case $N = U$. //

1.9.2. Lemma: If $M, N \in C(n)_{fp}$ [cf. 1.2], the homomorphism $\mathrm{Ext}^*_{C(n)}(N;M)[\phi_n^{-1}] \xrightarrow{\sim} \mathrm{Ext}^*_{C(n)}(N, M[v_n^{-1}])$ is an isomorphism.

Proof: Here ϕ_n is induced by multiplication $v_n : M \to M$. The universal mapping property of a localization defines a homomorphism $\mathrm{Hom}_{C(n)}(N;M)[\phi_n^{-1}] \to \mathrm{Hom}_{C(n)}(N;M[v_n^{-1}])$. If N is finitely-generated over U, this map is surjective; if M is finitely presented over U, the map is injective. Since the Ext group can be computed by a resolution in $C(n)_{fp}$, the lemma follows.

1.9.3. Lemma: If $N \in C(n)$, $M \in \mathcal{C}(n)$, the homomorphism

$$\mathrm{Ext}^*_{\mathcal{C}(n)}(N[v_n^{-1}];M) \xrightarrow{\sim} \mathrm{Ext}^*_{C(n)}(N, M)$$

is an isomorphism.

Proof: The universal mapping property of a localization implies an isomorphism $\mathrm{Hom}_{\mathcal{C}(n)}(N \otimes_U U(n);M) \xrightarrow{\sim} \mathrm{Hom}_{C(n)}(N, M)$, which gives rise to a spectral sequence $\mathrm{Ext}^*_{\mathcal{C}(n)}(\mathrm{Tor}_*^{U/I_{p,n-1}}(N;U(n));M) \Rightarrow \mathrm{Ext}^*_{C(n)}(N;M)$. Since $U(n)$ is flat over $U/I_{p,n-1}$, the lemma follows. //

The isomorphism of 1.9 is now defined by

$$\mathrm{Ext}^*_C(U, M)[\phi_n^{-1}] \xleftarrow{\sim} \mathrm{Ext}^*_{C(n)}(U/I_{p,n-1};M)[\phi_n^{-1}] \xrightarrow{\sim}$$

$$\ldots \xrightarrow{\sim} \mathrm{Ext}^*_{C(n)}(U/I_{p,n-1};M[v_n^{-1}]) \xleftarrow{\sim} \mathrm{Ext}^*_{\mathcal{C}(n)}(U(n);M[v_n^{-1}]).$$

§2. Local Arithmetic.

2.1. Proposition: Let D be a division algebra of finite rank $[D : \mathbb{Q}_p]$ over the p-adic field.

If center $D = \mathbb{Q}_p$, then $[D : \mathbb{Q}_p] = n^2$ for some n, and D has a $\mathbb{Q}_p$-basis $\omega^{p^s} F^t$, $0 \leq s, t \leq n-1$, where

ω is a primitive (p^n-1)-th root of 1 over $\mathbb{Q}_p$,

F satisfies $F^n = p^m$, $F\omega = \omega^p F$ (for some integer m with $(n, m) = 1$).

The class $\mathrm{inv}(D) = m/n \in \mathbb{Q}/\mathbb{Z}$ specifies D up to isomorphism.

Proof: cf., [4; II,§7,Theorem 49] and [2; (Serre) VI, §1, Appendix].

2.2. Definition: Let D^* denote the multiplicative group of nonzero elements of D. There is a canonical valuation

$$D^* \ni x \mapsto [D : \mathbb{Q}_p]^{-1} |\mathrm{Norm}\, x|_p = |x| \in \mathbb{Q}$$

normalized so $|p| = 1$. [Here $\mathrm{Norm} : D^* \to \mathbb{Q}_p^*$ is the determinant of left-multiplication, and $|\cdot|_p : \mathbb{Q}_p^* \to \mathbb{Z}$ is the standard p-adic valuation. Any unit $\delta \in D^*$ can be written uniquely in the form $\delta = \delta_0 \theta$, where $\theta = \omega^a F^b$ for some $a, b \in \mathbb{Z}$ and δ_0 is a strict unit, i.e., $\delta_0 = 1+x$ with $|x| > 0$.

2.3. Definition: Let $\bar{\mathbb{F}}_p = \bigcup_{n \geq 1} \mathbb{F}_{p^n}$ be the algebraic closure of $\mathbb{F}_p$; fix a primitive (p^n-1)-th root $\bar{\omega} \in \bar{\mathbb{F}}_p$ of unity.

If $k \in \mathbb{Z}$, we write $\bar{\mathbb{F}}_p^{\otimes k}$ for the one-dimensional $\bar{\mathbb{F}}_p$-module with generator γ_k; we define a D^*-action on $\bar{\mathbb{F}}_p^{\otimes k}$ by

$$\delta \in D^*, \ \alpha\gamma_k \in \bar{\mathbb{F}}_p^{\otimes k} \mapsto \omega^{-ak} \alpha^{p^b} \gamma_k \in \bar{\mathbb{F}}_p^{\otimes k}$$

where $\delta = \delta_0 \omega^a F^b$, $\alpha \in \bar{\mathbb{F}}_p$. [A strict unit $\delta_0 \in D^*$ thus acts trivially on $\bar{\mathbb{F}}_p^{\otimes k}$.]

2.4. Proposition: There is a natural isomorphism

$$\mathrm{Ext}_C^{i,2k}(U;U/I_{p,n-1})[\phi_n^{-1}] \xrightarrow{\sim} H^i(D_n^*;\bar{\mathbb{F}}_p^{\otimes k})$$

[where D_n is the p-adic division algebra of rank n^2 with $\mathrm{inv}(D) = 1/n$, and the cohomology is defined by continuous cochains in the natural topology of D_n^*].

Proof: We require the structure theorem for $\mathcal{C}(n)$:

2.4.1. Definition: There is a unique ring homomorphism $x_n : U \to \mathbb{Z}_{(p)}$ defined by $\sum_{m \geq 1} x_n(\mathbb{C}P(m-1))m^{-s} = (1-p^{n(1-s)-1})^{-1}$ [5; §5]. If $M \in C$, we write $M(x_n) = M \otimes_U \mathbb{Z}_{(p)}$ with $\mathbb{Z}_{(p)}$ regarded as a U-module via x_n. If $pM = 0$, we write $M(\bar{x}_n) = M(x_n) \otimes_{\mathbb{F}_p} \bar{\mathbb{F}}_p$.

2.4.2. Definition: If $\delta \in U(D_n) = \{\delta \in D_n^* \mid |\delta| = 0\}$ (i.e., $\delta = \delta_0 \omega^a$) then there exists [5: 5.4] a power series $[\delta](T) = \sum_{i \geq 0} \delta_i T^{i+1}$ with $\delta_i \in \bar{\mathbb{F}}_p$ such that $\delta \mapsto s_{\underline{\delta}} = \sum s^\alpha \delta^\alpha \in S^* \otimes \bar{\mathbb{F}}_p$ (where $\delta^\alpha = \delta_0^{\alpha_0} \ldots \delta_n^{\alpha_n} \in \bar{\mathbb{F}}_p$) defines a representation of the compact group $U(D_n)$ on $M(\bar{x}_n)$ with the following properties [5:5.6, 5.8]:

i) The action of $U(D_n)$ is continuous (in the discrete topology on $M(\bar{x}_n)$).

ii) If $\sigma : M(\bar{x}_n) = M(x_n) \otimes_{\mathbb{F}_p} \bar{\mathbb{F}}_p \to M(x_n) \otimes_{\mathbb{F}_p} \bar{\mathbb{F}}_p = M(\bar{x}_n)$ is defined by $\sigma(m \otimes a) = m \otimes a^p$, then $(F\delta F^{-1})(\sigma v) = \sigma(\delta v)$ $[v \in M(\bar{x}_n), \ \delta \in U(D_n)]$.

2.4.3. Example: If $M = U_*(S^{2k};\mathbb{F}_p)$ then $M(\bar{x}_n)$ is a 1-dimensional $\bar{\mathbb{F}}_p$-module, and $\delta = \delta_0\omega^a \in \{\delta \in D^* \mid |\delta| = 0\}$ acts by multiplication with $\bar{\omega}^{ak}$.

Proof: If $\gamma_k \in U_{2k}(S^{2k};\mathbb{F}_p)$ is the fundamental class, then $s_{\underline{t}}(\gamma_k) = t_0^k\gamma_k$ [5: 0.2]. Use [5: 5.4, 5.5]: $\omega \in D^*$ corresponds to $[\omega](T) = \bar{\omega}T$.

2.4.4. Definition: An algebraic representation of $U(D_n)$ is an $\mathbb{F}_p$-module V, together with an action of $U(D_n)$ on $\bar{V} = V \otimes_{\mathbb{F}_p} \bar{\mathbb{F}}_p$ satisfying 2.4.2, i), ii).

We can finally state

2.5. Structure Theorem: The functor $M \mapsto M(x_n) : \mathcal{C}(n) \to$ (Algebraic $U(D_n)$-Representations) is an equivalence of categories.

Proof: Cf., [5: §7] for an explicit inverse functor. //

2.6. Corollary: If $M \in C(n)_{fp}$ then there is a natural isomorphism

$$\mathrm{Ext}^*_C(U;M)[\phi_n^{-1}] \xrightarrow{\sim} \mathrm{Ext}^*_{\mathrm{Alg.}\, U(D_n)\text{-Rep's}}(\mathbb{F}_p, M(x_n)).$$

Proof: By 1.9, it suffices to show that $\mathrm{Ext}^*_{\mathcal{C}(n)}(U(n), M[v_n^{-1}])$ has the proper form. But by 2.4.3, $U(n) \ni \mathcal{C}(n)$ corresponds to the trivial $U(D_n)$-representation on $\mathbb{F}_p$, and $(M[v_n^{-1}])(x_n) = M(x_n)$ since $x_n(v_n) = 1$. //

Completion of the Proof of Proposition 2.4:

2.7. Definition: Let V be an algebraic $U(D_n)$-representation. Define an action of D_n^* on $\bar{V} = V \otimes_{\mathbb{F}_p} \bar{\mathbb{F}}_p$ by $(\delta_1 F^b)(v) = (\delta_1)(\sigma^b(v))$ where $\delta_1 = \delta_0\omega^a \in U(D_n)$.

[Evidently, $(\delta_1' F^c)((\delta_1 F^b)(v)) = \delta_1' \sigma^c(\delta_1 \sigma^b v) = \delta_1'(F^c \delta_1 F^{-c})\sigma^{b+c}(v)$ (by 2.4.2, ii) $= (\delta_1'(F^c \delta_1 F^{-c}))F^{b+c}(v)$, so the action makes sense.] //

Proposition 2.4 now follows immediately from standard facts about derived composite functors: for $\mathrm{Ext}^*_{\mathrm{Alg.}\,U(D_n)\text{-Rep's}}(\mathbb{F}_p, V)$ is the right-derived functor of $V \mapsto V^{U(D_n)}$, while $H^*(D_n^*; \bar{V})$ is the derived functor of $\bar{V} \mapsto \bar{V}^{D_n^*}$. But $V^{U(D_n)} = \bar{V}^{D_n^*}$, and the forgetful functor $V \mapsto \bar{V}$ is exact. Hence $\mathrm{Ext}^*_{\mathrm{Alg.}\,U(D_n)\text{-Rep's}}(\mathbb{F}_p; V) \xrightarrow{\sim} H^*(D_n^*; \bar{V})$.

To complete the proof, we take $V = \mathbb{F}_p^{\otimes k}$; by 2.4.3, this corresponds to $U_*(S^{2k}; \mathbb{F}_p)$. //

2.7. Definition: If D is a division algebra over $\mathbb{Q}_p$, we write $\mathbb{S}(D)$ for its group of strict units, i.e., $\delta \in \mathbb{S}(D)$ iff $\delta = 1+x \in D^*$ with $|x| > 0$. By Proposition 2.1, there is a semidirect product decomposition $D^* = \mathbb{S}(D) \cdot \theta$ where $\theta = \mu_{p^n-1} \cdot \mathbb{Z}$ has generators $[\omega^a F^b]$, $a, b \in \mathbb{Z}$, acting on $\mathbb{S}(D)$ by $(\omega^a F^b, \delta) \mapsto \omega^a F^b \delta F^{-b} \omega^{-a}$.

2.8. Proposition: The Hochschild-Serre spectral sequence for D^* degenerates to an isomorphism $\bigoplus_{k \in \mathbb{Z}} H^i(D^*; \bar{\mathbb{F}}_p^{\otimes k}) \cong H^i(\mathbb{S}(D); \mathbb{F}_p)[\phi_n, \phi_n^{-1}]$.

Proof: It is convenient to break θ into μ_{p^n-1} and $\mathbb{Z}$. Since μ_{p^n-1} has order prime to p, the spectral sequence

$$H^*(\mu_{p^n-1}; H^*(\mathbb{S}(D); \bar{\mathbb{F}}_p^{\otimes k})) \Rightarrow H^*(U(D); \bar{\mathbb{F}}_p^{\otimes k})$$

degenerates to an isomorphism

$$H^i(U(D); \bar{\mathbb{F}}_p^{\otimes k}) \simeq H^{i,2k}(\mathbb{S}(D); \bar{\mathbb{F}}_p)$$

with the right-hand side defined as follows: If μ_{p^n-1} acts on

$H^i(\mathbb{S}(D);\bar{\mathbb{F}}_p) = H^i(\mathbb{S}(D);\mathbb{F}_p) \otimes_{\mathbb{F}_p} \bar{\mathbb{F}}_p$ by $[\omega]c(g_0,\ldots,g_i) = c(\omega^{-1}g_0\omega,\ldots,\omega^{-1}g_i\omega)$, then $H^{i,2k}(\mathbb{S}(D);\bar{\mathbb{F}}_p) = \{c \in H^i(\mathbb{S}(D);\bar{\mathbb{F}}_p) \mid [\omega]c = \bar{\omega}^k c\}$ is the $\bar{\omega}^k$-eigenspace of $[\omega]$. [Note that $H^{i,2k}(\mathbb{S}(D);\bar{\mathbb{F}}_p)$ is naturally a bigraded algebra, with $k \in \mathbb{Z}/(p^n-1)\mathbb{Z}$, i.e., the second grading is cyclic. We introduce a formal indeterminate ϕ_n and identify

$$H^i(D_n^*;\bar{\mathbb{F}}_p^{\otimes k}) \cong H^{i,2k}(\mathbb{S}(D);\mathbb{F}_p) \cdot \phi_n^k.$$

2.8.1. Lemma: $[F] \in \mu_{p^n-1} \cdot \mathbb{Z}$ maps $H^{i,2k}(\mathbb{S}(D);\bar{\mathbb{F}}_p)$ to $H^{i,2k}(\mathbb{S}(D);\bar{\mathbb{F}}_p)$, and is a Frobenius linear ring homomorphism.

Proof: Suppose $c \in H^i(U(D);\bar{\mathbb{F}}_p^{\otimes k})$ is identified with $\Sigma c_j \otimes a_j\phi_n^k \in H^{i,2k}(\mathbb{S}(D);\mathbb{F}_p) \otimes \bar{\mathbb{F}}_p[\phi_n,\phi_n^{-1}]$. Evidently $[F]c = \Sigma c_j^F \otimes a_j^p\phi_n^k$ with $c^F(g_0,\ldots,g_i) = c(F^{-1}g_0F,\ldots,F^{-1}g_iF)$ [cf., 2.3]; thus $[F](\lambda c) = \lambda^p[F](c)$ if $\lambda \in \bar{\mathbb{F}}_p$, i.e., $[F]$ is Frobenius-linear.

Now suppose $[\omega]c = \bar{\omega}^k c$. Then $[\omega^p][F]c = [F][\omega]c = [F](\bar{\omega}^k c) = \bar{\omega}^{pk}[F]c$; that is, $[F]c \in H^{i,2k}(\mathbb{S}(D);\bar{\mathbb{F}}_p)$ also. //

2.8.2. Definition: $H^{i,k}(\mathbb{S}(D);\mathbb{F}_p)$ is the $\mathbb{F}_p$-vector space of $[F]$-invariant elements in $H^{i,k}(\mathbb{S}(D);\bar{\mathbb{F}}_p)$. Indeed, $H^{i,k}(\mathbb{S}(D);\bar{\mathbb{F}}_p) = H^{i,k}(\mathbb{S}(D);\mathbb{F}_p) \otimes_{\mathbb{F}_p} \bar{\mathbb{F}}_p$ as $[F]$-modules.

Completion of Proof of 2.8: The spectral sequence for $U(D) \cdot \mathbb{Z}$ collapses to $H^*(\mathbb{Z};H^*(U(D);\bar{\mathbb{F}}_p^{\otimes k})) = H^*(\mathbb{Z};H^{*,2k}(\mathbb{S}(D);\mathbb{F}_p) \otimes \bar{\mathbb{F}}_p) = H^{*,2k}(\mathbb{S}(D);\mathbb{F}_p) \cong H^*(D^*;\bar{\mathbb{F}}_p^{\otimes k})$.

From 2.8.2 it is clear that

$$\bigoplus_{k \in \mathbb{Z}/(p^n-1)\mathbb{Z}} H^*(D^*;\bar{\mathbb{F}}_p^{\otimes k}) \cong H^*(\mathbb{S}(D);\bar{\mathbb{F}}_p)^{[F]\text{-inv.}}.$$

It now suffices to show that $H^*(\mathbb{S}(D);\mathbb{F}_p)$ is isomorphic to $H^*(\mathbb{S}(D);\bar{\mathbb{F}}_p)^{[F]\text{-inv.}}$. Since $[F^n] = \sigma^n$ on $H^*(\mathbb{S}(D);\mathbb{F}_p)$, we may assume that

$c \in H^*(\mathbb{S}(D);\bar{\mathbb{F}}_p)^{[F]\text{-inv.}}$ is of the form $\sum_{j=0}^{n} c_j \otimes \bar{\omega}^{p^j}$, since the $\bar{\omega}^{p^j}$ form an $\mathbb{F}_p$-basis of $\mathbb{F}_{p^n}$. Such a class is [F]-invariant iff $c_j^F = c_{j+1}$; thus

$$H^*(\mathbb{S}(D);\mathbb{F}_p) \ni c \mapsto \sum c^{F^i} \otimes \bar{\omega}^{p^i} \in H^*(\mathbb{S}(D);\bar{\mathbb{F}}_p)^{[F]\text{-inv.}}$$

defines an isomorphism, and 2.8 is proved. //

2.9. <u>Completion of Proof of Theorem 1.7</u>: In view of 1.9, 2,4 and 2.8, we have an isomorphism $\mathcal{E}^{i,2k}(n) \xrightarrow{\sim} H^{i,2k}(\mathbb{S}(D_n);\mathbb{F}_p)\phi_n^k$ with the right-hand bigrading defined in 2.8.2. We recall from [5: 8.3]

2.9.1. <u>Finiteness Theorem</u>: The ring $H^*(\mathbb{S}(D);\mathbb{F}_p)$ is

i) a Poincaré duality algebra of formal dimension n^2 over $\mathbb{F}_p$ if p-1 does not divide n; while

ii) if $n = m(p-1)$, there exists $\tau \in H^*(\mathbb{S}(D);\mathbb{F}_p)$ which generates a polynomial subalgebra $\mathbb{F}_p[\tau]$ such that $H^*(\mathbb{S}(D);\mathbb{F}_p)$ is a <u>finite</u> $\mathbb{F}_p[\tau]$-module.

Theorem 1.7 is an immediate consequence. //

We note that 2.9.1 can be made slightly more precise:

2.9.2. <u>Proposition</u>: If $(p-1)\nmid n$, then $H^{*,*}(\mathbb{S}(D);\mathbb{F}_p)$ is a Poincaré algebra of formal dimension $(n^2,0)$; while if $n = m(p-1)$, the bidegree of τ has the form $(2i, 2k(p^m-1))$.

<u>Proof</u>: To prove the first assertion, we show that $\theta = \mu_{p^n-1} \cdot \mathbb{Z}$ acts trivially on the top-dimensional class in $H^*(\mathbb{S}(D);\mathbb{F}_p)$. Recall [8: I, Prop. 30, step 4] if G is a p-Poincaré duality group of formal dimension d, and G_0 is an open subgroup,

then res : $H^d(G;\mathbb{F}_p) \to H^d(G_0;\mathbb{F}_p)$ is an isomorphism. We fix $r \gg 0$, and apply this to $G = \mathbb{S}(D)$, $G_0 = \mathbb{S}_r(D) = \{1+p^r x \in D^* \mid |x| > 0\}$. By a theorem of Lazard, $H^*(\mathbb{S}_r(D);\mathbb{F}_p)$ is an exterior algebra on $H^1(\mathbb{S}_r(D);\mathbb{F}_p) = \mathrm{Hom}(\mathbb{S}_r(D);\mathbb{F}_p)$. We define $\psi_{i,j}(1+p^r x) = \bar{x}_{i,j}$ where $x = \Sigma x_{s,t}\omega^{p^s}F^t$ with $x_{s,t} \in \hat{\mathbb{Z}}_p$; thus the fundamental class Ψ of $\mathbb{S}(D)$ restricts to $\prod \psi_{i,j}$ $(0 \leq i, j \leq n-1)$. Since $[F]\psi_{i,j} = \psi_{i+1,j}$, we have $[F]\Psi = \Psi$ (the associated permutation is of deg $n^2(n-1) \equiv 0 \bmod 2$). To evaluate $[\omega]\Psi$, we regard $\mathbb{S}_r(D)^{ab}$ as the $\mathbb{F}_{p^n}$-vector space with basis $\{F^i\}$, $0 \leq i \leq n-1$; since $[\omega]\{F^i\} = \bar{\omega}^{p^i-1}\{F^i\}$, we have $\mathrm{determinant}_{\mathbb{F}_{p^n}} [\omega] \mid \mathbb{S}_r(D)^{ab} = \frac{N\bar{\omega}}{\bar{\omega}^n}$, where $N : \mathbb{F}^*_{p^n} \ni x \mapsto x^\nu \in \mathbb{F}^*_p$, $\nu = 1+p+\ldots+p^{n-1}$ is the norm. Consequently, $[\omega]\Psi = \det_{\mathbb{F}_p}[\omega] \cdot \Psi = N(\frac{N\bar{\omega}}{\bar{\omega}^n}) \cdot \Psi = \Psi$. The first assertion is now immediate from the definition of Poincaré duality algebra.

To prove the second assertion, recall [5: 8.6] that (for appropriate b) the field $K/\mathbb{Q}_p$ generated by $\omega^b F^m \in D_n$ contains a primitive p^{th} root of unity, and that the restriction homomorphism $H^*(\mathbb{S}(D_n);\mathbb{F}_p) \to H^*(\mu(K);\mathbb{F}_p)$ maps $\tau \in H^*(\mathbb{S}(D_n);\mathbb{F}_p)$ to a non-nilpotent element of $H^*(\mu(K);\mathbb{F}_p)$. [Here $\mu(K)$ is the group of roots of unity in K.] Since K is left fixed by conjugation with ω^r, $r = \frac{p^n-1}{p^m-1}$, it follows that $\mathrm{res}[\omega^r](\tau) = \mathrm{res}\,\tau$; in particular, if $[\omega]\tau = \bar{\omega}^{\ell}\tau$, then $\ell = 2k(p^m-1)$. //

§3. Example.

In this §, we specialize 1.7, 2.4 and 2.8 to the case $n = 2$; the isomorphism

$$\mathcal{E}^i(2)[\phi_2^{-1}] \simeq H^i(\mathbb{S}(D_2);\mathbb{F}_p) \otimes \mathbb{F}_p[\phi_2, \phi_2^{-1}]$$

leads to the study of the division algebra D_2 with $[D_2 : \mathbb{Q}_p] = 4$, $\mathrm{inv}(D_2) = \frac{1}{2}$, i.e., the p-adic quaternion algebra.

3.1. Definition: We write $\mathbb{Q}_p(\omega)$ for the field generated over $\mathbb{Q}_p$ by a primitive (p^2-1)th root ω of unity; we have $[\mathbb{Q}_p(\omega) : \mathbb{Q}_p] = 2$, with Galois group generated by the involution σ, $\sigma(\omega) = \omega^p$. If α_0 is an integer of $\mathbb{Q}_p(\omega)$, we write $\bar{\alpha}_0 \in \mathbb{F}_{p^2}$ for its residue class, modulo p.

If $\alpha \in D_2$, then by 2.1 we can write $\alpha = \alpha_0 + \alpha_1 F$ for unique $\alpha_0, \alpha_1 \in \mathbb{Q}_p(\omega)$; the conjugate α^c of α will be defined by $\alpha^c = \alpha_0^\sigma - \alpha_1 F$.
The reduced norm $N_{red}(\alpha) = \alpha \cdot \alpha^c = |\alpha_0|^2 - p|\alpha_1|^2$ (where $|\alpha_0|^2 = \alpha_0 \cdot \alpha_0^\sigma$) lies in $\mathbb{Q}_p$, and explicit computation shows that

$$N_{red} : D_2^* \to \mathbb{Q}_p^*$$

is a (surjective) homomorphism. The group $\mathbb{S}L(D_2)$ is defined by the exact sequence

$$1 \to \mathbb{S}L(D_2) \to \mathbb{S}(D_2) \xrightarrow{N_{red}} (1+p\hat{\mathbb{Z}}_p)^\times \to 1.$$

If p is odd, this sequence is split by $(1+p\hat{\mathbb{Z}}_p)^\times \ni (1+pa) \mapsto (1+pa)^{\frac{1}{2}} \in \mathbb{S}(D_2)$, and we have

$$\mathbb{S}(D_2) \simeq \hat{\mathbb{Z}}_p \times \mathbb{S}L(D_2).$$

In particular, if $p > 3$ then $\mathbb{S}L(D_2)$ is a p-Poincaré group of cohomological dimension 3; it is the analogue of the group S^3 of unit real quaternions.

3.2. Proposition: $$H^1(\mathbb{S}L(D_2);\mathbb{F}_p) \cong \mathbb{F}_{p^2}.$$

Proof: An element of $H^1(\mathbb{S}L(D_2);\mathbb{F}_p)$ is a homomorphism $\rho : \mathbb{S}L(D_2) \to \mathbb{F}_p$. We construct $\rho_0 : \mathbb{S}L(D_2) \to \mathbb{F}_{p^2}$ and show that it is universal among homomorphisms to $\mathbb{F}_p$-vector spaces.

If $\alpha \in \mathbb{S}L(D_2)$, then $|\alpha_0|^2 = 1+p|\alpha_1|^2$; it follows that $\alpha_0 \equiv 1 \bmod p$. Hence $\rho_0(\alpha) = \bar{\alpha}_1 \in \mathbb{F}_{p^2}$ is a homomorphism, for $\rho_0(\alpha\beta) = \alpha_1\beta_0^\sigma + \alpha_0\beta_1 \equiv \bar{\alpha}_1+\bar{\beta}_1 \bmod p$.

If ρ_0 does not possess the desired universal property, then there exists a nontrivial homomorphism $\rho_1 : \mathbb{S}L(D_2) \to \mathbb{F}_p$ which vanishes on $\{\alpha_0+\alpha_1F \in \mathbb{S}L(D_2) | \alpha_1 \equiv 0 \bmod p\} = \ker\rho_0$. Any homomorphism such as ρ_1 vanishes on commutators $\alpha\beta\alpha^{-1}\beta^{-1}$; hence

$$\rho_1(1+[\alpha,\beta]) = \rho_1(\alpha)+\rho_1(\beta)\ ,\ [\alpha,\beta] = \alpha\beta-\beta\alpha.$$

But by direct computation, $[\alpha,\beta] = [\alpha,\beta]_0 + [\alpha,\beta]_1F$ with $[\alpha,\beta]_1 = (\alpha_0-\alpha_0^\sigma)\beta_1-(\beta_0-\beta_0^\sigma)\alpha_1 \equiv 0 \bmod p$, i.e., $1+[\alpha,\beta] \in \ker\rho_0$, and $\rho_1 = 0$. //

3.3. Corollary: If $p > 3$,

$$\sum_{i\geq 0} T^i \dim H^i(\mathbb{S}L(D_2);\mathbb{F}_p) = 1+2T+2T^2+T^3$$

$$\sum_{i\geq 0} T^i \dim H^i(\mathbb{S}(D_2);\mathbb{F}_p) = 1+3T+\ T^2+3T^3+T^4\ .$$

Proof: By Poincare duality,

$$H^2(\mathbb{S}L(D_2);\mathbb{F}_p) \simeq H^1(\mathbb{S}L(D_2);\mathbb{F}_p) \simeq \mathbb{F}_{p^2}$$

$$H^3(\mathbb{S}L(D_2);\mathbb{F}_p) \simeq H^0(\mathbb{S}L(D_2);\mathbb{F}_p) \simeq \mathbb{F}_p. \ /\!/$$

It remains to describe the bigrading on $H^{*,*}(\mathbb{S}L(D_2);\mathbb{F}_p)$. [Note that the splitting $\mathbb{S}(D_2) = \hat{\mathbb{Z}}_p \times \mathbb{S}L(D_2)$ is invariant under the group $\theta(D_2) = \mu_{p^2-1} \cdot \mathbb{Z}$; hence

$$H^{*,*}(\mathbb{S}(D_2);\mathbb{F}_p) \simeq H^{*,*}(\hat{\mathbb{Z}}_p;\mathbb{F}_p) \otimes H^{*,*}(\mathbb{S}L(D_2);\mathbb{F}_p)$$

as bigraded algebras].

3.4. Proposition: $H^{1,*}(\mathbb{S}L(D_2);\mathbb{F}_p) = H^1(\mathbb{S}L(D_2);\overline{\mathbb{F}}_p)^{[F]\text{-inv.}}$ has $\mathbb{F}_p$-basis ρ_0, ρ_0^σ with bidegree $(1,2(p-1))$, $(1,2p(p-1))$ respectively; the Poincaré dual classes in $H^{2,*}$ have bidegree $(2,2p(p-1))$, $(2,2(p-1))$ respectively.

Proof: We have $\rho_0(F^{-1}\alpha F) = \rho_0(\alpha_0^\sigma + \alpha_1^\sigma F) = \rho_0(\alpha)^p \in \mathbb{F}_{p^2}$; hence $[F](\rho_0) = \rho_0$. Similarly, $\rho_0^\sigma(\alpha) = \overline{\alpha}_1^p \in \mathbb{F}_{p^2}$ is $[F]$-invariant. But $\rho_0(\omega^{-1}\alpha\omega) = \overline{\omega}^{p-1}\rho_0(\alpha)$, so (cf., 2.8) $\operatorname{bideg}\rho_0 = (1,2(p-1))$, and analogously for ρ_0^σ. Since the fundamental class of $\mathbb{S}L(D_2)$ has bidegree $(3,0)$, the degrees in $H^{2,*}$ follow from 2.9.2. $/\!/$

Institute for Advanced Study, June 1974

REFERENCES

[1] H. Cartan, S. Eilenberg; Homological Algebra, Princeton, 1956.

[2] J. W. S. Cassels, A. Fröhlich (ed.'s); Algebraic Number Theory, Thompson, 1967.

[3] J. Cohen; Stable Homotopy Theory, Lecture Notes in Math. 165, Springer.

[4] I. Kaplansky; Fields and Rings, Univ. of Chicago, 1969.

[5] J. Morava; "Structure theorems for cobordism comodules", preprint.

[6] S. P. Novikov; "Methods of algebraic topology, from the viewpoint of cobordism theory", Math. USSR (Izvestija) (AMS transl.) 1 (1967), 827-913.

[7] D. G. Quillen; "Elementary proofs of some results of cobordism theory using Steenrod operations", Advances in Math. 7 (1971), 29-56.

[8] J. P. Serre; Cohomologie Galoisienne, Lecture Notes in Math. 5, Springer.

[9] L. Smith; "On realizing complex bordism modules I, II, III", Amer. J. Math. 92 (1970), 793-856; 93 (1971), 226-263; 94 (1972), 875-890.

[10] H. Toda; "On spectra V(n)", in Proc. AMS Colloquium on Algebraic Topology (Madison), 273-278.

[11] ________; "On spectra realizing exterior parts of the Steenrod algebra", Topology 10 (1971), 53-65.

[12] B. B. Venkov; "On the cohomology of the unit groups of division algebras", Proc. Steklov Inst. 80 (1965), AMS transl., p. 73.

[13] R. Zahler; "The Adams-Novikov spectral sequence for the spheres", Annals of Math. 96 (1972), 480-504.

HYPERCOHOMOLOGY OF TOPOLOGICAL CATEGORIES

by

Jack Morava

§0. INTRODUCTION. We'll present a topological category C as a semisimplicial topological space, $\ldots C[2] \substack{\rightarrow\\\rightarrow\\\rightarrow} C[1] \rightrightarrows C[0]$, so C[0] is the space of objects of C, C[1] denotes its space of morphisms, C[2] consists of composable pairs of morphisms, and so forth. The geometric realization of C is the quotient $BC = \coprod_{n\geq 0} \Delta^n \times C[n]/(\text{face relations})$ of the union of the Cartesian products of the C[n] with the n-simplex Δ^n; we write $B_{zar}C$ to indicate that Δ^n has the Zariski topology of Segal, in which the closed sets are the unions of faces.

0.1. EXAMPLE: Let G be a topological group, viewed as a category with a single object. Then [7,20]

$$H^*(B_{zar}G;\underline{\mathbb{R}}) \cong H^*_c(G;\mathbb{R}),$$

where H_c^* denotes the van Est cohomology [i.e."Eilenberg-MacLane cohomology based on continuous cochains"] of G, while [for any sheaf S on a space X] $H^*(X;S)$ denotes the right-derived functor of global sections on X, with coefficients in S [Grothendieck cohomology], and $\underline{A}$ denotes [for any locally contractible Hausdorff abelian topological group] the sheaf of germs of continuous A-valued functions.

0.2. EXAMPLE: Let X be a compact Hausdorff space, viewed as a "discrete" category [in which every morphism is an identity]. Then

$$H^*(B_{zar}X;\underline{\mathbb{R}}) = C(X;\mathbb{R})$$

is the real Banach algebra of continuous functions on X, concentrated in dimension 0.

0.3. PROPOSITION: $C \mapsto H^*(B_{zar}C;\underline{\mathbb{R}})$ is a contravariant functor from categories to graded algebras, sending natural transformations (of functors) to identities (of the induced homomorphisms). In particular, an equivalence of categories induces an isomorphism of cohomology [19].

0.4. EXAMPLE: Let $G \times X \to X$ be a transformation group; let $[X/G]$ denote the category whose objects are elements $x,x' \in X$, with $\mathrm{Map}_{X/G}(x,x') = \{g \in G | gx = x'\}$; Thus $[X/G][1] = G \times X$, $[X/G][0] = X$, etc. The (equivariant) map $X \to pt$ induces a fibration $X \to B_{zar}[X/G] \to B_{zar}[pt/G] = B_{zar}G$

whose Leray spectral sequence collapses [following 0.2], if X is Hausdorff, to an isomorphism

$$H^*(B_{zar}[X/G];\mathbb{R}) \cong H^*_c(G;C(X;\mathbb{R})).$$

If G acts freely on X, then [X/G] is equivalent to the discrete category X/G and [by 0.3] $H^*(B_{zar}[X/G];\mathbb{R}) = 0$ if $* > 0$. Consequently the higher cohomology groups of $B_{zar}C$ are an estimate of the obstructions to regarding C as a Hausdorff space.

0.5. EXAMPLE: Let G be a real Lie group, and let $X = \underline{g}$ be its Lie algebra, with G acting by the adjoint representation. If G is compact and semisimple, then the topological quotient $\underline{g}/G$ is homeomorphic [11] to the quotient $\underline{h}/W$, in which $\underline{h}$ is a Cartan subalgebra [corresponding to a maximal torus T of G] and W is the Weyl group [= normalizer/centralizer] of T. Consequently $H^0(B_{zar}[\underline{g}/G];\mathbb{R})$ is isomorphic to the algebra of W-invariant real-valued continuous functions on $\underline{h}$, which is a completion of the ring of W-invariant polynomials in the weights [=linear functionals on $\underline{h}$] of G.

0.6. EXAMPLE: Let B be a closed subgroup of G, and let X be the homogeneous space G/B. The topological category [X/G] is then equivalent [by Shapiro's lemma] to [pt/B], whence

$$H^*_c(G;C(G/B;\mathbb{R})) \cong H^*_c(B;\mathbb{R}) \qquad [5].$$

For instance, G could be reductive, and B arithmetic [2].

0.7. EXAMPLE: Let G be a real Lie group, acting smoothly on a manifold X, with Lie algebra $\underline{g}$. If K is a maximal compact subgroup of G [whose Lie algebra $\underline{\underline{k}}$ is a maximal subalgebra of $\underline{g}$ with definite Killing form] then $G\times_K X \to X$ is a bundle whose fiber is the [contractible] symmetric space of maximal compact subgroups of G.

According to van Est, $H^*_c(G;A)$ is isomorphic to the relative Lie algebra cohomology of $\underline{g}$ mod $\underline{\underline{k}}$, with coefficients in the [contractible] module A [3]. More generally, the G-invariant differential forms along the fiber of the symmetric space bundle over X define a sheaf of differential graded algebras [or Sullivan real homotopy types] over the maximal Hausdorff quotient X/G of the group action, and the cohomology of the complex of global sections of this sheaf is $H^*(B_{zar}[X/G];\underline{\mathbb{R}})$. [Consequently one can at best hope that this cohomology is "locally" finitely-generated, i.e. as $H^0(B_{zar}[X/G];\underline{\mathbb{R}})$ - module.].

0.8 EXAMPLE: If Γ is the groupoid of germs of elements of a Lie pseudogroup G, then $H^*(B_{zar}\Gamma;\underline{\mathbb{R}})$ is the ring of continuous characteristic classes of G-foliations [9].

0.9. EXAMPLE: Let $T : \mathbb{R}^2 \to \mathbb{R}^2$ be an analytic volume-preserving diffeomorphism of the plane, with 0 as a fixed point. If an eigenvalue λ of T'(0) is real, T is conjugate to the linear transformation $\begin{bmatrix} \lambda & 0 \\ 0 & \lambda^{-1} \end{bmatrix}$, with 0 an isolated fixed point; thus if $X = \mathbb{R}^2 - \{0\}$, $G = \mathbb{Z}$ [acting by T] then $H^*(B_{zar}[X/G];\underline{\mathbb{R}})$ is isomorphic to the algebra of continuous functions on the real line.

If an eigenvalue λ of T'(0) is imaginary, then it lies on the unit circle, and T is called an elliptic diffeomorphism; its orbits, according to Poincaré, look something like this:

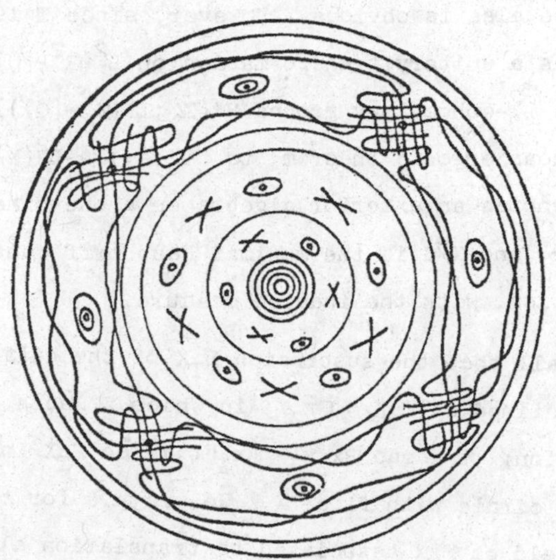

If λ is a primitive q^{th} root of 1, then T is conjugate to the normal form [1,23]

$$T(z) = z \exp(2\pi i a(|z|^2)) + P_q(z,\overline{z})$$

in which $z = x + iy \in \mathbb{C}$, $P_q \equiv 0$ modulo terms of degree $\geq q$, and $a(t) = \sum_{\frac{1}{2}q-1\geq i\geq 0} a_i(T)t^i$, the real numbers $a_i(T)$ being the Birkhoff invariants of T.

If $G = \mathbb{Z}$ acts as above $X = \mathbb{R}^2 - \{0\}$, then

$$\begin{aligned} H^*(B_{zar}[X/G];\underline{\mathbb{R}}) &= \ker(T-1) \text{ on } C^2(\mathbb{R}^2-\{0\};\mathbb{R}) && \text{If } * = 0 \\ &= \operatorname{coker}(T-1) && \text{If } * = 1 \\ &= 0 && \text{If } * \geq 2 \end{aligned}$$

and little else is obvious. However, since T is measure-preserving it induces a unitary transformation on $L^2(\mathbb{R}^2-\{0\})$ and the analogous L^2-cohomology groups $H^*(\mathbb{Z};L^2(\mathbb{R}^2-\{0\}))$ are isomorphic [by the mean ergodic theorem] to $E^*(e_1) \otimes L^2(X/\!\!/G)$, where $E^*(e_1)$ denotes an exterior algebra on a one-dimensional generator, and $X/\!\!/G$ is the maximal Hausdorff quotient of the group action, with the induced measure.

Recall that the suspension $\Sigma_T X$ of the diffeomorphism T may be defined as $\Sigma_T X = \mathbb{R} \times_{\mathbb{Z}} X$, in which $\mathbb{Z}$ acts on $\mathbb{R}$ by translation; the map $\mathbb{R}\times_{\mathbb{Z}} X \to \mathbb{R}\times_{\mathbb{Z}} pt$ makes $\Sigma_T X$ into a bundle over the circle with fiber X. We write θ for the canonical vectorfield on $\Sigma_T X$ associated to translation along $\mathbb{R}$.

If α is a 1-form on $\Sigma_T X$, we write $w(\alpha)$ for the class of the function $X \ni x \mapsto \langle \alpha(0,x), \theta(0,x)\rangle \in \mathbb{R}$ in $H^1(B_{zar}[X/G];\mathbb{R})$. If α is closed, then $w(\alpha)$ depends only on its class in $H^1_{DR}(\Sigma_T X;\mathbb{R})$ and $w : H^1_{DR}(\Sigma_T X;\mathbb{R}) \to H^1(B_{zar}[X/G];\mathbb{R})$ is a homomorphism. Dually we can define the function $w^* = X/\!\!/G \to H_1(\Sigma_T X;\mathbb{R})$ [which assigns to an orbit its winding number] by $w^*([x])([\alpha]) = (w[a])/e_1)(x)$, in which the homology of $\Sigma_T X$ is identified with the vectorspace dual to $H^1_{DR}(\Sigma_T X)$; it may be computed from the Wang sequence in the form of a short exact sequence

$$0 \to \operatorname{coker}(T^*-1) \text{ on } H_1(X;\mathbb{R}) \to H_1(\Sigma_T X;\mathbb{R}) \to H_0(X;\mathbb{R}) \to 0$$

[at least if X is connected].

The image of the invariant measure on X under w^* is a kind of probability measure on $H_1(\Sigma_T X;\mathbb{R})$, which is a conjugacy invariant of T if eg T is homotopic to the identity.

If the perturbation term P_q above vanishes identically, then X/G is parametrized by $|z|$, and it is easy to see that $w^*(|z|) = a(|z|^2)$.

§1. A GENERALISATION. We extend the cohomology defined above to coefficients which need not be locally contractible. Suppose that C is a semisimplicial object in the category of ringed spaces, with structure sheaf $\underline{O}_C$; thus $O_{C[n]}$ is the structure sheaf of the ringed space C[n]. [In the examples above, let $O_{C[n]}$ be the sheaf of germs of continuous real-valued functions.] Let E be a sheaf of semisimplicial $\underline{O}_C$-modules, as for example in §0.6; thus E[n] will be an $\underline{O}_{C[n]}$-module. The category of such $\underline{O}_C$- modules possesses enough injectives [12]; let $0 \to E \to I_0 \to I_1 \to \cdots$ be a resolution of E by such injectives. The global sections associated to such a resolution from a double complex where $(i,j)^{th}$ term is the module of sections of the sheaf $I_i[j]$. We write $\underline{H}^*(C;E)$ for the total cohomology of this complex, which we call the hypercohomology of C with coefficients in E. If E* is a graded sheaf of $\underline{O}_C$-modules, we define

$$\underline{H}^{i,j}(C;E) = \underline{H}^i(C,E^j).$$

1.1. EXAMPLE: If C is a semisimplicial topological space ring/ed by continuous real-valued functions as suggested above, then

$$\underline{H}^*(C;\underline{O}_C) \cong H^*(B_{zar}C;\underline{\mathbb{R}}) \qquad [20]$$

1.2. EXAMPLE: Let X be a bounded symmetric domain in $\mathbb{C}^n$, and let G be an arithmetic group of analytic automorphisms acting totally discontinuously [which is to say that for $\forall x \in X\ \exists$ neighborhood $U \ni x$ such that for $\forall g \in G - \{1\}$ we have $gU \cap U = \emptyset$.] on X. The jacobian of $g \in G$ is an element of the multiplicative group of $\mathbb{C}$, defining a character $w : G \to \mathbb{C}^x$. We write $\underline{H}^{**}[X/G]$ for the hypercohomology of the analytic category [X/G] with coefficients in the structure sheaf $\underline{O}_{X/G}$ twisted by powers of w, i.e. $\underline{H}^{i,j}[X/G] = \underline{H}^i([X/G]; \underline{O}_{X/G} \otimes w^j)$; here [X/G] is the semi-simplicial analytic space $G \times G \times X \rightrightarrows G \times X \rightrightarrows X$ as above, the structure sheaf $\underline{O}_{X/G}$ being the appropriate product of the sheaves of analytic functions on X and constant functions on G. Then $\underline{H}^{*,*}[X/G]$ becomes a bigraded algebra, and $\underline{H}^{0,*}[X/G]$ is the graded ring of modular forms on the Satake compactification of the analytic space X/G.

1.3. EXAMPLE: We interpret the category $[\mathbb{C}^{n+1} - \{0\}/\mathbb{C}^x]$ as that of lines through the origin in $\mathbb{C}^{n+1}$, whose objects are elements $(z_0,\cdots,z_n)$ of $\mathbb{C}^{n+1}-\{0\}$ and morphisms are multiplications $(z_0,\cdots,z_n) \mapsto (\lambda z_0,\cdots,\lambda z_n)$ by $\lambda \in \mathbb{C}^x$. The identity map of $\mathbb{C}^x$ is a character, in the preceding sense, of this transformation group, and we have

$H^{*,*}$ [lines through the origin in $\mathbb{C}^{n+1}$] $\cong H^*(\mathbb{C}P(n); \underline{O}_{\mathbb{C}P(n)}^{\otimes *})$,

in which $\mathbb{C}P(n)$ is a projective algebraic variety of dimension n in the sense of Serre, provided $\mathbb{C}$ be algebraically closed. By the explicit calculations [21,§65] we have the algebra $E(e_n) \otimes \mathbb{C}[t_0,\cdots,t_n]$ with degree $e_n = (n,0)$, deg $t_i = (0,2)$.

1.4. EXAMPLE: The category of Borel subgroups of the multiplicative group $\mathbb{H}^x$ of quaternions has the projective space $\mathbb{H}^x/B \cong \mathbb{C}P(1)$ of such things as objects, and left-multiplication by elements of $\mathbb{H}^*$ as transformations. The reduced norm [24] $\nu : \mathbb{H}^x \to \mathbb{R}^x$ is a character, as above, and we may interpret $\underline{H}^{**}[\mathbb{C}P(1)/\mathbb{H}^x]$ as the hypercohomology of a semisimplicial real algebraic variety. By the Hopf fibration we have

$$\underline{H}^{**}[\mathbb{C}P(1)/\mathbb{H}^x] \cong \underline{H}^*\,[pt/\mathbb{C}^x] \otimes \mathbb{R}[\nu,\nu^{-1}],\ \deg \nu = (0,2)$$

as bigraded algebras. By van Est

$$\underline{H}^*\ \ [pt/\mathbb{C}^x] \cong H^*_{lie}(\mathbb{C},\mathbb{R};\mathbb{R}) \cong E(e_1) \text{ with degree } e_1 = (1,0).$$

1.5. DEFINITION: If $F : C \to C'$ is a morphism of semi-simplicial ringed spaces, we write $\underline{R}^*F_*(E)$ for the right hyperderived [4] direct image of F, with coefficients in the sheaf E of $\underline{O}_C$-modules.

The Grothendieck spectral sequence of the composition of direct image functors associated to the diagram $C \xrightarrow{F} C' \to pt$ of semisimplicial ringed spaces is a Leray spectral sequence with E_2-term $\underline{H}^*(C';\ \underline{R}^*F_*(E))$ converging to $\underline{H}^*\ (C;E)$.

1.6. EXAMPLE: If the discrete group G acts totally discontinuously on an analytic space X, the hyperderived direct image of the quotient functor $[X/G] \to X/G$ has stalk $H^*_c(ISO[x];E_x)$

at $[x] \in X/G$, where $ISO[x] = \{g \in G | gx = x\}$ is discrete. [8,22].

1.7. REMARK: The preceeding examples suggest as a mythicomathematical principle that the hypercohomology of categories extends the theory of automorphic forms to transformation groups with bad quotients. However, the preceeding example is the only case where an explicit description of the hyperderived direct image of the quotient map is available. For other bad actions , see [13].

§2. EXAMPLES OVER LOCAL FIELDS.

Let D be a locally compact complete nondiscrete nontrivially skew field, with center C of characteristic 0. [The unique archimedean example of such a D is the field of quaternions.] Let $\mathbb{C}$ be the completion of an algebraic closure of C, and let $\mathbb{C}_0$ be the maximal unramified subfield of $\mathbb{C}$ containing C; then $\mathbb{C}_0 \otimes_C D$ is isomorphic to the ring of linear transformations of $\mathbb{C}_0^d$, for some $d > 1$. [24].

Let S(D) be the maximal compact subgroup of the multiplicative group of D, acting on the space $\mathbb{C}_0P(d-1)$ of lines through the origin in $\mathbb{C}_0^d$ by left-multiplication. We write $[\mathbb{C}_0P(d-1)/S(D)]$ for the semisimplicial analytic space associated to this transformation group. If C is archimedian we recover example 1.4, so from now on we assume that C is nonarchimedian [thus topologized by a valuation] but unramified. The maximal compact subring of C [resp. $\mathbb{C}_0$] may be identified with the ring $W(k_C)$ [resp. W(k)] of Witt vectors of the residue field k_C [resp. its algebraic closure k] of C. Thus S(D) is an analytic scheme over $W(k_C)$, while $\mathbb{C}_0P(d-1)$ is analytic over W(k); but for the purposes of this paper, we regard $\mathbb{C}_0P(d-1)$ as an analytic scheme over $W(k_C)$, by restriction of scalars.

2.1. PROPOSITION: The analytic subspace X_D of lines $L \in \mathbb{C}_0P(d-1)$ which are quasiinvariant [in the sense that for $\forall \delta \in S(D)$ we have $\delta L \equiv L$ (modulo p)] is an S(D)-invariant scheme parameterized by $W(k)[[t_1,\cdots,t_{d-1}]]$.

PROOF. The first assertion is clear from the definition; to show the second, we apply the theory of division algebras [24] to choose a basis for the action of D on $\mathbb{C}_0^d$ in which the quasiinvariant lines are those with projective coordinates congruent to $[0,\cdots,0,1]$(modulo p).

Let C_0 be a maximal commutative subfield of D which is unramified over C, and fix an embedding of C_0 in $\mathbb{C}_0$; let $\sigma \in \mathrm{Gal}(C_0/C)$ be the Frobenius automorphism. An element π of positive valuation exists in D, with the following properties:

i) $\pi^d \in C$.

ii) if $c \in C_0$ then $\pi^{-1}c\pi = \sigma(c)$.

iii) $\{1,\pi,\cdots,\pi^{d-1}\}$ is a basis for D as left C_0-vectorspace.

We use iii) to identify $\mathbb{C}_0^d$ with $\mathbb{C}_0 \otimes_{C_0} D$; thus if $\delta \in S(D)$, $x = (x_1,\cdots,x_d) \in C_0^d$ we define δx by its coordinates $\sum_{d\geq i\geq 1} (\delta x)_i \pi^{i-1} = (\sum_{d\geq i\geq 1} x_i\pi^{i-1})\delta^{-1}$. The line spanned by x [which may be assumed to have all coordinates integral, with at least one a unit] is quasiinvariant if for $\forall\delta \in S(D)$ $\exists$ unit w of W(k) such that $w(\Sigma\, x_i\pi^{i-1}) \equiv (\Sigma\, x_i\pi^{i-1})\delta$ (mod p), and it is easy to see by induction that the first coordinate not congruent to zero mod p is x_{d-1}. Letting $t_i = x_i x_{d-1}^{-1}$ defines a parametrization of X_0. ∎

Let $\mathbb{S}(D)$ be the Sylow prop-p subgroup of S(D) [p being the residue characteristic of C] and let the reduced norm $\nu : S(D) \to S(C)$ bigrade the hypercohomology of $[X_D/S(D)]_{an}$.

2.2. PROPOSITION:

$$\underline{H}^{**}(X_D/S(D)]_{an} \cong H_c^{**}(S(D);\ W(k)[[t_1,\dots t_{d-1}]])$$

PROOF: Since X_D is affine, the Leray spectral sequence of the functor $[X_D/S(D)] \to [pt/S(D)]$ degenerates to an isomorphism $\underline{H}^{**}[X_D/S(D)]_{an} \cong H_{an}^{**}(S(D);W(k)[[t_1,\dots,t_{d-1}]])$, in which the second subscript signifies Eilenberg-MacLane cohomology based on cochains analytic over $W(k_C)$. That filtration of $W(k)[[t_1,\dots,t_{d-1}]]$ which assigns to the monomial $t^a = t_1^{a_1}\cdots t_{d-1}^{a_{d-1}}$ the weight $|a| = \Sigma\, a_i$ is $S(D)$-invariant, so that cohomology has coefficients in the limit of a family of torsion-free modules of finite rank over $W(k)$; by the comparison theorem of Lazard [14, $\underline{V}$§2.3.10] the inclusion of the complex of analytic cochains into the continuous ones induces a cohomological isomorphism. ∎

2.3. COROLLARY: If $p \nmid d$ then in an appropriate grading

$$Gr^*\underline{H}^{**}[X_D/S(D)\ ]_{an} \cong H_c^*(S\,(C);H_c^*(SL(D);W(k)[[t_1,\dots,t_{d-1}]]^{\otimes *})$$

with

$$SL(D) = \ker \nu .$$

PROOF: This is the Hochschild-Serre spectral sequence of

the short exact sequence $1 \to SL(D) \to S(D) \to S(C) \to 1$, which is split if $p \nmid d$ by the inclusion of the center in D. ∎

2.4. In this paragraph we invoke Cartier's results to show that when $C = \mathbb{Q}_p$ and $[D] = d^{-1} \in Br(\mathbb{Q}_p)$, the hypercohomology of $[X_D/S(D)]_{an}$ is isomorphic to the E_2-term of the Adams-Novikov spectral sequence of the Bousfield localisation $S^0(\chi)$ of the 0-sphere with respect to the completion U_χ of complex bordism at a Hirzebruch genus $\chi : U \to \mathbb{F}_p = k_C$ of height d. [We've shown in the preceeding paper [17] that the groupoid scheme $(U_\chi, U_\chi U)$ of cooperations on U_χ is equivalent over W(k) to the split Hopf algebroid $(E_\chi, E_\chi \otimes H_\chi)$ in which E_χ is the formal power series ring which parameterizes *-isomorphism classes of lifts of the formal group law defined by χ from $\mathbb{F}_p$ to W(k), and H_χ in the Hopf algebra of (suitably Galois-invariant) locally constant W(k)-valued functions on S(D). If M, N $\in C_\chi$ are affine representations of [= comodules over] $(U_\chi, U_\chi U)$ then there is a Grothendieck spectral sequence with

$$E_2 = H_c^*(S(D); \operatorname{Ext}^*_{E_\chi}(M(\chi), N(\chi)) \Rightarrow \operatorname{Ext}^*_{C_\chi}(M,N) \otimes W(k)$$

[where $M(\chi) = M \otimes_{U\chi} E_\chi$, etc.] coming from the composition $\operatorname{Hom}_{E_\chi \otimes H_\chi}(M(\chi),-) \cong H_c^0(S(D); \operatorname{Hom}_{E_\chi}(M(\chi),-))$ of functors [for any $E_\chi \otimes H_\chi$-comodule $N(\chi)$ can be imbedded in the acyclic comodule (E_χ-injective envelope of $N(\chi)) \otimes H_\chi$.]

2.4.1. COROLLARY: If $p \nmid d$ then

$$\operatorname{Ext}^{**}_{C_\chi}(U_\chi, U_\chi) \cong H_c^*(\mathbb{S}(\mathbb{Q}_p); H_c^*(\mathbb{S}L(D); E_\chi)^{\otimes *})$$

PROOF: $S(D)$ acts on the χ-adic completion of $U^*(CP(\infty)) \cong U_\chi[[T]]$ by $[\delta](T) \equiv \omega(\delta)\ T \bmod T^2$, defining a character $\omega : S(D) \to W(k)^x$ which may be identified with the composition $S(D) \to S(D)/\$L(D) \cong k_D^x \times \$(\mathbb{Q}_p)$. Restriction to S^2 in $CP(\infty)$ shows that ω induces the bigrading on $\operatorname{Ext}^*_{C_\chi}(U_\chi, U_\chi)$. The assertion follows from the preceeding proposition, with $H_c^{**}(\$L(D); E_\chi)$ given the $2(p^d-1)$- periodic bigrading associated to the action of the cyclic group k_D^x. ∎]

It remains to see that $E_\chi \cong W(k)[[t_1,\cdots,t_{d-1}]]$ as $S(D)$-modules. Let Φ be a formal group of finite height d over k, with Cartier module M_Φ of p-typical curves, and let M_G be the corresponding module of curves in a lift G of Φ to the ring $W(k)$. If $\sigma_\Phi : M_\Phi \to M_G$ is the canonical $(W(k),F)$ - linear Cartier section [15, VII §6.14] then the composition $\lambda_G : M_\Phi \to M_G \to M_G/VM_G \cong$ tangent space of G at 0 is a lift to $W(k)$ of the composition

$M_\Phi \to M_\Phi/VM_\Phi \cong$ tangent space of Φ at 0 [$\cong k$, if we study only one-dimensional formal groups].

Consequently the line spanned by λ_G in the Dieudonne module $\operatorname{Hom}_{W(k)}(M_\Phi, W(k))$ is quasi-invariant, in the sense of 2.1, under the natural action of $\operatorname{Aut} \Phi \cong S(D)$. Conversely, any element of the projectified Dieudonne module of Φ has a surjective representative $\lambda' : M_\Phi \to W(k)$ whose kernel L determines the *-isomorphism class of a lift of Φ provided $L \equiv VM_\Phi \pmod p$ [15, VII §7,17]. Now M_Φ is isomorphic to $W(k) \otimes_{W(F_q)} \mathrm{End}\,\Phi$, where $q = p^d$, with action of the Verschiebung defined by $Vx = x\pi^{d-1}$ with π as in 2.1 [15, VI §7.8, §7.39].

2.5. We conclude with Honda's [10] description of the topological quotient $X_D/S(D)$. We write $W(k)_\sigma[[F]]$ for the Hilbert power series ring, where σ is the Frobenius element of $\mathrm{Gal}(\overline{F}_p/F_p)$, and $Fa = \sigma(a)F$ for $a \in W(k)$. Let G be a lift of Φ as above, with logarithm $\log_G(T)$. Then there exists an "Eisenstein" polynomial $E_G(F) = \sum_{0\le i\le d} e_i(G)F^i$ [where $e_0(G) = p$, $e_d(G) \not\equiv 0 \pmod p$, and $e_i(G) \equiv 0 \pmod p$ otherwise] such that

i) $E_G(F)*\log_G(T) = \sum_{0\le i\le d} e_i(G)\log_G^{\sigma^i}(T^{p^i}) \equiv 0 \pmod p$

ii) lifts G, G' of Φ are isomorphic if and only if there exists $w \in W(k)^x$ such that $u^{-1}E_G u = E_{G'}$.

iii) Any such Eisenstein polynomial E defines a formal grouplaw with $\log_E(T) = \sum_{i\ge 0} b_i T^{p^i}$ by $pE^{-1} = \sum_{i\ge 0} b_i F^i$.

We write $[e_1(G),\cdots,e_d(G)]$ for the equivalence class of $(e_1(G),\cdots,e_d(G)) \in W(k)^d$ under the action of $W(k)^x$ defined by

$$u\cdot(e_1,\cdots,e_d) = (u^{\sigma-1}e_1,\cdots,u^{\sigma^d-1}e_d)\ ,\ \text{where}$$

$E_G(F) = \Sigma\, e_i(G)F^i$ is the Honda-Eisenstein polynomial of the lift G. Note that in a natural sense, $[e_1(G),\cdots,e_d(G)] \equiv [0,\cdots,0,1] \pmod p$, and that the full isomorphism class of a lift [i.e. element of $X_D/S(D)$] is determined by the projective coordinates [e(G)] of some representative.

Let $\Theta_i : W(k_D)^x \times W(k) \to W(k)$ denote the one-dimensional representation of $W(k_D)^x$ defined by $\Theta_i(\lambda)$ = multiplication by $\sigma^i(\lambda)\cdot\sigma^{i+1}(\lambda)\cdots\sigma^{d-1}(\lambda)$; then $\Theta = \bigoplus_{1\le i\le d-1} \Theta_i$ is a free $W(k)$-module of rank $d - 1$, and $W(k_D)^x$ acts freely by

$$\lambda(\theta_1,\dots,\theta_{d-1}) = (\text{Norm }\lambda\cdot\theta_1,\dots,\sigma^{d-1}(\lambda)\theta_{d-1}).$$

The exact sequence [6]

$$1 \to W(k)^x \xrightarrow{u \mapsto u^{-1}\sigma^{d-1}(u)} W(k)^x \to W(k_D)^x \to 1$$

shows that $\Theta/W(k_D)^x$ parameterizes the quotient $X_D/S(D)$; in fact [by iii) above] we define a functor

$$E : [\Theta/W(k_D)^x]_{an} \to [X_D/S(D)]_{an}$$

which induces an isomorphism of the quotient $\hat{Z}_p$-analytic spaces.

The E_2-term of the Leray spectral sequence for the quotient functor $Q : [X_D/S(D)]_{an} \to X_D/S(D)$ is thus $\underline{H}^*([\Theta/W(k_D)^x]_{an}; \underline{R}^*Q_*(\underline{O}_{X_D}/S(D)))$. It can be shown that the isotropy group of a point of X_D under the action of $\mathbb{S}(D)$ is a group of cohomological dimension d, at least if $(p-1) \nmid d$, and that the cohomological dimension of $\mathbb{S}(D)$ is d^2 [16]; this implies that the stalk of $\underline{R}^*Q_*$ is <u>not</u> the continuous cochain cohomology of the isotropy group. [I am indebted to Doug Ravenel for explaining this to me, in a conversation at the Northwestern conference].

Jack Morava
23 July 1977
SUNY at Stony Brook

References

1. V. I. Arnol'd, A. Avez, Ergodic problems of classical mechanics, [Benjamin, N.Y.], appendix 28.

2. A. Borel, Cohomology of arithmetic groups, Vancouver conference I (1974) p. 435-442.

3. R. Bott, Remarks on continuous cohomology, Manifolds conference, Tokyo (1973).

4. H. Cartan, S. Eilenberg, Homological Algebra [Princeton University Press].

5. W. Casselman, D. Wigner, Continuous cohomology and a conjecture of Serre, Inventiones Math. 25(1974) p. 199-211

6. M. Demazure, P. Gabriel, Groupes Algebriques I [North-Holland].

7. D. B. Fuks, I. M. Gelfand, Classifying space for principal fibrations over Hausdorff bases, DAN 181 (1968) p. 515-518.

8. A. Grothendieck, Tohoku 9 (1957) p.119-221

9. A. Haefliger, Seminaire Bourbaki, no. 412

10. T. Honda, On the theory of commutative formal groups, J. Math. Soc. Japan 22 (1970), p. 213-246.

11. Wu-Yi Hsiang, Cohomological theory of compact transformation groups, Ergebnisse Math, 1976 [Springer].

12. L. Illusie, Complex Cotangent..., Springer 239.

13. J. L. Koszul, Cohomologie des actions locales de groups de Lie, Symposia Math. XVI (1974) 399-407.

14. M. Lazard, Groups p-adiques analytiques, Publ. Math. IHES 26

15. ________, Commutative Formal Groups, Springer 443.

16. J. Morava, Extensions of cobordism comodules [to appear].

17. ________, Completions of complex cobordism [this conference].

18. S. Schwarzman, Asymptotic cycles, Ann. Math. 66 (1957) p. 270-284.

19. G. Segal, Categories and classifying spaces, Publ. Math. IHES 34 (1968) p. 105-112.

20. _______, On the classifying space of a topological group in the Gelfand - Fuks sense, Functs. Analiz 9 (1975) p. 48-50.

21. J. P. Serre, Faisceaux algebriques coherents, Ann. Math. 61 (1955) p. 197-278.

22. B. B. Venkov, On the cohomology of unit groups in division algebras, Proc. Steklov Inst. 80 (1965) p. 73-100.

23. E. Zehnder , Comm. Pure and Applied Maths. 26 (1973) p. 131-182.

24. A. Weil, Basic Number Theory, Springer.

A NOVICE'S GUIDE TO THE ADAMS-NOVIKOV SPECTRAL SEQUENCE

Douglas C. Ravenel*
University of Washington
Seattle, Washington 98195

Ever since its introduction by J. F. Adams [8] in 1958, the spectral sequence that bears his name has been a source of fascination to homotopy theorists. By glancing at a table of its structure in low dimensions (such have been published in [7], [10] and [27]; one can also be found in §2) one sees not only the values of but the structural relations among the corresponding stable homotopy groups of spheres. It cannot be denied that the determination of the latter is one of the central problems of algebraic topology. It is equally clear that the Adams spectral sequence and its variants provide us with a very powerful systematic approach to this question.

The Adams spectral sequence in its original form is a device for converting algebraic information coming from the Steenrod algebra into geometric information, namely the structure of the stable homotopy groups of spheres. In 1967 Novikov [44] introduced an analogous spectral sequence (formally known now as the Adams-Novikov spectral sequence, and informally as simply the Novikov spectral sequence) whose input is algebraic information coming from MU^*MU, the algebra of cohomology operations of complex cobordism theory (regarded as a generalized cohomology theory (see [2])). This new spectral sequence is formally similar to the classical one. In both cases, the E_2-term is computable (at least in principle) by purely algebraic methods and the E_∞-term is the bigraded object associated to some filtration of the stable homotopy groups of spheres (the filtrations are <u>not</u> the same for the

*Partially supported by NSF

two spectral sequences). However, it became immediately apparent, for odd primes at least, that the Novikov spectral sequence has some striking advantages. Its E_2-term is smaller and there are fewer differentials, i.e. the Novikov E_2-term provides a better approximation to stable homotopy than the Adams E_2-term. Most of the groups in the former are trivial for trivial reasons (the sparseness phenomenon to be described in Corollary 3.17) and this fact places severe restrictions on when nontrivial differentials can occur. It implies for example that $E_2^{**} = E_{2p-1}^{**}$. For $p = 3$, the entire Novikov spectral sequence through dimension 80 can be legibly displayed on a single page (hopefully this will be done in [52]; see [75] for a table through dimension 45), whereas the Adams spectral sequence through a comparable range requires 4 pages (see [36]).

In the Adams spectral sequence for $p > 2$, the first nontrivial differential originates in dimension $pq - 1$ (where $q = 2p - 2$) and is related to the odd primary analogue of the nonexistence of elements of Hopf invariant one (see §2). The latter result is, in the context of the Novikov spectral sequence (even for $p = 2$), a corollary of the structure of the 1-line $E_2^{1,*}$, which is isomorphic to the image of the J-homomorphism (see [1]).

In the Novikov spectral sequence for $p > 2$, the first nontrivial differential does not occur until dimension $p^2q - 2$ and is a consequence of Toda's important relation in stable homotopy $\alpha_1 \beta_1^p = 0$ (see [70], [71] and [56]). An analogous differential occurs in the Adams spectral sequence as well.

The situation at the prime 2 is quite different. At first glance (see Zahler's table in [75]) the Novikov spectral sequence appears to be less efficient than the Adams spectral sequence. The first nontrivial differential in the former originates in dimension 5 whereas the first nontrivial Adams differential does not originate until dimension 15. In looking at Zahler's table one is struck by the abundance of differentials, and also by the nontrivial group extensions occuring in dimensions 3 and 11 (the table stops at dimension 17).

These apparent drawbacks have been responsible for public apathy toward the 2-primary Novikov spectral sequence up until now. An object of this paper, besides providing a general introduction to the subject, is to convince the reader that the Novikov spectral sequence at the prime 2 is a potentially powerful (and almost totally untested) tool for hacking one's way through the jungles of stable homotopy. In particular in §7 we will show how it can be used to detect some interesting new families of elements recently constructed by Mahowald.

The plan of the rest of the paper is as follows:

In §2, we will discuss the classical Adams spectral sequence and some of the questions it raised about the stable homotopy.

In §3, we will set up the Novikov spectral sequence.

In §4, we will discuss the relation between the two spectral sequences and show how comparing the two E_2-terms for $p = 2$ leads to a complete determination of stable homotopy through dimension 17.

In §5, we discuss what we call 'first order' phenomena in the Novikov spectral sequence, i.e. we show how it detects the image of the J-homomorphism and related elements.

In §6 and §7, we discuss second order phenomena, i.e. certain possible new families of homotopy elements which are difficult if not impossible even to conceive of without the Novikov spectral sequence.

In §8, we will discuss some recent theoretical developments which have led to some unexpected insights into the nature of stable homotopy and (most interestingly) the relation between it and algebraic number theory. In other words, we will discuss the theory of Morava stabilizer algebras and the chromatic spectral sequence, in hopes of persuading more people to read (or at least believe) [37], [58], [51] and [39].

I have tried to write this paper in the expository spirit of the talk given at the conference. Naturally, I have expanded the lecture considerably in order to make the paper more comprehensive and useful to someone wishing to begin research in this promising area. At two points however, I have been unable to resist giving some fairly detailed proofs which have not appeared (and probably will not appear) elsewhere. In §5, you will find a new partial proof of Theorem 5.8, which describes the image of the J-homomorphism and related phenomena at the prime 2. The proof uses techniqes which can be generalized to higher order phenomena (such as those described in §6 and §7) and it makes no use of the J-homomorphism itself. In §7 are derivations of some consequences of certain hypotheses concerning the Arf invariant elements and Mahowald's η_j's.

I am painfully aware of the esoteric nature of this subject and of the difficulties faced by anyone in the past who wanted to become familiar with it. I hope that this introduction will make the subject more accessible and that there will be greater activity in what appears to be a very fertile field of research.

The E_2-term can be written either as $\mathrm{Ext}_A^{**}(\mathbb{F}_p, \mathbb{F}_p)$ (Ext in the category of A-modules) or $\mathrm{Ext}_{A_*}^{**}(\mathbb{F}_p, \mathbb{F}_p)$ (Ext in the category A_*-comodules). The distinction here is didactic, but in the case $E = BP$ (the Novikov spectral sequence) the formulation in terms of comodules leads to a substantial simplification.

The identification of the E_2-term can be carried out for general E provided that E is a ring spectrum and $E \wedge E$ is a wedge of suspensions of E. This is the case when E = MU, BP or MSp, but not if E = bo or bu. (For the homotopy type of $E \wedge E$ in these two cases, see [35] and [6] §III 17 respectively.)

We now specialize to the case $p = 2$. Table 1, which displays the behavior of the spectral sequence through dimension 19 is provided for the reader's amusement. Before commenting on it, we will discuss $\mathrm{Ext}_A^{1,*}(\mathbb{F}_2, \mathbb{F}_2)$, the Adams "1-line".

Proposition 2.5

$$\mathrm{Ext}_A^{1,t}(\mathbb{F}_2, \mathbb{F}_2) = \begin{cases} \mathbb{F}_2 & \text{if } t = 2^i \\ 0 & \text{otherwise} \end{cases}$$

The generator of $\mathrm{Ext}_A^{1,2^i}(\mathbb{F}_2, \mathbb{F}_2)$ is denoted by h_i and represented by $\xi_1^{2^i}$ in the cobar complex (2.3).

Proof. In (2.3), there are no coboundaries in $\bar{A}_*$, so all cocylces in that group are nontrivial. An element is a cocycle iff its image in A_* is primitive, i.e. if it is dual to a generator of A. A is generated by the elements Sq^{2^i} [66], so the result follows. □

The first 4 of these generators detect well-known elements in stable homotopy: h_o detects 2ι, where ι generates the zero stem, while h_1, h_2, and h_3 detect the suspensions of the 3 Hopf fibrations $S^3 \to S^2$, $S^7 \to S^4$ and $S^{15} \to S^8$ respectively.

§2. The Classical Adams Spectral Sequence

In this section, we discuss the outstanding features of the classical mod 2 Adams spectral sequence. Readers who are already knowledgeable in this area will lose very little by skipping this section.

A general formulation of the Adams spectral sequence is the following. We have a diagram of spectra

$$X = X_o \leftarrow X_1 \leftarrow X_2 \leftarrow X_3 \leftarrow \cdots \qquad (2.1)$$
$$\downarrow \qquad \downarrow \qquad \downarrow \qquad \downarrow$$
$$Y_o \qquad Y_1 \qquad Y_2 \qquad Y_3$$

where $X_{s+1} \to X_s \to Y_s$ is a cofibration for each s. Then from the theory of exact couples (see [7]) we have

Theorem 2.2 Associated to the diagram (2.1) there is a spectral sequence $\{E_r^{s,t}\}$ with differentials $d_r : E_r^{s,t} \to E_r^{s+r,t+r-1}$ such that:

(a) $E_1^{s,t} = \pi_{t-s} Y_s$;

(b) $d_1 : E_1^{s,t} \to E_1^{s+1,t}$ is induced by the composite $Y_s \to \Sigma X_{s+1} \to \Sigma Y_{s+1}$;

(c) the spectral sequence converges to $\pi_* \tilde{X}$ where $\tilde{X}$ is the cofibre of $\varprojlim X_i \to X \to \tilde{X}$. □

The diagram (2.1) is called an Adams resolution if $\varprojlim X_i$ is weakly contractible after localizing at some prime p. In this case, the spectral sequence will converge to the p-localization of $\pi_* X$.

Needless to say, the spectral sequence is useful only if one knows $\pi_* Y_s$. This is often the case if we set $Y_s = X_s \wedge E$, where E is the representing spectrum for some familiar homology theory, such as ordinary mod p homology theory. In that case, we have the E_*-homology Adams spectral sequence for π_*X. For a more detailed discussion, see [6] §III 15. The case $E = MU$ or BP is that of the Novikov spectral sequence.

If X is connective and $E = H\mathbb{F}_p$ (the mod p Eilenberg-MacLane spectrum) or BP (the Brown-Peterson spectrum), then $\tilde{X}$ is the p-adic completion of X or the p-localization of X respectively (see [11] or [12]). If either X or E fail to be connective (e.g. if E is the spectrum representing K-theory) then the relation between X and $\tilde{X}$ (which Bousfield calls the E-nilpotent completion of X) is far from obvious.

Theorem 2.2 yields the classical mod p Adams spectral sequence if we set $X = S^o$, $E = H\mathbb{F}_p$, and $Y_s = X_s \wedge E$. If we denote X_1 by $\bar{E}$, we have $X_s = \bar{E}^{(s)}$ (the s-fold smash product of $\bar{E}$ with itself) and $Y_s = E \wedge \bar{E}^{(s)}$ for $s > 0$. It follows that each Y_s is a wedge of mod p Eilenberg-MacLane spectra and that for $s > 0$, $\pi_*\Sigma^{-s}Y_s = \bar{A}_*^{\otimes s}$ where $\bar{A}_*$ is the agumentation ideal of the dual mod p steenrod algebra A_*. One can show further that the Adams E_1-term in this case is isomorphic to the normalized cobar complex

$$(2.3) \quad \mathbb{F}_p \xrightarrow{\delta_o} \bar{A}_* \xrightarrow{\delta_1} \bar{A}_* \otimes \bar{A}_* \xrightarrow{\delta_2} \bar{A}_* \otimes \bar{A}_* \otimes \bar{A}_* \xrightarrow{\delta_3} \cdots$$

that one uses to compute the cohomology of the Steenrod algebra. Specifically, we have

$$\delta_s(a_1 \otimes a_2 \cdots \otimes a_s) = \sum_{i=1}^{s} (-1)^i a_i \otimes \cdots a_{i-1} \otimes \Delta(a_i) \otimes a_{i+1} \cdots a_s$$

where $a_i \in \bar{A}^*$ and $\Delta\colon \bar{A}_* \to \bar{A}_* \otimes \bar{A}_*$ is the coproduct. In this way, we arrive at Adams' celebrated original theorem.

<u>Theorem 2.4</u> (Adams [8]). There is a spectral sequence converging to the p-component of π_*S^o with $E_2^{s,t} = \mathrm{Ext}_A^{s,t}(\mathbb{F}_p, \mathbb{F}_p)$, where A is the mod p Steenrod algebra. □

(These elements are customarily denoted by η, ν and σ respectively.)

The question then arises as to whether h_i for $i > 3$ is a permanent cycle in the spectral sequence and therefore detects a homotopy element. This question has some interesting implications.

Theorem 2.6 The following statements are equivalent:

(a) h_i is a permanent cycle in the Adams spectral sequence.

(b) There is a 2-cell complex $X = S^n \cup e^{n+2^i}$ such that Sq^{2^i} is nontrival in $H^*(X; \mathbb{F}_2)$.

(c) $\mathbb{R}^{2^i}$ can be made into a division algebra over $\mathbb{R}$.

(d) S^{2^i-1} is parallelizable. □

A proof can be found in [4].

In one of the more glorious moments of algebraic topology, Adams answered the question in the following spectacular way.

Theorem 2.7 (Adams [4]). For $i > 3$, h_i is not a permanent cycle in the Adams spectral sequence. More precisely, $d_2\ h_i = h_0 h_{i-1}^2 \neq 0$. □

We now comment on Table 1. A similar table showing $E_2^{s,t}$ for $t - s \leq 70$ (but not showing any differentials) can be found in [67], where the method for computing it developed by May [32] [33] is discussed. Differentials up to $t - s = 45$ have been computed and published in [10] and [31].

The vertical axis s is filtration or cohomological degree. The horizontal axis is $t - s$, so all elements in the same topological dimension will have the same horizontal co-ordinate. Each small circle represents a basis element of the vector space $E_2^{s,t} = \mathrm{Ext}_A^{s,t}(\mathbb{F}_2, \mathbb{F}_2)$. When a space is empty, the corresponding vector space is trivial. $\mathrm{Ext}_A^{**}(\mathbb{F}_2, \mathbb{F}_2)$ has a commutative algebra structure, as does $E_r^{s,t}$ for $r > 2$, and the differentials are

TABLE I. THE CLASSICAL ADAMS SPECTRAL SEQUENCE FOR P=2

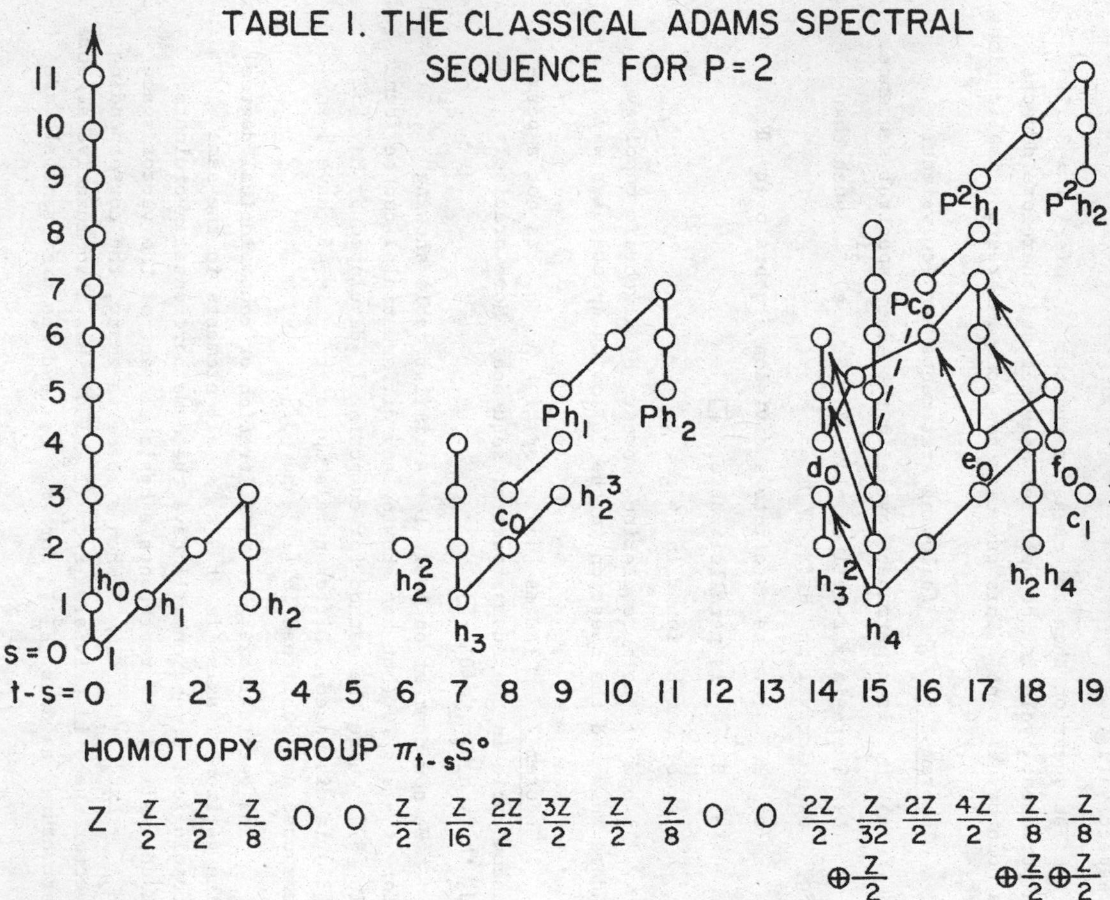

HOMOTOPY GROUP $\pi_{t-s}S^\circ$

t-s	0	1	2	3	4	5	6	7	8	9	10	11	12	13	14	15	16	17	18	19
$\pi_{t-s}S^\circ$	Z	$\frac{Z}{2}$	$\frac{Z}{2}$	$\frac{Z}{8}$	0	0	$\frac{Z}{2}$	$\frac{Z}{16}$	$\frac{2Z}{2}$	$\frac{3Z}{2}$	$\frac{Z}{2}$	$\frac{Z}{8}$	0	0	$\frac{2Z}{2}$	$\frac{Z}{32}\oplus\frac{Z}{2}$	$\frac{2Z}{2}$	$\frac{4Z}{2}$	$\frac{Z}{8}$	$\frac{Z}{8}\oplus\frac{Z}{2}\oplus\frac{Z}{2}$

derivations, i.e. the spectral sequence is one of commutative algebras. Hence many of the elements of $E_2^{s,t}$ are products of elements in lower filtration (i.e. lower values of s). The vertical lines represent multiplication by h_o (all powers of h_o are nontrivial), and the solid diagonal lines going up and to the right indicate multiplication by h_1. Certain multiplicative relations are built into the table, e.g. $h_1^3 = h_o^2h_2$, $h_1^3d_o = h_o^3e_o$, $h_2^3 = h_1^2h_3$, etc. Differentials in the spectral sequence are indicated by solid arrows going up and to the left, e.g. $d_2h_4 = h_oh_3^2$ (by Theorem 2.7) and $d_3h_oh_4 = h_od_o$. The broken line going from $h_o^3h_4$ to Pc_o indicates a nontrivial extension in the multiplicative structure, i.e. if ρ is the element of $\pi_{15}S^o$ detected by $h_o^3h_4$, then $\eta\rho \in \pi_{16}S^o$ is detected by Pc_o. The elements which are not products of h_i's can be expressed as Massey products (see [34]), e.g. $c_o = \langle h_1, h_o, h_2^2\rangle$, $c_1 = \langle h_2, h_1, h_3^2\rangle$, $d_o = \langle h_o, h_2^2, h_o, h_2^2\rangle$, and $f_o = \langle h_o^2, h_3^2, h_2\rangle$. The letter P denotes a periodicity operator $P: E_2^{s,t} \cap \ker h_o^4 \to E_2^{s+4,t+12}$ (see [5]), $Px = \langle x, h_o^4, h_3\rangle$. (In particular, $Ph_3 = h_o^3h_4$ and $Ph_2^2 = h_o^2d_o$.) Its analogue in the Novikov spectral sequence will be discussed in some detail in §5.

The corresponding homotopy groups are listed on the lower part of the table. They can be read off from the spectral sequence with the help of

<u>Proposition 2.8</u> If a and h_oa are nonzero permanent cycles in the Adams spectral sequence then the homotopy element detected by the latter is twice that detected by the former, i.e. multiplication by h_o in $E_2^{s,t}$ corresponds to multiplication by 2 in π_*S^o.

<u>Proof</u>. The statement is certainly true in dimension zero since we know $\pi_oS^o = Z$ and $\mathrm{Ext}_A^{*,*-*}(\mathbb{F}_2, \mathbb{F}_2) = \mathbb{F}_2[h_o]$. The statement in higher dimensions follows from the multiplicative properties of the spectral sequence. □

Note that all differentials in the spectral sequence decrease $t - s$ by 1 and increase s by at least 2. Hence elements that are low enough in the table cannot be targets of nontrivial differentials, while those that are high enough cannot be sources of same. In §4, we will show how all differentials and group extensions through dimension 17 can be determined by comparing the Adams and Novikov E_2-terms.

We conclude this section with a discussion of the Adams 2-line.

Theorem 2.9 (Adams [4]) $\mathrm{Ext}_A^{2,*}(\mathbb{F}_2, \mathbb{F}_2)$ has as a basis the elements $h_i h_j$ with $i \leq j$ and $i \neq j - 1$. □

As in the case of the 1-line, we can ask what happens to these elements in the spectral sequence. There is no possibility of any of them being the target of a differential, as such a differential would have to originate on the 0-line, which is trivial in positive dimensions. Hence any of these elements which is a permanent cycle will detect a nontrivial homotopy element. In the range of our table, we see that all such elements except $h_o h_4$ are permanent cycles. A big step forward in answering this question is

Theorem 2.10 (Mahowald-Tangora [30]) With the exceptions $h_o h_2$, $h_o h_3$, $h_2 h_4$ and possibly $h_2 h_5$ and $h_3 h_6$, the only elements of $\mathrm{Ext}_A^{2,*}(\mathbb{F}_2, \mathbb{F}_2)$ which can possibly be permanent cycles are h_j^2 and $h_1 h_j$. □

The elements h_j^2 are commonly known as the Arf (or Kervaire) invariant elements, due to the following result.

Theorem 2.11 (Browder [13]) There is a framed $(2^{j+1} - 2)$-manifold with nontrival Kervaire (or Arf) invariant iff the element h_j^2 is a permanent cycle in the Adams spectral sequence. □

These elements are known to survive (i.e. to be permanent cycles) for $j \leq 5$. They have been the object of intense investigation by Barratt, Mahowald, and others (see [29]). The corresponding homotopy element is commonly known as θ_j. Barratt and Mahowald have privately expressed the belief that if θ_j can be shown to exist for all of stable homotopy will follow with relative ease.

The survival of the elements h_ih_j is closely related to that of h_j^2. If θ_j exists and $2\theta_j = 0$, then the Toda bracket $\eta_{j+1} = \langle \theta_j, 2, \eta \rangle$ is detected by h_1h_{j+1}.

Mahowald has recently devised an extremely ingenious construction to prove

Theorem 2.12 (Mahowald [28]) The element $\eta_j \in \pi_{2^j}S^o$ exists for all $j \geq 3$ i.e. h_1h_j is a permanent cycle. []

In §7, we will indicate how θ_j and η_j appear in the Novikov spectral sequence and how the latter produces a new family of homotopy elements.

A computation of the Adams 3-line can be found in [74].

§3 Setting up the Novikov Spectral Sequence

The Novikov spectral sequence for the p-localization of the stable homotopy groups of spheres is obtained from Theorem 2.2 by setting $X = S^0$ and $Y_s = X_s \wedge BP$, where BP is the Brown-Peterson spectrum. If we replace BP by MU (the Thom spectrum associated with the unitary group; its homotopy is the complex cobordism ring) we obtain a 'global' Novikov spectral sequence which converges to all of π_*S^0, not just the p-component. Novikov [44] knew that the p-localizaton of the MU spectral sequence is isomorphic to the BP spectral sequence but he did not know how to compute with the latter. In either case, the identification of the E_2-term is as in the classical case and we have

Theorem 3.1 (Novikov [44], Adams [6] §III 15) There are spectral sequences

$$E_2^{s,t} = \mathrm{Ext}_{MU_*MU}^{s,t} (MU_*, MU_*) \Rightarrow \pi_*S^0$$

and

$$E_2^{s,t} = \mathrm{Ext}_{BP_*BP}^{s,t} (BP_*, BP_*) \Rightarrow (\pi_*S^0)_{(p)}. \quad \square$$

Novikov [44] and Zahler [75] used MU^*MU instead of MU_*MU. Since the former is not of finite type, this approach leads to certain technical difficulties, as one can see by reading [75].

In order to make Theorem 3.1 more explicit (see Proposition 3.16), we will describe BP_*BP. MU_*MU was determined by Novikov [44], and BP_*BP by Quillen [50]. Both are described lucidly by Adams [6], §II 16. An illuminating functorial description of BP_*BP has been given by Landweber [22], and will be discussed briefly in §8.

The structure of MU_*MU is easier to describe than that of BP_*BP. Nevertheless, the latter object, being much smaller, is easier to compute with, even if it takes some time to convince oneself of this. (It took me about four years.)

It would be a disservice to the reader not to begin the description of BP_*BP with a brief discussion of formal group laws. It is safe to say that every major conceptual advance in this subject since Quillen's work [50] has been connected directly or indirectly with the theory of formal group laws.

Definition 3.2 A one dimensional commutative formal group law over a commutative ring with unit R (hereafter and herebefore referred to simply as a formal group law) is a power series $F(x, y) \in R[[x, y]]$ such that

(i) $F(0, x) = F(x, 0) = x$ (identity element)

(ii) $F(x, y) = F(y, x)$ (commutativity)

(iii) $F(F(x, y), z) = F(x, F(y, z))$ (associativity).

Examples 3.3

(i) $F(x, y) = x + y$, the additive formal group law.

(ii) $F(x, y) = x + y + xy$, the multiplicative formal group law so named because $1 + F(x, y) = (1 + x)(1 + y)$.

(iii) $$F(x, y) = \frac{x\sqrt{1-y^4} + y\sqrt{1-x^4}}{1 + x^2 y^2}.$$

This is a formal group law over $Z[1/2]$, originally discovered by Euler in his investigation of elliptic integrals (see [62] pp. 1-9).

There are notions of homomorphisms and isomorphisms of formal groups over R, whose easy definitions we leave to the reader. A comprehensive and down-to-earth account of the theory of formal group laws has recently been provided by Hazewinkel [18]. Fröhlich's book [17] is also useful.

There is also a change of rings homomorphism. If $F(x, y)$ is a formal group law over R and $\theta: R \to S$ is a ring homomorphism, then $\theta(F(x, y))$ is a formal group law over S. With this in mind, we can ask for a universal formal group law $F_U(x, y)$ over a certain ring L (named for its discoverer, Lazard [25]) such that for any formal group law F over any ring R there is a unique ring homomorphism $\theta: L \to R$ such that $F(x, y) = \theta(F_U(x, y))$.

The Lazard ring L and the universal formal group law $F_U(x, y)$ are easy to construct. Simply write $F_U(x, y) = \sum a_{i,j} x^i y^i$ and regard the coefficients $a_{i,j}$ as indeterminates and set $L = Z[a_{i,j}]/(\sim)$, where $\sim$ denotes the relations among the $a_{i,j}$ imposed by Definition 3.2. It is obvious then that L and $F_U(x, y)$ have the desired properties. However, it was not easy to determine L explicitly. After Lazard [25] did so, Quillen made the following remarkable observation. Before stating it, recall that $MU^* = Z[x_1, x_2 \cdots]$ where $\dim x_i = -2i$ and $MU^* CP^\infty \simeq MU^*[[t]]$ where $t \in MU^2 CP^\infty$ is canonically defined (see [6] §II 2). Then we have

<u>Theorem 3.4</u> (Quillen [50]). The complex cobordism ring MU^* is isomorphic to the Lazard ring L, and under this isomorphism, $\Delta(t) = F_U(t \otimes 1, 1 \otimes t)$, where $\Delta: MU^* CP^\infty \to MU^*(CP^\infty x CP^\infty)$ is the map induced by the tensor product (of complex line bundles) map $CP^\infty x CP^\infty \to CP^\infty$. □

Proofs can be found in [6] §II 8, and [14]. This result establishes an intimate connection between complex cobordism and formal group laws. Most of the advances in the former since 1969 (the date of Quillen's theorem) have ignored complex manifolds entirely. It would be nice in some sense to have a description of the spectrum MU which is rooted entirely in formal group laws and which makes no mention of Thom spectra or complex manifolds. A recent result of Snaith [65] appears to be a step in this direction. Also the results of [59] imply that $MU^* CP^\infty$ as a Hopf algebra actually characterizes MU.

To proceed with the narrative, we have

Proposition 3.5 Let $F(x, y)$ be a formal group law over a torsion free ring R and define $f(x) \in R\otimes Q[[x]]$ by

$$f(x) = \int_0^x \frac{dt}{f_2(t, 0)}$$

where

$$f_2(x, y) = \frac{\partial F}{\partial y}.$$

Then $f(F(x, y)) = f(x) + f(y)$, i.e. $f(x)$ is an isomorphism over $R\otimes Q$ between F and the additive formal group law. □

Definition 3.6 The power series $f(x)$ above is the logarithm $\log_F(x)$ of the formal group law F.

The word logarithm is used because in the case of the multiplicative formal group law (Example 3.3(ii)),

$$\log_F x = \log(1 + x) = \sum_{n\geq 1} (-1)^{n+1} \frac{x^n}{n}$$

Theorem 3.7 (Mischenko [44]) $\log_{F_U} x = \sum_{n\geq 0} [CP^n] \frac{x^{n+1}}{n+1}$, where $[CP^n]$ denotes the element of MU^* represented by the complex manifold CP^n. □

Definition 3.8 A formal group law F over a torsion free ring is p-typical if $\log_F x = \sum_{i\geq 0} \ell_i x^{p^i}$.

This definition is due to Cartier [16] and can be generalized to rings with torsion (e.g. finite fields). See [18], [23] or [9].

Theorem 3.9 (Cartier [16]) Every formal group law F over a torsion free $Z_{(p)}$-algebra R is canonically isomorphic to a p-typical formal group law F_T, such that if $\log_F x = \sum a_i x^i$, then $\log_{F_T} x = \sum a_{p^i} x^{p^i}$. □

Proofs can be found in [18], [23] and [9]. It is possible to define a universal p-typical formal group law $F_T(x, y)$ over a p-typical Lazard ring L_T and we have

Theorem 3.10 (Quillen [50]). The p-typical Lazard ring L_T is isomorphic to the Brown-Peterson coefficient ring $BP_* = Z_{(p)}[v_1, v_2 \cdots]$ where $\dim v_n = 2(p^n - 1)$ in such a way that in $BP^*(CP^\infty \times CP^\infty)$, $\Delta(t) = F_T(t\otimes 1, 1\otimes t)$, where Δ is as in Theorem 3.4. □

Quillen's proof translates Cartier's canonical isomorphism (theorem 3.9) into a canonical retraction λ of $MU_{(p)}$ (the localization of MU at the prime p) onto BP. Its action on homotopy is determined by

$$\lambda_*[CP^n] = \begin{cases} [CP^n] & \text{if } n = p^i - 1 \\ 0 & \text{otherwise.} \end{cases}$$

It follows that $\ell(x) = \sum_{i\geq 0} [CP^{p^i-1}] \frac{x^{p^i}}{p^i}$ is the logarithm for the universal p-typical formal group law. We let $\ell_i = [CP^{p^i-1}]/p^i \in BP_*\otimes Q$, so $\ell(x) = \sum \ell_i x^{p^i}$, and $BP_*\otimes Q = Q[\ell_i]$. We define the formal sum $\sum^F x_i$ by $\ell(\sum^F x_i) = \sum \ell(x_i)$ or equivalently, $\sum^F x_i = F(x_1, F(x_2, F(x_3, \cdots)\cdots)$.

We are now ready to describe BP_*BP. Since it is $\pi_*BP\wedge BP$, there are two maps $\eta_L, \eta_R\colon BP_* \to BP_*BP$ (the left and right units) induced by $BP = BP \wedge S^0 \to BP \wedge BP$ and $BP = S^0 \wedge BP \to BP \wedge BP$ respectively. The latter map $\eta_R\colon \pi_*BP \to BP_*BP$ is the Hurewicz map in BP_*-homology.

Since BP_*BP is the dual of BP^*BP, the algebra of BP^*-cohomology operations, it has a coproduct $\Delta\colon BP_*BP \to BP_*BP\otimes_{BP_*} BP_*BP$, where the tensor product is with respect to the bimodule structure given by the maps η_R and η_L.

The maps η_R and Δ are related by the commutative diagram

$$
(3.11)\qquad
\begin{array}{ccc}
BP_* & \xrightarrow{\eta_R} & BP_*BP = BP_*\otimes_{BP_*} BP_* BP \\
\downarrow{\scriptstyle \eta_R} & & \downarrow{\scriptstyle \eta_R \otimes 1} \\
BP_*BP & \xrightarrow{\Delta} & BP_*BP\otimes_{BP_*} BP_* BP
\end{array}
$$

Again, we refer the reader to [22] for novel and illuminating interpretation of this structure, which admittedly seems a bit peculiar at first.

Theorem 3.12 (Quillen [50], Adams [6] §II 16). As an algebra, $BP_*BP = BP_*[t_1, t_2, \cdots]$ where $\dim t_i = 2(p^i - 1)$. The structure maps $\Delta\colon BP_*BP \to BP_*BP\otimes_{BP_*} BP_*BP$ and $\eta_L, \eta_R\colon BP \to BP_*BP$ are given by

$$\sum_{i\le 0} \ell(\Delta(t_i)) = \sum_{i,j\ge 0} \ell(t_i \otimes t_j^{p^i})$$

(or equivalently $\sum_{i\ge 0}^{F} \Delta(t_i) = \sum_{i,j\ge 0}^{F} t_i\otimes t_j^{p^i}$) where $t_o = 1$, $\eta_L(\ell_n) = \ell_n$, and $\eta_R \ell_n = \sum_{0\le i\le n} \ell_i t_{n-i}^{p^i}$. □

The reader can easily verify that these formulae satisfy (3.11), and moreover that using (3.11), η_R determines Δ.

Example $\Delta(t_1) = t_1\otimes 1 + 1\otimes t_1$, $\Delta(t_2) = t_2\otimes 1 + t_1\otimes t_1^p + 1\otimes t_2 - \ell_1 \sum_{0<i<p} \binom{p}{i} t_1^i \otimes t_1^{p-i}$, and $\Delta(t_3)$ is too messy to write down in public. Partial simplification of these formulae is achieved in [57] and [54].

We are not finished yet. In order to use Theorem 3.12 in practice, we need to relate the generators ℓ_1 of $BP_*\otimes Q$ to generators v_i of BP_*, which as yet have not been defined.

Theorem 3.13 (Hazewinkel [18], [19]. Generators v_i of BP can be defined recursively by the formula

$$p\ell_n = \sum_{0\le i<n} \ell_i v_{n-i}^{p^i} . \quad \square$$

Example

$$\ell_1 = \frac{v_1}{p}, \quad \ell_2 = \frac{v_1^{1+p}}{p^2} + \frac{v_2}{p},$$

and

$$\ell_3 = \frac{v_1^{1+p+p^2}}{p^3} + \frac{v_1 v_2^p + v_2 v_1^{p^2}}{p^2} + \frac{v_3}{p}.$$

The following formula for $\eta_R(v_n)$ is useful

Theorem 3.14 [57]

$$\sum_{\substack{i>0\\ j\ge 0}}^{F} v_i t_j^{p^i} \equiv \sum_{\substack{i>0\\ j\ge 0}}^{F} \eta_R(v_i)^{p^j} t_j \mod (p) . \quad \square$$

Example. $\eta_R v_2 \equiv v_2 + v_1 t_1^p - v_1^p t_1 \mod p$.

Araki [9] has defined another set of generators v_i by the slightly different formula $p\ell_n = \sum_{0\le i\le n} \ell_i v_{n-i}^{p^i}$ where $v_o = p$. This gives messier expressions for the ℓ_i, e.g. $\ell_1 = v_1/(p - p^p)$, but the analogue of Theorem 3.14 is true on the nose, not just mod (p). Araki's and Hazewinkel's generators are the same mod (p).

We conclude this section by exhibiting a complex whose cohomology is $\mathrm{Ext}_{BP_*BP}(BP_*, BP_*)$, i.e. the E_2-term of the Novikov spectral sequence. Let $\Omega^*(BP_*)$ be the complex

$$\Omega^o BP_* \xrightarrow{d^o} \Omega^1 BP_* \xrightarrow{d^1} \Omega^2 BP_* \xrightarrow{d^2} \Omega^3 BP_* \longrightarrow \cdots$$

where

$$\Omega^s BP_* = BP_* \otimes_{BP_*} BP_*BP \otimes_{BP_*} \cdots \otimes_{BP_*} BP_*BP$$

(s factors of BP_*BP) with

(3.15) $$d^s(v \otimes x_1 \otimes \cdots x_s) = \eta_R(v) \otimes x_1 \otimes \cdots x_s$$
$$+ \sum_{i=1}^{s} (-1)^i \, v \otimes x_1 \otimes \cdots x_{i-1} \otimes \Delta(x_i) \otimes x_{i+1} \otimes \cdots x_s$$
$$- (-1)^s v \otimes x_1 \otimes \cdots x_s \otimes 1 .$$

This is the <u>cobar complex</u> for BP_*. Let $\overline{BP_*BP} = (t_1, t_2 \cdots) \subset BP_*BP$ and define the <u>normalized cobar complex</u> $\tilde{\Omega}^s(BP_*)$ by $\tilde{\Omega}^o BP_* = BP_*$ and $\tilde{\Omega}^s(BP_*) = BP_* \otimes_{BP_*} \overline{BP_*BP} \otimes_{BP_*} \cdots \overline{BP_*BP}$ (with s factors $\overline{BP_*BP}$). Then (3.15) gives $d^s : \tilde{\Omega}^s BP_* \longrightarrow \tilde{\Omega}^{s+1} BP_*$, e.g. $d^o\, v = \eta_R(v) - \eta_L(v)$. Then we have

<u>Proposition 3.16</u>

$$H^* \Omega^* BP_* \cong H^* \tilde{\Omega}^* BP_* \cong \mathrm{Ext}_{BP_*BP}(BP_*, BP_*). \; \square$$

<u>Corollary 3.17</u> (Sparseness) In the Novikov spectral sequence $E_r^{s,t} = 0$ if $q \nmid t$, where $q = 2p - 2$, and $E_{qm+2}^{s,t} = E_{qm+q-1}^{s,t}$ for all $m \geq 0$. $\square$

Two systematic methods of computing Ext BP_* through a range of dimensions have been developed [54] [55], and we hope they will be applied soon.

§4 Comparing the Adams and Novikov Spectral Sequences

In this section, we make some general remarks about the relation between the two spectral sequences, and then we make a specific comparison in low dimensions (≤ 17) at $p = 2$.

To begin with, the natural map $BP \to H\mathbb{F}_p$ of spectra induces a map of Adams resolutions (2.1) and hence a spectral sequence homomorphism (i.e. one which commutes with differentials) Φ from the Novikov to the Adams spectral sequence. This implies the following

Proposition 4.1 If a homotopy element is detected in the Adams spectral sequence by an element in $E_\infty^{s,t}$, then it is detected in the Novikov spectral sequence by an element in some $E_\infty^{s',t'}$ with $t' - s' = t - s$ and $s' \le s$. If a homotopy element is detected in the Novikov spectral sequence by an element in $E_\infty^{s,t}$, then it is detected in the Adams spectral sequence by an element in some $E_\infty^{s',t'}$ with $t' - s' = t - s$ and $s' \ge s$.

This fact, along with knowledge of the behavior of Φ on the 1-line (see §5) leads to an easy proof of the first part of Theorem 2.7 and its odd primary analogue. The latter was first proved by other means by Liulevicius [26] and Shimada-Yamanoshita [61]. In §9 of [39] we calculated the image of Φ on the 2-line for $p > 2$ and thereby proved an odd primary analogue of Theorem 2.10. Our ignorance of the Novikov 2-line (see §6) for $p = 2$ prevented us from giving a similar proof of Theorem 2.10 itself.

Next, we will describe some spectral sequences which indicate a certain relationship between the two E_2-terms. We begin with a Cartan-Eilenberg ([15] p. 349) spectral sequence for the Adams E_2-term. Recall that for $p > 2$ $A_* = \mathbb{F}_p[\xi_1, \xi_2 \cdots] \otimes E(\tau_0, \tau_1 \cdots)$ and for $p = 2$ $A_* = \mathbb{F}_2[\xi_1, \xi_2 \cdots]$ (see [66]). Define extension of Hopf algebras $E \to A \to P$ by $P_* = \mathbb{F}_p[\xi_i]$ for $p > 2$ and $P_* = \mathbb{F}_2[\xi_i^2]$ for $p = 2$.

Theorem 4.2 (a) There is a spectral sequence converging to $\mathrm{Ext}_A(\mathbb{F}_p, \mathbb{F}_p)$ with $E_2^{s,t} = \mathrm{Ext}_P(\mathbb{F}_p, \mathrm{Ext}_E(\mathbb{F}_p, \mathbb{F}_p))$

(b) $\mathrm{Ext}_E(\mathbb{F}_p, \mathbb{F}_p) = \mathbb{F}_p[a_o, a_1, \cdots]$

where $a_i \in \mathrm{Ext}^{1,2p^i-1}$ is represented by the image of τ_i (or ξ_{i+1} if $p = 2$) in the cobar complex for E.

(c) The spectral sequence collapses from E_2 for $p > 2$.

Proof: (a) is a special case of Theorem XVI 6.1 of [15]. (b) follows from the fact that E is an exterior algebra. For (c) observe that for $p > 2$, we can give A_* a second grading based on the number of τ's which is preserved by both the coproduct and product. (The coproduct does not preserve this grading for $p = 2$). The fact that differentials must respect this grading implies that the spectral sequence collapses. []

Next, we construct the so-called algebraic Novikov spectral sequence ([44], [36]) which converges to the Novikov E_2-term and has itself the same E_2-term (indexed differently) as that of that Cartan-Eilenberg spectral sequence above.

Let $I = (p, v_1, v_2 \cdots) \subset BP_*$. This ideal is independent of the choice of generators v_i. If we filter BP_* by powers of I, the associated bigraded ring $E^o BP_*$ is isomorphic to $\mathbb{F}_p[a_o, a_1, \cdots]$ where a_i has dimension $2(p^i - 1)$ and filtration 1 and corresponds to the generators v_i (where $v_o = p$). This filtration can be extended to BP_*BP and to the normalized cobar complex $\tilde{\Omega}^*(BP_*)$. We have $E^o BP_*BP = E^o BP_*[t_i]$ and Theorem 3.12 implies that $\Delta t_n = \sum_{0 \le i \le n} t_i \otimes t_{n-i}^{p^i} \in E^o BP_*BP$. It follows that $BP_*BP/I \cong P_*$ as Hopf algebras. To describe the coboundary operator in $E^o \tilde{\Omega}^*(BP_*)$, it remains to determine $d^o = E^o BP_* \to E^o \overline{BP_*BP}$. It follows from (3.15) and Theorem 3.14 that $d^o a_n = \sum_{0 \le i < n} a_i t_{n-i}^{p^i}$. This agrees, via the appropriate isomorphism, with the d_1 in the Cartan-Eilenberg spectral sequence of Theorem 4.2. Combining all these remarks we get

Theorem 4.3 (Novikov [44], Miller [36]) The filtration of $\tilde{\Omega}^*BP_*$ by powers of I leads to a spectral sequence converging to $\mathrm{Ext}_{BP_*BP}(BP_*BP_*)$ whose E_2-term is isomorphic to $\mathrm{Ext}_{P_*}(\mathbb{F}_p, \mathbb{F}_p[a_i])$. □

Theorems 4.2 and 4.3 give algebraic spectral sequences having the same E_2-term (up to reindexing) and converging to the Adams and Novikov E_2-terms respectively. For $p > 2$ the former collapses, so in that case the spectral sequence of Theorem 4.3 can be regarded as passing from the Adams E_2-term (reindexed) to the Novikov E_2-term. Presumably (but this has not been proved) differentials in this spectral sequence correspond in some way to differentials in the Adams spectral sequence. For example, one can easily find the Hopf invariant differentials, i.e. those originating on the Adams 1-line, in this manner. Philosophically, Theorems 4.2 and 4.3 imply that for $p > 2$, any information that can be gotten out of the Adams spectral sequence can be obtained more efficiently from the Novikov spectral sequence.

Another way of describing this situation is the following. According to the experts (i.e. M. C. Tangora), all known differentials in the Adams spectral sequence for odd primes are caused by two phenomena. Each is a formal consequence (in some devious way possibly involving Massey products [34]) of either the Hopf invariant differentials or the relations described by Toda in [70] and [71]. In computing the Novikov E_2-term via Theorem 4.3 or any other method one effectively computes all the Hopf invariant differentials in one fell swoop and is left with only the Toda type differentials to contend with. Better yet, for $p = 3$, all known differentials in the Novikov spectral sequence are formal consequences of the first one in dimension 34 (see [52]). One is tempted to conjecture that this is a general phenomenon, i.e. that if one knows the Novikov E_2-term and the first nontrivial differential, then one knows all of the stable homotopy groups of spheres. However, apart from limited empirical evidence, we have no reason to believe in such an optimistic conjecture.

At the prime 2, the relation between the Adams and Novikov E_2-terms is more distant since the spectral sequence of Theorem 4.2 does not collapse. In this case, the Adams spectral sequence does yield some information more readily than the Novikov spectral sequence, and the use of the two spectral sequences in concert provides one with a very powerful tool which has, as yet, no odd primary analogue. We will illustrate by comparing the two through dimension 17, the limit of Zahler's computation [75].

Table 2 is a reproduction of Zahler's table, with the added feature that all elements are named. We will explain this notation in the next two sections. Each $E_2^{s,t}$ is finite except for $E_2^{o,o} = Z_{(2)}$. Each circle in the table represents an element of order 2 and each square represents an element of higher order. Specifically, $\alpha_{2i/j}$ has order 2^j. The diagonal lines going up and to the right indicate multiplication by $\alpha_1 = \eta$, and an arrow pointing in this direction indicates that multiplication by all powers of α_1 is nontrivial. The arrows going up and to the left indicate differentials, and the broken vertical lines indicate nontrivial group extensions.

We now show all the differentials and extensions in the two tables can be deduced by purely algebraic arguments, i.e. without resorting to any geometric considerations.

First, observe that there is no room for any nontrivial differentials in the Adams spectral sequence below dimension 14. (The multiplicative structure precludes nontrivial differentials on h_1 and $h_1\ h_3$.) There are also no nontrivial group extensions in this range other than those implied by proposition 2.8. (The fact that $2\eta = 0$ precludes nontrivial extensions in dimensions 8 and 9.) One also knows that $\eta^3\sigma = 0$ because $\eta^3\sigma = \eta\nu^3 = (\eta\nu)\nu^2 = 0$. One can deduce that $\eta^3\sigma = 0$, instead of the element detected by Ph_1^2, by comparing the filtrations of the corresponding elements in the Novikov spectral sequence. The former $\alpha_1\ ^3\alpha_{4/4}$ has filtration 4, while the latter, $\alpha_1\ \alpha_5$, has filtration 2.

TABLE 2. THE NOVIKOV SPECTRAL SEQUENCE FOR P = 2

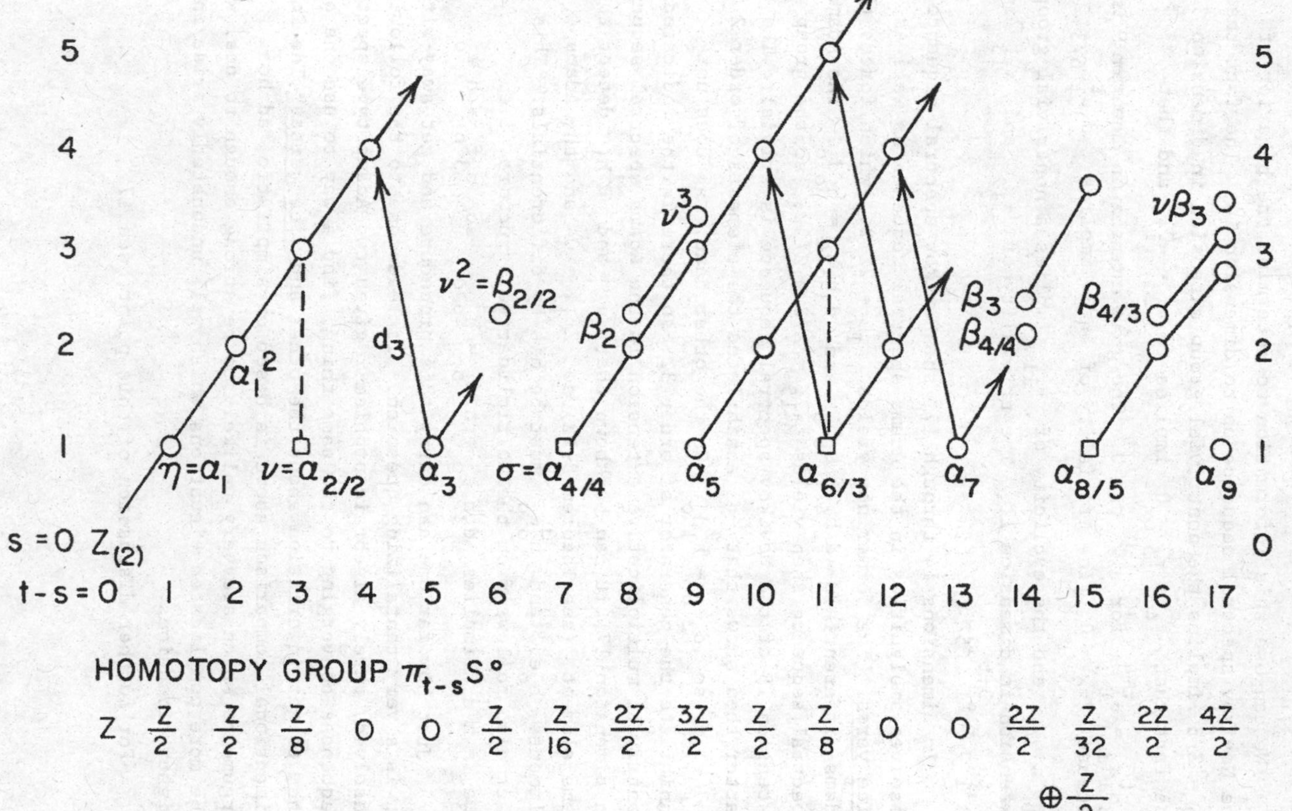

We can use this information to determine the behavior of the Novikov spectral sequence up to dimension 14. The fact that $\pi_3 = Z/8$ implies the nontrivial group extension in dimension 3. The fact that $\pi_4 = \pi_5 = 0$ implies $d_3\alpha_3 = \alpha_1^4$, and that $d_3\alpha_1^t\alpha_3 = \alpha_1^{t+4}$ for all $t \geq 0$. The group extension in dimension 9 is trivial because $2\pi_9 = 0$. The triviality of $\eta^3\sigma$ implies $d_3\alpha_1^t\alpha_{6/3} = \alpha_1^{3+t}\alpha_{4/4}$, and the cyclicity of π_{11} implies a nontrivial group extension in dimension 11. The triviality of π_{12} and π_{13} imply $d_3\alpha_1^t\alpha_7 = \alpha_1^{3+t}\alpha_5$.

In dimensions 14 through 17, the Novikov spectral sequence resolves ambiguities in the Adams spectral sequence as well as vice versa. The former now yields $\pi_{14} = 2Z/2$, which forces the Adams differentials $d_2h_4 = h_oh_3^2$ and $d_3h_oh_4 = h_od_o$. The Adams spectral sequence then yields $\pi_{15} = Z/2 \oplus Z/32$, so the group extension 15 of the Novikov spectral sequence is trivial. The latter then shows that η annihilates the elements of order 2 in π_{15}, so $d_2e_o = h_1^2d_o$. On the other hand, η does not annihilate the generator of order 32, so there is the indicated nontrivial multiplicative extension in the Adams spectral sequence. In dimension 17, it can be shown that α_9 and P^2h_1 detect the same element, (see Theorem 5.12) so $2\pi_{12} = 0$ and the Adams elements $h_o^2e_1$ and $h_o^3e_o$ must be hit by differentials. This last fact also follows from the multiplicative structure, i.e. $d_3e_o = h^2d_o$ implies $d_3h_1e_o = h_1^3d_o = h_o^3e_o$, so $d_3f_o = h_oe_o$.

Just how far one can carry this procedure and get away with it is a very tantalizing question. It leads one to the following unsolved, purely algebraic problem: given two Adams type spectral sequences converging to the same thing, find a way to use one of them to get information about the other and vice versa. The low dimensional comparison above is based on simplistic, ad hoc arguments which are very unlikely to be strong enough to deal with the more complicated situations which will undoubtedly arise in higher dimensions.

For further discussion of this point, see §7.

§5 First Order Phenomena in the Novikov Spectral Sequence

We will not say exactly what we mean by nth order phenomena until §8. Roughly speaking, first order phenomena consist of Im J and closely related homotopy elements as described by Adams in [1]. The manner in which the Novikov spectral sequence detects these elements was apparently known to Novikov [44] and was sketched by Zahler [75]. Most of the detailed computations necessary were described in §4 of [39] but some of the proofs we present here are new.

We begin by computing the Novikov 1-line. First, we need some notation. For a BP_*BP-comodule M, $\mathrm{Ext}_{BP_*BP}(BP_*, M)$ will be denoted simply by Ext M. If $M = BP_*X$, then Ext M is the E_2-term of the Novikov spectral sequence for π_*X.

Proposition 5.1 If M is a cyclic BP_*-module, Ext M = $H^*(M \otimes_{BP_*} \tilde{\Omega} BP_*)$. □

A proof can be found in §1 of [39].

Now $\mathrm{Ext}^1 BP_*$, the Novikov 1-line, is a torsion group, so we begin by finding the elements of order p. Consider the short exact sequence

$$(5.2) \qquad 0 \to BP_* \xrightarrow{p} BP_* \to BP_*/(p) \to 0.$$

The image of the connecting homomorphism $\delta_o : \mathrm{Ext}^o BP_*/(p) \to \mathrm{Ext}^1 BP_*$. is, by elementary arguments, the subgroup of elements of order p. The following result was first published by Landweber [21] and can be derived easily from Theorem 3.14.

Theorem 5.3 Let $I_n = (p, v_1, \cdots v_{n-1}) \subset BP_*$. Then BP_*/I_n is a BP_*BP-comodule and $\mathrm{Ext}^o BP_*/I_n \cong \mathbb{F}_p[v_n]$. □

Corollary 5.4 $\mathrm{Ext}^o BP_*/(p) = \mathbb{F}_p[v_1]$ and $\delta_o v_1^t = \alpha_t \neq 0 \in \mathrm{Ext}^1 BP_*$ for all $t > 0$.

Proof The nontriviality of α_t follows from the long exact sequence in Ext associated with (5.2), in which we have

$$\mathrm{Ext}^0 BP_* \to \mathrm{Ext}^0 BP_*/(p) \xrightarrow{\delta_o} \mathrm{Ext}^1 BP_* .$$

In positive dimensions, δ_o is monomorphic because $\mathrm{Ext}^0 BP_*$ is trivial. □

In [75] α_t denotes the generator of $\mathrm{Ext}^{1,2t}\ BP^*$ for $p = 2$, but our α_t is an element of order 2 in that group.

All that remains in computing $\mathrm{Ext}^1 BP_*$, the Novikov 1-line, is determining how many times we can divide α_t by p. From §4 of [39] we have

Theorem 5.5

(a) For $p > 2$, $\alpha_t \in \mathrm{Ext}^{1,qt}\ BP_*$ is divisible by t but not by pt, i.e. $\mathrm{Ext}^{1,qt} BP_* \simeq Z/(p^{1+\nu(t)})$ where $p^{\nu(t)}$ is the largest power of p which divides t.

(b) For $p = 2$, $\alpha_t \in \mathrm{Ext}^{1,2t} BP_*$ is divisible by

$$\begin{cases} t \quad \text{but not by} \quad 2t \quad \text{if} \quad t \quad \text{is odd or} \quad t = 2 \\ 2t \quad \text{but not by} \quad 4t \quad \text{if} \quad t \quad \text{is even and} \quad t > 2; \end{cases}$$

i.e.

$$\mathrm{Ext}^{1,2t} BP_* = \begin{cases} Z/(2) & \text{if} \quad t \quad \text{is odd} \\ Z/(4) & \text{if} \quad t = 2 \\ Z/(2^{2+\nu(t)}) & \text{if} \quad t \quad \text{is even} \quad \text{and} \quad t > 2. \end{cases}$$ □

It is easy to see that α_t is divisible by $p^{\nu(t)}$. From the fact that $\eta_R v_1 = v_1 + pt_1$ (using Hazewinkel's v_1 (Theorem 3.13) and Theorem 3.12), one computes $\delta_o v_1^t = \frac{1}{p}[(v_1 + pt_1)^t - v_1^t]$ which is easily seen to be divisible by $p^{\nu(t)}$.

We can now explain part of the notation of Table 2. $\alpha_{t/i}$ denotes a certain element (defined precisely in [39]) of order p^i in $\mathrm{Ext}^{1,qt}BP_*$. In particular, $\alpha_{t/1} = \alpha_t$ and $p^{i-1}\alpha_{t/i} = \alpha_t$.

As in §2, one can ask which of these elements are permanent cycles.

Theorem 5.6 (Novikov [44]) For $p > 2$, Im J maps isomorphically to $\mathrm{Ext}^1 BP_*$, i.e. each element of $\mathrm{Ext}^1 BP_*$ is a nontrivial permanent cycle and in homotopy $p\alpha_t = 0$ for all $t > 0$. □

The homotopy elements $\alpha_t \in \pi_{qt-1} S^0$ can also be constructed inductively by Toda brackets, specifically $\alpha_t = \langle \alpha_{t-1}, p\iota, \alpha_1 \rangle$ [72].

As Table 2 indicates, the situation at $p = 2$ is not so simple. Let $x_t \in \mathrm{Ext}^{1,2t}BP_*$ be a generator. Then from [39] §4 we have

Theorem 5.7 For all $s > 0$ and $t \neq 2$, $\alpha_1^s x_t$ generates a nontrivial summand of order 2 in $\mathrm{Ext}^{1+2,2s+2t}BP_*$. □

(This is a consequence of Theorem 5.10 below.) Note that this says that for $t > 2s + 2$ all the groups $E_2^{s,t}$ which are not trivial by sparseness (Corollary 3.17) are in fact nontrivial.

The behavior of these elements in the spectral sequence and in homotopy is as follows.

Theorem 5.8 In the Novikov spectral sequence for $p = 2$

(a) $d_3\, \alpha_1^s\, \alpha_{4t+3} = \alpha_1^{s+3}\, \alpha_{4t+1}$ and $d_3\, \alpha_1^s\, x_{4t+6} = \alpha_1^{3+s}\, x_{4t+4}$ for all $s, t \geq 0$.

(b) For $t > 0$, the elements x_{4t}, $\alpha_1 x_{4t}$, $\alpha_1^2 x_{4t}$, α_{4t+1}, $\alpha_1\alpha_{4t+1}$

$\alpha_1^2\alpha_{4t+1}$ and $2x_{4t+2} = \alpha_{4t+2/2}$ are all nontrivial permanent cycles, as are α_1, α_1^2, α_1^3 and $\alpha_{4/2} = x_2$. In π_*S^o, we have $2\alpha_{4t} = 2\alpha_{4t+1} = 0$ and $2\alpha_{4t+2} = \alpha_1^2\alpha_{4t+1}$, i.e. there is a nontrivial group extension in dimension $8t + 3$.

(c) The image of the J-homomorphism is the group generated by x_{4t}, $\alpha_1 x_{4t}$, $\alpha_1^2 x_{4t}$ and $\alpha_{4t+2/2}$ (which generates a Z/8 summand with $4\alpha_{4t+2/2} = \alpha_1^2\alpha_{4t+1}$).

This result says that the following pattern occurs in the Novikov E_∞-term as a direct summand for all $k > 0$.

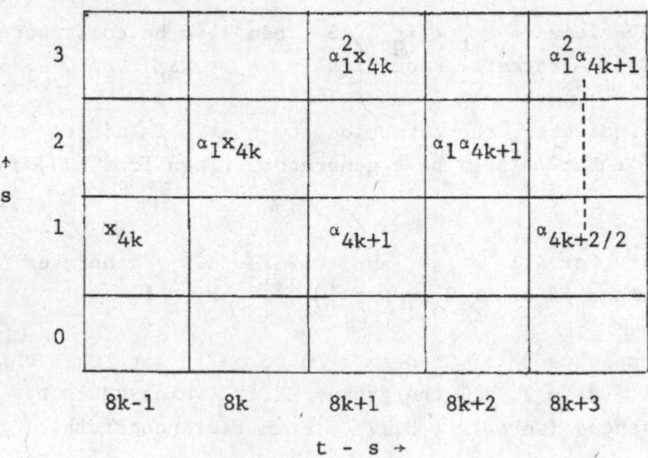

where all elements have order 2 except $\alpha_{k+2/2}$ which has order 4 and x_{4k} which has order $2^{\nu(k)+4}$, and the broken vertical line indicates a nontrivial group extension.

In [27] the elements x_{4t}, α_{4t+1} and $\alpha_{4t+2/2}$ are denoted by ρ_t, μ_t and ξ_t respectively, while Adams [1] denotes α_{4t+1} and $\alpha_1\alpha_{4t+1}$ by μ_{8t+1} and μ_{8t+2} respectively.

Parts (a) and (b) seem to have been known to Novikov [44] as was the fact that Im J maps onto the groups indicated in (c). The fact that this map from Im J is an isomorphism requires the Adams Conjecture [1], [49]. We will prove (a) and a weaker form of (b), namely we will only show that the elements said to have order 16 or less are permanent cycles. Another proof of this fact, based on a comparison of the Adams and Novikov spectral sequences can be derived from Theorem 5.12. The J homomorphism can be used to show that x_{4t} is a permanent cycle.

Our proof is based on an analysis of the mod 16 Moore spectrum, which we denote by M(16). As it is somewhat involved, the reader may want to proceed directly to §6.

We begin with Table 3, which displays the Novikov spectral sequence for M(16) through dimension 13. The notation is the same as in Table 2, from which Table 3 can be easily deduced. Circles represent elements or order 2, and squares represent elements of higher order. The orders of 1, ν, b, σ, v, $\alpha_{6/3}$ and d are 16, 4, 4, 16, 16, 8 and 8 respectively. There are various multiplicative relations among these elements, e.g. $2d = vb$, $v\alpha_{2t+1} = \alpha_{2t+5}$, and $v\alpha_{4t+2/3} = \alpha_{4t+6/3}$ which are easy to find.

The element $v \in \mathrm{Ext}^{0,8}BP_*/16$ has the property that $v^t \neq 0$ for all $t > 0$. Since v is a permanent cycle and M(16) is a ring spectrum, nontrivial differentials and group extensions respect multiplication by powers of v. We wish to describe which elements of Ext $BP_*/16$ are not annihilated by any power v, i.e. to describe Ext $BP_*/16$ mod 'v-torsion'. The methods of [39] (also sketched in §8) make this possible. Let $R = (Z/16)[v,\alpha_1]/(2\alpha_1)$. Then we have

<u>Theorem 5.9</u> In dimensions ≥6, Ext $BP_*/16$ mod v-torsion is the R-module generated by c, v, va, d, σ, $a\sigma$, $\alpha_{6/3}$ and α_7 with relations $2c = 2va = 8d = 2a\sigma = 8\alpha_{6/3} = 2\alpha_7 = 0$. (More precisely,

TABLE 3. THE NOVIKOV SPECTRAL SEQUENCE FOR THE MOD 16 MOORE SPECTRUM M(16)

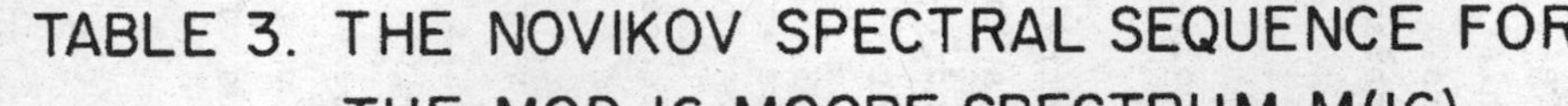

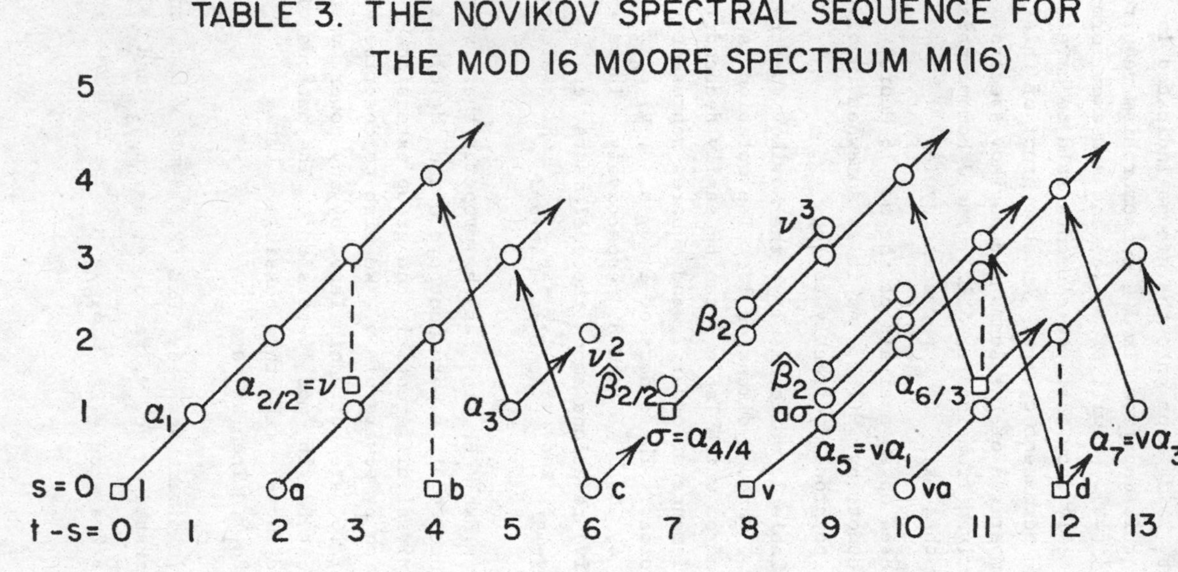

$a = 8v_1$ $b = 4v_1^2$ $c = 8v_1^3$ $v = v_1^4 + 8v_1v_2$ $d = 2v_1^6 + 8v_1^3v_2$

$\widehat{\beta}_{2/2} = 8(t_1^4 + v_1^2t_1^2)$ $\widehat{\beta}_2 = v_1\widehat{\beta}_{2/2}$

this describes the image of $\mathrm{Ext}\ BP_*/16$ in $v^{-1}\ \mathrm{Ext}\ BP_*/16$ in dimensions ≥ 6. In dimensions 0 through 5 one also has the elements $1, a, \nu$ (note $v\nu = 2\alpha_{6/3}$), b ($vb = 2d$) and α_3 ($v\alpha_3 = \alpha_7$).) □

This can be deduced from the corresponding statement about $\mathrm{Ext}\ BP_*/2$, namely

Theorem 5.10 $\mathrm{Ext}\ BP_*/2$ modulo v_1-torsion (i.e. the image of $\mathrm{Ext}\ BP_*/2$ in $v_1^{-1}\ \mathrm{Ext}\ BP_*/2$) is $\mathbb{F}_2[v_1, \alpha_1, \sigma]/(\sigma^2)$. □

The method of proof for this result will be discussed in §8.

In order to relate the behavior of the spectral sequence for M(16) to that for the sphere, we need the Geometric Boundary Theorem.

Theorem 5.11 (Johnson-Miller-Wilson-Zahler [20]) Let $W \xrightarrow{f} X \xrightarrow{g} Y \xrightarrow{h} \Sigma W$ be a cofibre sequence of finite spectra such that $BP_*(h) = 0$, i.e. such that

$$0 \longrightarrow BP_*W \xrightarrow{f_*} BP_*X \xrightarrow{g_*} BP_*Y \longrightarrow 0$$

is exact, and let $\delta\colon \mathrm{Ext}^*BP_*Y \to \mathrm{Ext}^{*+1}BP_*W$ be the connecting homomorphism. Then if $\bar{x} \in \mathrm{Ext}\ BP_*Y$ is a permanent cycle detecting $x \in \pi_*Y$, then $\delta(\bar{x}) \in \mathrm{Ext}\ BP_*W$ is a permanent cycle detecting $h_*(x) \in \pi_*\Sigma W$. □

Now we can prove Theorem 5.8 a) In $\pi_*M(16)$ we have $\alpha_1^3 a = 0$ so $\delta(x_1^3 v^t a) = \alpha_1^3\alpha_{4t+1} = 0$. Hence by Theorem 5.11, a differential must hit $\alpha_1^3\alpha_{4t+1}$, and by Sparseness (Corollary 3.17) and our knowledge of $\mathrm{Ext}^1 BP_*$ (Theorem 5.5 (b)), the only possibility $d_3\alpha_{4t+3}$.

For the other family of d_3's, one can show that $\delta_o\ (a\sigma\alpha_1^2 v^t) = \alpha_1^3\ x_{4t+4}$, so by Theorem 5.11, $\alpha_1^3\ x_{4t+4}$ must be hit by a differential, and the only possible source is $\alpha_{4t+6/3}$.

For the group extensions in (b), we have $\alpha_1^2 v^t a = 4v^t b$ in $\pi_* M(16)$, so $\delta_o(\alpha_1^2 v^t a) = \alpha_1^2 \alpha_{4t+1}$ detects twice the element by $\delta_o(2v^t b) = \alpha_{4t+2}$. For the permanent cycles of order 16 or less, we have in $\pi_* S^o$, $\alpha_{4t+4/4} = \delta v^t \sigma$, $\alpha_{4t+1} = \delta v^t a$, and $\alpha_{4t+2/2} = \delta v^t \nu$. This concludes the proof of our weakened form of theorem 5.8. □

We draw the reader's attention to the basic idea of the above proof. Theorem 5.9 provides a lever with which we can extrapolate the low dimensional information of Table 3 to the infinite amount of information contained in Theorem 5.8. This kind of extrapolation is typical of applications of the Novikov sepctral sequence to stable homotopy; a finite amount of low dimensional information can often be made to yield an infinite number of nontrivial homotopy elements.

We conclude this section with a discussion of how the phenomena of Theorem 5.8 appear in the Adams spectral sequence. It follows from Corollary 5.4 that any element of order 2 in Ext BP_* can be 'multiplied' (modulo some indeterminacy) by v_1. In other words, the $\mathbb{F}_2[v_1]$-module structure of Ext $BP_*/2$ translates to a Massey product operator which sends an element x of order 2 to $\langle x, 2, \alpha_1 \rangle$. In a similar way, the fact that $\text{Ext}^o\, BP_*/16 \subset (Z/16)\,[v_1^4 + 8v_1v_2]$ leads to an 8-dimensional periodicity operator which sends an element x of order 16 to $\langle x, 16, \alpha_{4/4} \rangle$.

This is readily seen to correspond to the Adams periodicity operator $Px = \langle x, h_o^4, h_3 \rangle$ discussed in §2. The important difference is that the Novikov operator preserves filtration, while the Adams operator raises filtration by 4. With this in mind, one can prove

<u>Theorem 5.12</u> The Novikov elements $\alpha_{4t/4}$, $\alpha_1 x_{4t}$, $\alpha_1^2 x_{4t}$, α_{4t+1}, $\alpha_1\alpha_{4t+1}$ and $\alpha_{4t+2/2}$ correspond to (i.e. detect the same

homotopy elements as) the Adams elements $P^{t-1} h_3$, $P^{t-1} c_o$, $h_1 P^{t-1} c_o$, $P^t h_1$, $P^t h_1^2$ and $P^t h_2$ respectively for all $t \geq 1$ (the last three elements correspond for $t \geq 0$.) □

Hence the Adams spectral sequence shows that these elements are permanent cycles, since they lie along the vanishing line ([5]) and the Novikov spectral sequence shows that they are nontrivial.

For future reference, we mention the behavior of the elements of Theorem 5.10.

Theorem 5.13 In the Novikov spectral sequence for the mod 2 Moore spectrum $M(2)$

(a) $d_3 v_1^{4t+2} = \alpha_1^3 v_1^{4t}$, $d_3 v_1^{4t+3} = \alpha_1^3 v_1^{4t+1}$, $d_3 v_1^{4t+2}\sigma = \alpha_1^3 v_1^{4t}\sigma$ and $d_3 v_1^{4t+3}\sigma = \alpha_1^3 v_1^{4t+1}\sigma$.

(b) The elements $\alpha_1^i \sigma^\varepsilon v_1^{4t+j}$ are nontrivial permanent cycles for $0 \leq i \leq 2$, $\varepsilon = 0, 1$; $j = 0, 1$; $t \geq 0$.

(c) The homotopy element detected by $\alpha_1^2 \sigma^\varepsilon v_1^{4t}$ ($\varepsilon = 0, 1$; $t \geq 0$) is twice that detected by $\sigma^\varepsilon v_1^{4t+1}$.

All homotopy elements implied by (b) except $\sigma^\varepsilon v_1^{4t+1}$ have order 2. □

This can be proved by using Theorem 5.10 and comparing the Adams and Novikov spectral sequences for $M(2)$ through dimension 8.

We will need the following odd primary analogue

Theorem 5.14(a) For $p \geq 3$, Ext BP_*/p modulo v_1-torsion (i.e. the image of Ext BP_*/p in v_1^{-1} Ext $BP_*(p)$ is $\mathbb{F}_p[v_1, \alpha_1]/(\alpha_1^2)$.

(b) Each element v_1^t and $\alpha_1 v_1^t$ a $(t \geq 0)$ is a nontrivial permanent cycle and the corresponding homotopy element has order p. □

§6 Some Second Order Phenomena in the Novikov Spectral Sequence for Odd Primes

As remarked at the beginning of §5, we postpone our definition of nth order phenomena until §8. Unlike the first order phenomena, which was essentially described by Adams in [1], second and higher order phenomena are still largely unexplored. The infinite families of elements discovered in recent years by Larry Smith [63], [64], Oka [45], [46], [47] and Zahler [76] are examples of what we call second order families. The γ-family of Toda [73] is an example of a third order family. We will comment on this family and the unusual publicity received by its first member at the end of the section. The elements η_j recently constructed by Mahowald [28] presumably fit into not one but a series of second order families, as we shall describe in §7.

We will treat the odd primary case first because it is simpler. We begin by considering the computation of $Ext^2 BP_*$. (See the beginning of §5 for the relevant notation.) As in the case of $Ext^1 BP_*$, it is a torsion group of finite type, and the subgroup of order p is precisely the image of $\delta_o: Ext^1 BP_*/p \to Ext^2 BP_*$. By Theorem 5.3, $Ext^1 BP_*/p$ is a module over $\mathbb{F}_p[v_1]$. Its structure modulo v_1-torsion is given by Theorem 5.14. It can also be shown that for each $t \geq 0$, $v_1^t \alpha_1 \in Ext^1 BP_*/p$ is the mod p reduction of an element and $Ext^1 BP_*$, and therefore in ker δ_o. Hence we are interested in elements of $Ext^1 BP_*/p$ which are v_1-torsion, i.e. which are annihilated by some power of v_1. To get at the v_1-torsion submodule of $Ext^1 BP_*/p$, we first study the elements which are killed by v_1 itself.

To this end, consider the short exact sequence

$$(6.1) \qquad 0 \to \Sigma^q BP_*/p \xrightarrow{v_1} BP_*/p \to BP_*/I_2 \to 0 ,$$

and let $\delta_1: Ext^* BP_*/I_2 \to Ext^{1+*} BP_*/I_1$ be the connecting homomorphism. By Theorem 5.3, $Ext^o BP_*/I_2 = \mathbb{F}_p[v_2]$, so we get

Proposition 6.2 There are nontrivial elements $\beta_t \equiv \delta_1(v_2^t) \in \mathrm{Ext}^{1,(p+1)tq-q} BP_*/p$ for all $t > 0$. □

To finish computing $\mathrm{Ext}^1 BP_*/p$, one needs to determine how many times one can divide β_t by v_1. We let $\beta_{t/i}$ denote an element (if such exists) such that $v_1^{i-1}\beta_{t/i} = \beta_{t/1} \equiv \beta_t$. It is clear then that the v_1-torsion submodule of $\mathrm{Ext}^1 BP_*/p$ is generated over $\mathbb{F}_p$ by such elements.

Theorem 6.3 For all primes p, $0 \neq \beta_{sp^i/j} \in \mathrm{Ext}^{1,(sp^i(p+1)-j)q} BP_*/p$ exists for all $s > 0$, $i \geq 0$ and $0 < j \leq p^i$. (Precise definitions of these elements are given in the proof below.)

Proof: The basic fact that we need is that $\eta_R(v_2) \equiv v_2 + v_1 t_1^p - v_1^p t_1 \mod p$ (Theorem 3.14). From this, we get $\eta_R\, v_2^{p^i} \equiv v_2^{p^i} \mod (p, v_1^{p^i})$, so $v_2^{sp^i} \in \mathrm{Ext}^0 BP_*/(p, v_1^{p^i})$ for $s \geq 0$. Let $\tilde{\delta}$ be the connecting homomorphism for

$$(6.4) \qquad 0 \to \Sigma^{qp^i} BP_*/p \xrightarrow{v_1^{p^i}} BP_*/p \to BP_*/(p, v_1^{p^i}) \to 0 .$$

Then we can define $\beta_{sp^i/p^i} = \tilde{\delta}_1(v_2^{sp^i})$, and for $j < p^i$, $\beta_{sp^i/j} = v_1^{p^i-j}\beta_{sp^i/p^i}$. □

The interested reader can verify that this method of defining elements in $\mathrm{Ext}^1 BP_*/p$ is far easier than writing down explicit cocycles in $\tilde{\Omega} BP_*/p$.

Note that Theorem 6.3 says that $\mathrm{Ext}^1 BP_*/p$ contains $\mathbb{F}_p$-vector spaces of arbitrarily large finite dimension. For example β_{p^2/p^2} and β_{p^2-p+1} are both in $\mathrm{Ext}^{1,p^3q} BP_*/p$.

Unfortunately, Theorem 6.3 is not the best result possible. Further v_1 divisibility does occur, e.g. one can define β_{2p^2/p^2+p-1}. The complete computation of $\mathrm{Ext}^1 BP_*/p$ for $p > 2$ (and of $\mathrm{Ext}^1 BP_*/I_n$ for all p and $n > 1$) was first done by Miller-Wilson in [40] and redone (including the case $p = 2$, $n = 1$) in §5 of [39]. However, the elements of Theorem 6.4 will suffice for our purposes here.

The next and final step in the computation of $\mathrm{Ext}^2 BP_*$ is to determine how much $\delta_o(\beta_{i/j})$ (which will also be denoted by $\beta_{i/j}$) can be divided by p. This was done for $p > 2$ in §6 of [39] and announced in [38]. The computational difficulties encountered there are formidable. The problem is still open for $p = 2$, but it is certain that the methods of [39], if pushed a little further, will yield the answer.

We denote by $\beta_{i/(j,k)}$ a certain element with $p^{k-1}\beta_{i/(j,k)} = \beta_{i/(j,1)} \equiv \beta_{i/j}$. Then along the lines of Theorem 6.3 we have

<u>Theorem 6.5</u> For all primes p

$0 \neq \beta_{sp^i/(tp^j,1+j)} \in \mathrm{Ext}^{2,(sp^i(p+1)-tp^j)q} BP_*$ exists for all $s > 0$ and $0 < t < p^{i-2j}$ (an will be defined in the proof below), except $\beta_1 = 0$ for $p = 2$.

<u>Proof</u> From $\eta_R v_2 \equiv v_2 + v_1 t_1^p - v_1^p t_1 \mod p$ (Theorem 3.14) we obtain $\eta_R(v_2^{p^{2j+k}}) \equiv v_2^{p^{2j+k}} \mod (p^{1+j}, v_1^{p^{j+k}})$, and since $\eta_R v_1 = v_1 + pt_1$, we have $\eta_R v_1^{p^j} \equiv v_1^{p^j} \mod p^{1+j}$. It follows that $v_2^{p^{2j+k}} \in \mathrm{Ext}^o BP_*/(p^{1+j}, v_1^{p^{j+k}})$ and $v_1^{p^j} \in \mathrm{Ext}^o BP_*(p^{1+j})$. Let $\tilde{\delta}_o$ and $\tilde{\delta}_1$ be the connecting homomorphisms for the short exact sequences

$$0 \to BP_* \xrightarrow{p^{1+j}} BP_* \longrightarrow BP_*/(p^{1+j}) \longrightarrow 0$$

and

$$0 \to \Sigma^{qp^{j+k}} BP_*/(p^{1+j}) \xrightarrow{v_1^{p^{j+k}}} BP_*/(p^{1+j}) \to BP_*/(p^{1+j}, v_1^{p^{j+k}}) \to 0$$

respectively. Then we can define

$$\beta_{sp^{2j+k}/(tp^j,1+j)} = \tilde{\delta}_o(v_1^{p^j(p^k-t)}\tilde{\delta}_1(v_2^{sp^{2j+k}})).$$

The nontriviality of these elements can be seen by looking at the long exact sequences in Ext associated with the short exact sequences above. The one nontrivial fact that is needed is that the image of the mod (p^{1+j}) reduction $\mathrm{Ext}^1 BP_* \to \mathrm{Ext}^1 BP/(p^{1+j})$ consists of elements which are not annihilated by any power of $v_1^{p^j}$. □

Again, this is not the best result possible, but these elements will suffice for our purposes.

Note that Theorem 6.5 says that $\mathrm{Ext}^2 BP_*$ contains elements of arbitrarily high order, but that they occur very infrequently. For example, $\beta_{p^2/(p,2)}$ is the first element of order p^2, and and it is in dimension 130 for $p = 3$, and $\beta_{p^4/(p^2,3)}$, the first element of order p^3, is in dimension 1258 for $p = 3$.

Theorem 6.5 gives most of the additive generators of $\mathrm{Ext}^2 BP_*$ for $p > 2$. This group is much more complicated then $\mathrm{Ext}^1 BP_*$.

As the reader might guess, the question to ask now is which elements in this group are permanent cycles in the Novikov spectral sequence. This problem is far from being solved. Some progress has been made for $p \geq 5$. The current state of the art is

Theorem 6.6 For $p \geq 5$, the following elements in $\mathrm{Ext}^2 BP_*$ are permanent cycles, and the nontrivial homotopy elements they detect have the same order as the corresponding elements in the E_2-term.

(a) (Smith [63]) β_t for $t > 0$.

(b) (Smith [64], Oka [45], Zahler [76]) $\beta_{pt/j}$ for $t > 0$ and $0 < j < p$.

(c) (Oka [46]) $\beta_{tp/p}$ for $t \geq 2$.

(d) (oka [45]) $\beta_{tp^2/j}$ for $t > 0$ and $1 \leq j \leq 2p - 2$

(e) (Oka [47]) $\beta_{tp^2/j}$ for $t \geq 2$. and $1 \leq j \leq 2p$

(f) (Oka [47]) $\beta_{tp^2/(p,2)}$ for $t \geq 2$. □

Some of the elements in (b) - (e) were initially denoted by ε or ρ with various subscripts. On the other hand, we have

Theorem 6.7 [56] For $p \geq 3$ and $i \geq 1$ the element $\beta_{p^i/p^i} \in \mathrm{Ext}^2 BP_*$ is not a permanent cycle; in fact $d_{2p-1}\beta_{p^i/p^i} \equiv \alpha_1 \beta^p_{p^{i-1}/p^{i-1}}$ modulo certain indeterminacy. □

The special case $i = 1$ was first proved by Toda [70], [71] and it gives the first nontrivial differential in the Novikov spectral sequence for $p \geq 3$.

Theorem 6.6 (a) is definitely false for $p = 3$, for we have e.g. $d_5\beta_4 = \pm\alpha_1\beta_1^2\beta_{3/3} \neq 0$. We hope to have more to say about this in [52]. Tentative computations indicate for example that (for $p = 3$) β_t is a permanent cycle iff $t \not\equiv 4, 7$ or 8 mod 9.

We will now sketch the proof of Theorem 6.6 (a), as the proofs of (b) - (f) are all based on the same idea. Let $M(p)$ denote the mod p Moore spectrum. Then applying BP homology to the cofibration

$$(6.8) \qquad S^0 \xrightarrow{p} S^0 \longrightarrow M(p)$$

yields the short exact sequence (5.2). In this instance, we say 6.8 realizes (5.2). In [63] Smith shows that for $p \geq 3$ there is a map $\alpha\colon \Sigma^q M(p) \to M(p)$ which in BP homology realizes multiplication by v_1. We denote the cofibre of α by $M(p, v_1)$, so the cofibration

$$\Sigma^q M(p) \xrightarrow{\alpha} M(p) \longrightarrow M(p, v_1)$$

realizes the sequence 6.1. (It is not hard to see that this cannot be done for $p = 2$, but one can construct the spectrum $M(2, v_1^4)$. Our proof of Theorem 5.8 is based on the existences of $M(16, v_1^4 + 8v_1v_2)$.)

Next, Smith shows that for $p \geq 5$ there is a map $\beta\colon \Sigma^{(p+1)q} M(p, v_1) \to M(p, v_1)$ which realizes multiplication by v_2, so the cofibration

$$\Sigma^{(p+1)q} M(p, v_1) \xrightarrow{\beta} M(p, v_1) \to M(p, v_1, v_2)$$

realizes the short exact sequence

$$(6.9) \quad 0 \to \Sigma^{(p+1)q} BP_*/I_2 \xrightarrow{v_2} BP_*/I_2 \to BP_*/I_3 \to 0 .$$

(The map β does not exist for $p = 3$.)

Then it is not hard to show (with two applications of Theorem 5.11) that the composite

$$S^{t(p+1)q} \to \Sigma^{t(p+1)q} M(p) \to \Sigma^{t(p+1)q} M(p, v_1) \xrightarrow{\beta^t} M(p, v_1) \to \Sigma^{q+1} M(p) \to S^{q+2}$$

(where the first two maps are inclusions of low dimensional skeleta, and the last two maps are projections obtained by pinching low dimensional skeleta) is a homotopy element detected by $\beta_t \in \mathrm{Ext}^2 BP_*$.

In other words, the existence of $\beta_t \in \pi_* S^0$ is based on the existence of the map $\beta: \Sigma^{(p+1)q} M(p, v_1) \to M(p, v_1)$. As in §5, the low dimensional information required to construct this map can be extrapolated by the Novikov spectral sequence into an 'infinite amount' of information, i.e. the existence and nontriviality of β_t for all $t > 0$.

Parts (b) and (d) of Theorem 6.6 are based in a similar manner on the existence of maps

$$\Sigma^{p(p+1)q} M(p, v_1^{p-1}) \to M(p, v_1^{p-1}) \quad \text{and} \quad \Sigma^{p^2(p+1)q} M(p, v_1^{2p-2}) \to M(p, v_1^{2p-2})$$

realizing multiplication by v_2^p and $v_2^{p^2}$ respectively. For (c) the complex $M(p, v_1^p, v_2^p)$ does not exist (its existence would contradict Theorem 6.7 for $i = 1$), but Oka [46] constructs $M(p, v_1^p, v_2^{2p})$ and $M(p, v_1^p, v_2^{3p})$ from self-maps of $M(p, v_1^p)$ which yield the indicated elements. Parts (e) and (f) are proved in a similar manner.

We should point out that the 4-cell and 8-cell complexes M( ) above are not necessarily unique, i.e. a complex whose BP-homology is a cyclic BP_*-module is not in general characterized by that module. What is essential to the argument above is the existence of a self-map of the appropriate 4-cell complex which realizes multiplication by the appropriate power of v_2.

In a similar spirit, Theorem 6.7 implies

<u>Theorem 6.10</u> [56] For $p \geq 3$ and $i \geq 1$ there is no connective spectrum X such that $BP_*X = BP_*/(p, v_1^{p^i}, v_2^{p^i})$. □

In [73] Toda considers the existence of complex $M(p, v_1, v_2 \cdots v_n)$ which he calls $V(n)$ and which he characterizes in terms of their cohomology as modules over the Steenrod algebra. (Such a description of the $M(\ )$'s considered above will not work unless one is willing to resort to (much) higher order cohomology operations. We regard this fact as another advantage of BP-homology.) He proves

Theorem 6.11 (Toda [73]) For $p \geq 7$ the complex $V(3) = M(p, v_1, v_2, v_3)$ exists and is the cofibre of a map $\gamma: \Sigma^{(p^2+p+1)q} M(p, v_1, v_2) \to M(p, v_1, v_2)$. □

Let δ_2 be the connecting homomorphism for the short exact sequence (6.9). (Recall that δ_o and δ_1 are the connecting homomorphisms of (5.2) and (6.1) respectively.) Then we can define

$$(6.12) \quad \gamma_t = \delta_0 \delta_1 \delta_2 (v_3^t) \in \mathrm{Ext}^{3,(t(p^2+p+1)-(p+2))q} BP_* .$$

From Theorem 6.10, we derive

Corolary 6.13 For $p \geq 7$ the elements $\gamma_t \in \mathrm{Ext}^3 BP_*$ are permanent cycles for all $t > 0$. □

However, the nontriviality of these elements is far from obvious. The status of γ_1 was the subject of a controversy [48], [68], [3] which attracted widespread attention [43], [60]. In order to settle the question for all t one must know $\mathrm{Ext}^2 BP_*$ in all of the appropriate dimensions. Having determined the latter, we proved

Theorem 6.14 [38] [39] For $p \geq 3$, the element $\gamma_t \in \mathrm{Ext}^3 BP_*$ is nontrivial for all $t > 0$. □

The γ's are an example of what we call third order phenomena.

§7. Some Second Order Phenomena in the Novikov Spectral Sequence for the Prime 2.

We must assume that the reader is familiar with the notation introduced in the previous two sections. Our current knowledge of 2-primary second order phenomena is in some sense even sketchier than in the odd primary case. Nevertheless, the situation is quite tantalizing, especially in light of Mahowald's recent result (Theorem 2.12) on the existence of the elements η_j. We will see below that the Novikov spectral sequence provides a very suitable setting for understanding these elements and the families of elements that could possibly derive from them.

As in §6, we begin with a discussion of $\mathrm{Ext}^2 BP_*$, this time for $p = 2$. It is a torsion group, so to get at the elements of order 2, we look at $\mathrm{Ext}^1 BP_*/2$. It is a module over $\mathbb{F}_2[v_1]$ (Theorem 5.3) and its structure modulo v_1-torsion is given by Theorem 5.10. Unlike the odd primary case, not all of the v_1-torsion free part of $\mathrm{Ext}^1 BP_*/2$ is in the kernel of the connecting homomorphism $\delta_o\colon \mathrm{Ext}^1 BP_*/2 \to \mathrm{Ext}^2 BP_*$. Indeed, the summand of $\mathrm{Ext}^2 BP_*$ indicated in Theorem 5.7 is precisely the image under δ_o of the summand of $\mathrm{Ext}^1 BP_*/2$ given by Theorem 5.10. We call the former summand the <u>first order part</u> of $\mathrm{Ext}^2 BP_*$. (For $p > 2$, the first order part of $\mathrm{Ext}^2 BP_*$ is trivial.) The <u>second order part</u> of $\mathrm{Ext}^2 BP_*$ is that summand associated (via division by powers of 2) with the image under δ_o of the v_1-torsion submodule of $\mathrm{Ext}^1 BP_*/2$. This submodule contains all the elements provided by Theorem 6.4 (which is valid for all primes) as well as some more exotic elements which are described in §5 of [39].

Similarly $\mathrm{Ext}^2 BP_*$ itself contains the summand of Theorem 5.7, the subgroup (which is not a summand) provided by Theorem 6.5 (which is also valid for all primes) and some additional elements which have not yet been determined. For emphasis, we repeat that the determination of $\mathrm{Ext}^2 BP_*$ for $p = 2$ is still an open problem, but the methods of [39] are surely adequate for solving it.

We now wish to relate certain elements of Theorem 6.5 to elements in the Adams E_2-term. The manner in which elements of the two E_2-terms correspond to each other is difficult to define precisely, although in many cases it is easy enough to see in practice. Proposition 4.1 gives a correspondance only between nontrivial permanent cycles, and the homomorphism $\Phi\colon \mathrm{Ext}\, BP_* \to \mathrm{Ext}_A(\mathbb{F}_2, \mathbb{F}_2)$ is nontrivial only on a very small number of elements. Most of the elements in $\mathrm{Ext}^2 BP_*$ correspond in some way to elements of higher filtration in the Adams E_2-term.

A working (but not completely precise) procedure for matching elements in the two E_2-terms is the following. Theorem 4.2 and 4.3 give us two spectral sequences having essentially the same E_2-term and converging to the Adams and Novikov E_2-terms respectively. Hence we can take an element in the Novikov E_2-term, represent it (not uniquely in general) by some permanent cycle in the E_2-term of Theorem 4.3, and then see what happens to the corresponding element in the E_2-term of Theorem 4.2. The latter may fail to be a permanent cycle in the spectral sequence of Theorem 4.2. This would probably mean that the element we started with is not a permanent cycle in the Novikov spectral sequence (and that it supports a differential in some way related to a differential of Theorem 4.2), but this assertion has not been proved. It could also happen that the element we get is the target of a differential in 4.2. This might mean either that our original element is the target of some Novikov differential or that it 'corresponds' to an element having higher Adams filtration than originally estimated.

Of course, this procedure could be reversed (i.e. we could start with Adams elements and try to get Novikov elements), and the same remarks would apply. As we tried to indicate at the end of §4, making all of this more precise, especially nailing down the possible method of computing both Adams and Novikov differentials is an important unsolved algebraic problem.

With the above reservations in mind, we make

<u>Assertion 7.1</u> For $p = 2$

(a) $\Phi\left(\beta_{2^i/2^i}\right) = h_{i+1}^2$ and $\Phi\left(\beta_{2^i/2^i-1}\right) = h_1 h_{i+2}$. (This is a theorem.)

Under the procedure outlined above,

(b) for $j > i \geq 0$, $\beta_{2^j/2^j-2^i}$ corresponds to $h_o^{2^i-1} h_{1+i} \; h_{2+j}$ and $\beta_{2^{1+i}/(2^i,2)}$ corresponds to $h_o^{2^i-2} h_{1+i} \; h_{3+i}$;

(c) for $t \geq 0$, $\beta_{2^j/2^j-2^i-4t}$ corresponds to $P^t \; h_o^{2^i-1} \; h_{1+i} \; h_{2+j}$ and $\beta_{2^{1+i}/(2^i-4t,2)}$ corresponds to $P^t \; h_o^{2^i-2} \; h_{1+i} \; h_{3+i}$.

<u>Argument</u>: An odd primary analogue of (a) is proved in §9 of [39].

For (b) and (c), recall the definition of $\beta_{2^j/2^j-2^i}$. Let δ_o and $\tilde{\delta}_1$ be the connecting homomorphisms for the short exact sequences $0 \to BP_* \xrightarrow{2} BP_* \to BP_*/2 \to 0$ and

$$0 \to \Sigma^{2^{j+1}} BP_*/2 \xrightarrow{v_1^{2^j}} BP_*/2 \to BP_*/\left(2,\, v_1^{2^j}\right) \to 0$$

respectively. Then $v_2^{2^j} \in \operatorname{Ext}^o BP_*/\left(2,\, v_1^{2^j}\right)$ and $\beta_{2^j/2^j-2^i} = \delta_o\left(v_1^{2^i} \; \tilde{\delta}_1\left(v_2^{2^j}\right)\right)$ (see Theorem 6.4).

The spectral sequence of Theorem 4.3 has obvious analogues converging to $\operatorname{Ext} BP_*/\left(2,\, v_1^{2^j}\right)$, and $\operatorname{Ext} BP_*/2$, and we can compute $\beta_{2^j/2^j-2^i}$ in the E_2-terms of those spectral sequences.

We have $\tilde{\delta}_1 v_2^{2^j} = t_1^{2^{j+1}}$ modulo terms with higher I-filtration, and $\delta_o\left(v_1^{2^i}\tilde{\delta}_1\left(v_2^{2^j}\right)\right) \equiv \delta_o v_1^{2^i} t_1^{2^{j+1}} \equiv 2^{2^i-1} t_1^{2^i} \mid t_1^{2^{1+j}}$. Since 2 corresponds to h_o and $t_1^{2^i}$ corresponds to h_{1+i}, we get the element $h_o^{2^i-1} h_{1+i} h_{2+j}$ as desired. The argument for $\beta_{2^{1+i}/(2^i,2)}$ is similar.

For (c) we use the fact (see the discussion preceeding Theorem 5.12) that multiplication by v_1^4 in Ext $BP_*/2$ and Ext $BP_*/4$ corresponds to the Adams periodicity operator P. □

The discussion that follows will be of a more hypothetical nature. We will see how various hypotheses relating to the Arf invariant elements and Mahowald's η_j (Theorem 2.12) imply the existence of new families of homotopy elements. We list our hypothesis in order of decreasing strength.

Hypothesis $\underline{7.2}_i$ $(i \geq 2)$ $\beta_{2^i/2^i}$ is a permanent cycle and and the corresponding homotopy element can be factored $S^{6\cdot 2^i} \to \Sigma^{6\cdot 2^i} M\left(2, v_1^{2^i}\right) \xrightarrow{\beta} M\left(2, v_1^{2^i}\right) \to S^{2+2\cdot 2^i}$, i.e. the map β realizes multiplication by $v_2^{2^i}$.

Hypothesis $\underline{7.3}_i$ $\beta_{2^i/2^i}$ is a nontrivial permanent cycle and the corresponding homotopy element has order 2.

Hypothesis 7.4_i $\beta_{2^i/2^i-1}$ is a permanent cycle and the corresponding homotopy element has order 4 and is annihilated by ν.

Theorem 2.12 and 7.1(a) imply that there is a permanent cycle equal to $\beta_{2^i/2^i-1}$ modulo ker Φ. It appears unlikely that the error term in ker Φ would affect any of the arguments that follow, so we assume for simplicity that it is zero.

Similarly, if the Arf invariant element θ_{i+1} exists it is detected by $\beta_{2^i/2^i}$ modulo ker Φ.

Hypothesis 7.2 is known to be false for $i = 2$, and we have included it mainly to illustrate the methodology in as simple a way as possible. The statement that $\beta_{2^i/2^i}$ extends to $M(2, v_1^i)$ or, by duality that it coextends, is equivalent to the Toda bracket [69]. $\langle \beta_{2^i/2^i}, 2_\iota, \alpha_{2^i}, 2_\iota \rangle$ being defined and trivial. Mahowald has shown the following substitute for it.

Theorem 7.5 There is a map $\beta: \Sigma^{48} M(4, v_1^4) \to M(4, v_1^4)$ which realizes multiplication by v_2^8. $\square$

Corollary 7.6 The elements $\beta_{8t/(4,2)} \in \mathrm{Ext}^{2,48t-8}\, BP_*$ for all $t > 0$ are nontrivial permanent cycles and the corresponding homotopy elements have order 4.

Proof: The argument is similar to that of Theorem 6.6 (which is discussed following Theorem 6.7). We use Theorem 5.11 twice the show that the composite

$$S^{48t} \to \Sigma^{48t} M(4, v_1^4) \xrightarrow{\beta^t} M(4, v_1^4) \to S^{10}$$

is detected by $\beta_{8t/(4,2)}$. □

From 7.1 we see that $\beta_{8/(4,2)} \in \pi_{38}S^o$, $\pi_{16/(4,2)} \in \pi_{86}S^o$ and $\beta_{32/(4,2)} \in \pi_{182}S^o$ are detected in the Adams spectral sequence by $h_o^2 h_3 h_5$, $Ph_o^6 h_4 h_6$, and $P^3 h_o^{14} h_5 h_7$ respectively.

<u>Proposition 7.7</u>

(a) Hypothesis 7.2_i implies 7.3_i.

(b) Hypothesis 7.3_i implies 7.4_i.

<u>Proof</u>: (a) If $\beta_{2^i/2^i}$ extends to $M\left(2, v_1^{2^i}\right)$, it certainly extends to $M(2)$ and so has order 2.

(b) If $\beta_{2^i/2^i}$ has order 2, then it extends to a map $f: \Sigma^{4\cdot 2^i-2} M(2) \to S^o$. By Theorem 5.13, $v_1 \in \pi_2 M(2)$ has order 4, $\nu v_1 = 0$ and $f_*(v_1) = \beta_{2^i/2^i-1}$ by Theorem 5.11 and an easy calculation. □

Note that the proof of (b) shows $2\beta_{2^i/2^i-1} = \alpha_1^2\beta_{2^i/2^i} \in \pi_{2^{i+2}}S^o$ if 7.3_i holds.

The Hypothesis 7.2 - 7.4 provide homotopy elements as follows

<u>Theorem 7.8</u>

(a) If 7.2_i holds, then the following elements are (not necessarily nontrivial) permanent cycles: $\beta_{s\cdot 2^i/2^i-4t-a}\sigma^\varepsilon \alpha_1^j$ for $s > 0$; $a = 0, 1$; $\varepsilon = 0, 1$; $j = 0, 1, 2$.

(b) If 7.3_i holds, then the elements of (a) with $s = 1$ are permanent cycles.

(c) If 7.4_i holds, all of the elements of (b) except $\beta_{2^i/2^i}$, and $\alpha_1\beta_{2^i/2^i}$, are permanent cycles.

Proof:

(a) Let $\beta: \Sigma^{6\cdot 2^i} M\left(2, v_1^{2^i}\right) \to M\left(2, v_1^{2^i}\right)$ be the map of 7.2, Then $\beta_{s\cdot 2^i/2^i}$ is the composition

$$S^{6\cdot 2^i} \to \Sigma^{6\cdot 2^i\cdot s} M\left(2, v_1^{2^i}\right) \xrightarrow{\beta^s} M\left(2, v_1^{2^i}\right) \to S^{2+2^{i+1}} .$$

The other elements are obtained by composing the elements of $\pi_* M(2)$ given by Theorem 5.13 with the map

$$\Sigma^{6s\cdot 2^i} M(2) \to \Sigma^{6s\cdot 2^i} M\left(2, v_1^{2^i}\right) \xrightarrow{\beta^s} M\left(2, v_1^{2^i}\right) \to S^{2+2^{i+1}} .$$

(b) Compose the elements of Theorem 5.13 with the extension of $\beta_{2^i/2^i}$ to $M(2)$.

(c) The indicated elements with $a = 1$ can be obtained by composing the extension of $\beta_{2^i/2^i-1}$ with the appropriate elements given by the mod 4 analogue of Theorem 5.13. The element $\alpha_1^2\beta_{2^i/2^i}$ is $2\beta_{2^i/2^i-1}$ by the proof of Proposition 7.7(b); $\sigma\beta_{2^i/2^i}$ and $\beta_{2^i/2^i-4}$ can be realized as homotopy elements of order 2 by the Toda brackets [69] $\langle \beta_{2^i/2^i-1}, \nu, \eta\rangle$ and $\langle \beta_{2^i/2^i-1}, \eta^3, 2, \eta\rangle$ respectively. (The latter bracket is defined because $\eta^3\beta_{2^i/2^i-1} = 4\nu\beta_{2^i/2^i-1} = 0$.) Then the remaining elements can be obtained by composing the extensions of these two to $M(2)$ and composing with the elements of Theorem 5.13. □

The above theorem does not assert that the indicated elements are nontrivial, and some of them are likly to be trivial, such as $\alpha_1\beta_{s\cdot 2^i/2^i-4t}$, $\alpha_1^2\beta_{s\cdot 2^i/2^i-4t}$ with $2^{i-3} \le t < 2^{i-2}$ (since $\beta_{s\cdot 2^i/2^i-4t}$ in this is divisible by 2 by Theorem 6.5). The possible nontriviality is the subject of work in progress which will be reported elsewhere. At the moment, we can offer the following.

Theorem 7.9 The elements $\alpha_1\beta_{s\cdot 2^i/2^i-4t-1} \in \mathrm{Ext}^3\, BP_*$ and $\alpha_1^2\beta_{s\cdot 2^i/2^i-4t-1} \in \mathrm{Ext}^4\, BP_*$ for $0 \le t < 2^{i-2}$ (as well as $\beta_{s\cdot 2^i/2^i-4t-1}$, $\beta_{s\cdot 2^i/2^i-4t} \in \mathrm{Ext}^2\, BP_*$) are nontrivial in the Novikov E_2-term. □

Corollary 7.10 If the elements of Theorem 7.9 are permanent cycles, then the corresponding homotopy elements are nontrivial.

Proof: By sparseness (Corollary 3.17) a Novikov differential hitting any of these elements would have to originate on the 0-line or the 1-line. The former is trivial in positive dimensions, and all differentials originating on the latter were accounted for in theorem 5.8. □

We cannot resist commenting on how hard it would be to prove similar results using only the Adams spectral sequence. The proof of Theorem 7.9 is based on methods (see §8) which have no counterpart in the Adams spectral sequence. Even if somehow one could prove that the corresponding elements are nontrivial in the Adams E_2-term, they would have such high filtration that it would be extremely difficult to show that they are not hit by nontrivial

Adams differentials. The low filtration of elements in the Novikov spectral sequence makes it a very effective detecting device.

We remind the reader that none of the Hypotheses 7.2 - 7.4 are currently known to be true for all i. This is unfortunate in view of the following

Theorem 7.11 If for some $i \geq 2$

(i) $M\left(4, v_1^{2^i}\right)$ is a ring spectrum,

(ii) $\beta_{2^{i+1}/(2^i,2)}$ is a permanent cycle and

(iii) the corresponding homotopy element has order 4, then the elements $\beta_{s\cdot 2^{i+1}/(4j,2)}$ and $\alpha_1^k \beta_{s\cdot 2^{i+1}/4j-1}$ for $s > 0$;

$k = 0, 1, 2;$ $i \geq 2$ and $0 < j \leq 2^{i-2}$ are nontrivial permanent cycles.

Proof: The nontriviality follows from Corollary 7.10. Let $\beta\colon \Sigma^{48} M(4, v_1^4) \to M(4, v_1^4)$ be the map of Theorem 7.5 and $\alpha\colon \Sigma^8 M(4) \to M(4)$ a map which realizes multiplication by v_1^4. Then consider the following commutative diagram.

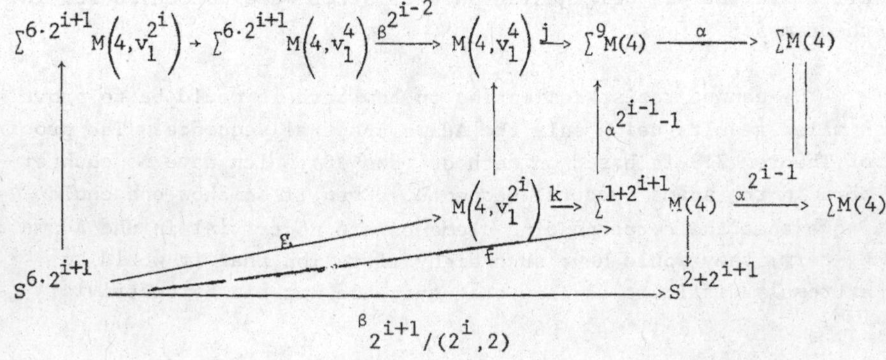

All of the maps except f and g are obvious; the last two maps of the two top rows are cofibre sequences, i.e. $\alpha \circ j = \alpha^{2^{i-1}} \circ k = 0$. The map f exists because $\beta_{2^{i+1}/(2^i,2)}$ has order 4. The commutativity of the diagram implies that $\alpha^{2^{i-1}} \circ f = 0$, so g exists. The multiplicative structure of $M\left(4, v_1^{2^i}\right)$ can be used to extend g to $\Sigma^{6\cdot 2^{i+1}} M\left(4, v_1^{2^i}\right)$. Thus we obtain a map

$$\Sigma^{6\cdot 2^{i+1}} M\left(4, v_1^{2^i}\right) \xrightarrow{\tilde{\beta}} M\left(4, v_1^{2^i}\right)$$

which realizes multiplication by $v_2^{2^{i+1}}$. We can then obtain the desired homotopy elements by composing

$$\Sigma^{6s\cdot 2^{i+1}} M(4) \to \Sigma^{6s\cdot 2^{i+1}} M\left(4, v_1^{2^i}\right) \xrightarrow{\tilde{\beta}^s} M\left(4, v_1^{2^i}\right) \to S^{2+2^{i+1}}$$

with the appropriate elements of $\pi_* M(4)$. □

Hence the hypotheses of Theorem 7.11 imply that a large collection of elements in $\mathrm{Ext}^2 BP_*$ are permanent cycles. Mahowald has an argument for the first hypothesis [77], but the status of the others is less clear. Theorem 7.11 has the following analogue.

<u>Theorem 7.12</u> Let $p \geq 5$. If for some $i \geq 1$

(i) $M\left(p, v_1^{p^i-1}\right)$ is a ring spectrum,

(ii) β_{p^i/p^i-1} is a permanent cycle and

(iii) the corresponding homotopy element has order p, then $\beta_{sp^i/j}$ is a permanent cycle (and the corresponding homotopy element of order p) for all $s > 0$, $i \geq 1$ and $0 < j < p^i$.

Proof: We argue as in Theorem 7.11, replacing Theorem 7.5 with the assertion that there is a map

$$\beta: \Sigma^{p(p+1)q} M\left(p, v_1^{p-1}\right) \to M\left(p, v_1^{p-1}\right)$$

realizing multiplication by v_2^p. This map has been constructed by Smith [64] and Oka [45] in the proof of Theorem 6.6(b). □

We hope to extend this result to $p = 3$ in [52]. Oka has recently announced [78] a proof of the first hypothesis for all i. The second is likely to follow from an odd primary analogue of Mahowald's Theorem 2.12. The third hypothesis, however, could be quite difficult to prove.

§8 Morava stabilizer Algebras and the Chromatic Spectral Sequence (the inner mysteries of the Novikov E_2-term)

The reader may well wonder how it is possible to prove results such as Theorems 5.7, 5.9, 5.10, 5.14(a) and 7.9, which state that various systematic families of elements in the Novikov E_2-term are nontrivial. The basic technique in each case is to study the map $\text{Ext}\, BP_* \to \text{Ext}\, v_n^{-1} BP_*/I_n$ (where $n = 1$ for the results of §5 and $n = 2$ for Theorem 7.9). The latter group is surprisingly easy to compute due to two startling isomorphisms (Theorems 8.4 and 8.7 below) originally discovered by Jack Morava [42]. It was this computability that motivated us to do the work that led to [39]. Morava's work implies that there is a deep, and previously unsuspected connection between algebraic topology and algebraic number theory. Where it will eventually lead to is anybody's guess.

After describing how to compute $\text{Ext}\, v_n^{-1} BP_*/I_n$, we will set up the chromatic spectral sequence (and explain why it is so named), which is a device for feeding this new found information into the Novikov E_2-term in a most systematic way. We will see that it reveals patterns of periodicity (which may carry over to stable homotopy itself; see [53]) hitherto invisible. In particular, we will define nth order phenomena in the Novikov spectral sequence.

In order to get at $\text{Ext}\, v_n^{-1} BP_*/I_n$, we need to define some auxillary objects. Let $K(n)_* = Q$ for $n = 0$ and $\mathbb{F}_p[v_n, v_n^{-1}]$ for $n > 0$, and make it a BP_*-module by sending v_i to zero for $i \neq n$ (where $v_o = p$). For $n > 0$, $K(n)_*$ is a graded field in the sense that every graded module over it is free. Next, define

$$(8.1) \qquad K(n)_* K(n) = K(n)_* \otimes_{BP_*} BP_*BP \otimes_{BP_*} K(n)\,,$$

where the tensor products on the left and right are with respect to the BP_*-module structures on BP_*BP induced by η_L and η_R respectively (see §3). $K(n)_* K(n)$ where a coassociative, non-cocommutative coproduct Δ from BP_*BP. Theorem 3.14 allows us to describe its algebra structure very explicitly.

Theorem 8.2 [58] $K(0)_* K(0) = Q$ and for $n > 0$, $K(n)_* K(n) = K(n)_*[t_1, t_2 \cdots]/\left(v_n t_i^{p^n} - v_n^{p^i} t_n\right)$. The coproduct Δ is given by

$$\sum_{i\geq 0}^F \Delta(t_i) = \sum_{i,j\geq 0}^F t_i \otimes t_j^{p^i},$$ where $t_o = 1$ and F is the formal group law over $K(n)_*$ given by the map $BP_* \to K(n)_*$ (see Theorem 3.10).

<u>Proof</u>: By definition (8.1) $K(n)_*K(n) = v_n^{-1} BP_*BP/(v_i, \eta_R v_i : i \neq n)$. In $K(n)_*K(n)$, $\eta_R v_n = v_n$ and Theorem 3.14 reduces to

(8.3) $$\sum_{i\geq 0}^F v_n t_i^{p^n} = \sum_{i\geq 0}^F v_n^{p^i} t_i .$$

Each side of (8.3) has at most one formal summand in each dimension, so we can formally cancel and get $v_n t_i^{p^n} = v_n^{p^i} t_i$ by induction on i. The formula for Δ follows from Theorem 3.12. □

Now $K(n)_*K(n)$ is a Hopf algebra over $K(n)_*$, so we can define its cohomology $\mathrm{Ext}_{K(n)_*K(n)}(K(n)_*, K(n)_*)$ in the usual manner. We now come to our first surprise.

<u>Theorem 8.4</u> [37]

$$\mathrm{Ext}\; v_n^{-1} BP_*/I_n \cong \mathrm{Ext}_{K(n)_*K(n)}(K(n)_*, K(n)_*).$$ □

Since $K(n)_* K(n)$ is much smaller than BP_*BP, this result simplifies the computation of $\mathrm{Ext}\; v_n^{-1} BP_*/I_n$ considerably. In §3 of [58] we filter $K(n)_*K(n)$ in such a way that the associated bigraded object is the dual of the universal enveloping algebra of a restricted Lie algebra. In [51] we use this filtration to construct a May spectral sequence [32] converging to the desired

Ext group. We use this device then to compute $\mathrm{Ext}\ v_n^{-1}BP_*/I_n$ for $n = 0, 1, 2$, and $\mathrm{Ext}^s\ v_n^{-1}\ BP_*/I_n$ for all n and $s = 0, 1, 2$.

However, deeper insight into the structure of $K(n)_*K(n)$ is gained as follows. Forgetting the grading, make $\mathbb{F}_p$ into a $K(n)_*$ module by sending v_n to 1, and let

(8.5) $$S(n)_* = K(n)_*K(n)\otimes_{K(n)_*} \mathbb{F}_p .$$

The $S(n)_*$ is a commutative, noncocommuative Hopf algebgra over $\mathbb{F}_p$ with algebra structure

(8.6) $$S(n)_* \cong \mathbb{F}_p[t_1, t_2 \cdots]/(t_i^{p^n} - t_i).$$

Its dual $S(n)$ (defined in the appropriate way in [58]) is called the <u>nth Morava stabilizer algebra</u> $S(n)$. This brings us to our second surprise.

<u>Theorem 8.7</u> [58] $S(n)\otimes\mathbb{F}_{p^n} \cong \mathbb{F}_{p^n}[S_n]$, the group algebra over $\mathbb{F}_{p^n}$ of a certain pro-p group S_n, to be described below. □

<u>Corollary 8.8</u> [58] $\mathrm{Ext}\ v_n^{-1}BP_*(I_n)\otimes_{K(n)_*} \mathbb{F}_p = H_c^*(S_n; \mathbb{F}_p)$, where the latter is the continuous cohomology of S_n with constant mod p coefficients. □

In §2 of [58] we also show how it is possible to recover the bigrading of $\mathrm{Ext}\ v_n^{-1}\ BP_*/I_n$ from $H_c^*(S_n; \mathbb{F}_p)$. For continuous cohomology of p-adic groups, see Lazard [24].

We will now describe the group S_n. Let Z_p denote the p-adic integers, and $\mathbb{F}_{p^n}$ the field with p^n elements. There is a complete local ring $W(\mathbb{F}_{p^n})$ (called the Witt ring of $\mathbb{F}_{p^n}$) which is a degree n extension of Z_p obtained by adjoining an element ω satisfying $\omega^{p^n-1} = 1$. The residue field of $W(\mathbb{F}_{p^n})$ is $\mathbb{F}_{p^n}$ and the extension $W(\mathbb{F}_{p^n}): Z_p$ is a lifting of the extension $\mathbb{F}_{p^n} : \mathbb{F}_p$. The Frobenius automorphism (which sends x to x^p) of the latter lifts to an automorphism of $W(\mathbb{F}_{p^n})$ over Z_p which sends ω to ω^p.

Let $E_n = W(\mathbb{F}_{p^n}) \langle\langle T \rangle\rangle/(T^n - p)$, i.e. the power series ring over $\mathbb{F}_{p^n}$ on one noncommuting variable T with $T^n = p$ and $T\omega = \omega^p T$. Then E_n is a noncommutative complete local ring with maximal ideal (T) and residue field $\mathbb{F}_{p^n}$. It is a simple algebra over Z_p with rank n^2 with Z_p-basis $\{\omega^i T^j : 0 \le i, j < n\}$. Tensoring it with the p-adic numbers Q_p (the field of fractions of Z_p) gives D_n which is a division algebra with center Q_p and Hasse invariant $\frac{1}{n}$. (The latter is an invariant in Q/Z which classifies such division algebras).

<u>Definition 8.9</u> The group S_n is the group of units in E_n which are congruent to 1 modulo (T).

Details of the above description can be found in [58].

<u>Examples 8.10</u> For $p = 2$ $S_1 \simeq Z_2^{\times}$, the group of units in Z_2. Hence $S_1 \simeq Z/2 \oplus Z_2$ and $H_c^*(S_1; \mathbb{F}_2) = \mathbb{F}_2[x, y]/(y^2)$ with $x, y \in H^1$. Corollary 8.8 leads to $\operatorname{Ext} v_1^{-1} BP_*/2 \simeq \mathbb{F}_2[v_1, v_1^{-1}, \alpha_1, \sigma]/(\sigma^2)$.

<u>8.11</u> For $p > 2$, $Z_p^{\times} \simeq \mathbb{F}_p^{\times} \oplus Z_p \simeq Z/(p-1) \oplus Z_p$ and $H_c^*(S_1; \mathbb{F}_p) = \mathbb{F}_p[X]/(x^2)$ with $x \in H^1$. Corollary 8.8 leads to $\operatorname{Ext} v_1^{-1} BP_*/p \simeq \mathbb{F}_p[v_1, v_1^{-1}, \alpha_1]/(\alpha_1^2)$.

8.12 For $p \geq 5$,

$$\dim H_c^i(S_2; \mathbb{F}_p) = \begin{cases} 1 & \text{for } i = 0, 4 \\ 3 & \text{for } i = 1, 3 \\ 4 & \text{for } i = 2 \\ 0 & \text{for } i > 4 \end{cases}$$

so $\operatorname{Ext} BP_*/I_2$ contains a free $\mathbb{F}_p[v_2]$ module on 12 generators.

8.13 For $p = 2$, S_2 contains the quaternion group G of order 8 and the restriction map $H_c^*(S_2; \mathbb{F}_2) \to H^*(G; S_2)$ is onto.

8.14 For all primes p, S_{p-1} contains a subgroup of order p and the restriction map $H_c^*(S_{p-1}; \mathbb{F}_p) \to H^*(Z/(p); \mathbb{F}_p)$ is onto. In [56] we use this map to show that for $p > 2$ all monomials in the elements β_{p^i/p^i} are nontrivial. This fact is used in the proof of Theorem 6.7.

Details of Examples 8.10 - 8.13 can be found in [51]. A useful reference for the continuous cohomology of p-adic Lie groups is Lazard [24], from which Morava has extracted

Theorem 8.15 [41] If $(p-1)\nmid n$, $H_c^*(S_n; \mathbb{F}_p)$ is a Poincare duality algebra of dimension n^2. If $(p-1)|n$, there is an element $b \in H_c^*(S_n^1; \mathbb{F}_p)$ such that $H_c^*(S_n; \mathbb{F}_p)$ is a finitely generated free module over $\mathbb{F}_p[b]$. □

You may well ask why Theorems 8.4 and 8.7 are true. We will try to give a heuristic explanation. Recall that a groupoid is a small category in which every morphism is an equivalence. The relevant example of such is the category $\underline{\underline{F(R)}}$ of p-typical form group laws (Definition 3.8) over a commutative $Z_{(p)}$-algebra R, and strict isomorphisms between them. (A strict isomorphism f is one with $f(x) \equiv x \bmod x^2$.) Landweber [22] has shown that the set of ring homomorphisms from BP_*BP to R is in one-to-one correspondance with the set of morphisms in $\underline{\underline{F(R)}}$, and the various structure maps of BP_*BP correspond to the various structures of the groupoid. Hence BP_*BP is a cogroupoid object in the algebras, and Haynes Miller [36] has christened such objects Hopf algebroids (a Hopf algebra being a co-group object).

Loosely speaking, Ext BP_* can be thought of as the cohomology of the groupoid of formal group laws over BP_* which are strictly isomorphic to the universal one. The map $BP_* \to v_n^{-1} BP_*/I_n$ induces a formal group law F_n over the latter by Theorem 3.10, and Ext $v_n^{-1} BP_*/I_n$ can be regarded as the cohomology of the groupoid of formal group laws over $v_n^{-1} BP_*/I_n$ which are strictly isomorphic to F_n. It can be shown (e.g. §19.4 of [18]) that any such formal law is canonically strictly isomorphic to the one F_n induced by $BP_* \to K(n)_* \to v_n^{-1} BP_*/I_n$. An argument similar to Landweber's shows that $\mathrm{Hom}_{\mathrm{Rings}}(K(n)_* K(n), R)$ is the groupoid of strict isomorphisms between formal group laws over the $Z(p)$-algebra R induced from the one over $K(n)_*$. It follows that Ext $v_n^{-1} BP_*/I_n$ and $\mathrm{Ext}_{K(n)_*K(n)}(K(n)_*, K(n)_*)$ are the cohomology groups of equivalent (in the sense of equivalence of categories) groupoids and are therefore isomorphic. This argument is the idea behind Theorem 8.4. The proof given in [37] is less abstract and less enlightening. *

For Theorem 8.7 we set $v_n = 1$ and get the Hopf algebra $S(n)_*$. (Note that $K(n)_*K(n)$ is a Hopf algebra over $K(n)_*$, but a Hopf algebroid over $\mathbb{F}_p$.) We are now dealing with the groupoid of strict isomorphisms of formal group laws over R induced by maps $\mathbb{F}_p \to R$.

* The heuristic proof of Theorem 8.4 described in the second paragraph will be made precise in a forthcoming paper by Jack Morava.

Since there is at most one such map (there are none unless R is an $\mathbb{F}_p$-algebra) our groupoid is actually the strict automorphism group of the induced formal group law over R. In the case $R = \mathbb{F}_{p^n}$, this group is known (§III. 2 of [17] or §20.4 of [18]) to be S_n, whence Theorem 8.7.

We now turn to the chromatic spectral sequence which was first introduced in §3 of [39]. For the reader's amusement, we will try to reconstruct the line of thought which led to its formulation. In §5, we observed that all elements of order p^i in $\mathrm{Ext}^1 BP_*$ are in the image of the connecting homomorphism for the short exact sequence

$$(8.16) \qquad 0 \to BP_* \xrightarrow{p^i} BP_* \longrightarrow BP_*/p^i \longrightarrow 0.$$

We would like to obtain all elements of finite order, and hence all of $\mathrm{Ext}^1 BP_*$ from a single short exact sequence. We have maps of short exact sequences

$$\begin{array}{ccccccc} 0 \to BP_* & \xrightarrow{p^i} & BP_* & \longrightarrow & BP_*/p^i & \to 0 \\ \| & & \downarrow p & & \downarrow & \\ 0 \to BP_* & \xrightarrow{p^{i+1}} & BP_* & \longrightarrow & BP_*/p^{i+1} & \to 0. \end{array}$$

Taking the direct limit over increasing i we get

$$(8.17) \qquad 0 \to BP_* \to p^{-1}BP_* \to BP_*/p^\infty \to 0$$

i.e. the tensor product of BP_* with the short exact sequence $0 \to Z \to Q \to Q/Z \to 0$. Since $\mathrm{Ext}^1 BP_*$ is a torsion group and $\mathrm{Ext}\, p^{-1}BP_* = Q$ concentrated in dimension 0, we see that the connecting homomorphism $\mathrm{Ext}^0 BP_*/p^\infty \to \mathrm{Ext}^1 BP_*$ is an isomorphism in positive dimensions.

In §6, we saw that all of $\mathrm{Ext}^2 BP_*$ (unless $p = 2$, in which case we get all of $\mathrm{Ext}^2 BP_*$ not accounted for in §5) comes from $\mathrm{Ext}^0 BP_*/(p^{1+i}, v_1^{p^{i+j}})$ by composing the connecting homomorphisms of (8.16) and

$$(8.18)\quad 0 \to \Sigma^{qp^{i+j}} BP_*/p^{1+i} \xrightarrow{v_1^{p^{i+j}}} BP_*/p^{1+i} \to BP_*/(p^{1+i}, v_1^{p^{i+j}}) \to 0$$

Moreover, there are maps

$$\begin{array}{ccccccc} 0 \to BP_*/p^{1+i} & \xrightarrow{v_1^{p^{i+j}}} & \Sigma^{-qp^{i+j}} BP_*/p^{i+1} & \to & \Sigma^{-qp^{i+j}} BP_*/(p^{1+i}, v_1^{p^{i+j}}) \to 0 \\ \Big\| & & \Big\downarrow v_1^{(p-1)p^{i+j}} & & \Big\downarrow \\ 0 \to BP_*/p^{1+i} & \xrightarrow{v_1^{p^{1+i+j}}} & \Sigma^{-qp^{1+i+j}} BP_*/p^{i+1} & \to & \Sigma^{-qp^{1+i+j}} BP_*/(p^{1+i}, v_1^{p^{1+i+j}}) \to 0. \end{array}$$

We can take the direct limit over increasing i and j and get

$$(8.19)\quad 0 \to BP_*/p^\infty \to v_1^{-1} BP_*/p^\infty \to BP_*/(p^\infty, v_1^\infty) \to 0$$

and it can be shown that for $p > 2$ the map (composition of connecting homomorphisms of 8.17 and 8.19 $\mathrm{Ext}^0 BP_*/(p^\infty, v_1^\infty) \to \mathrm{Ext}^2 BP_*$ is also an isomorphism in positive dimensions. Computing an Ext^0 is easier than computing an Ext^2 because there are no coboundaries to worry about.

We can splice 8.17 and 8.19 together to get a 4-term exact sequence

$$(8.20)\quad 0 \to BP_* \to p^{-1} BP_* \to v_1^{-1} BP_*/p^\infty \to BP_*/(p^\infty, v_1^\infty) \to 0.$$

Then Ext BP_* can be computed in terms of the Ext groups of the other comodules by means of a baby spectral sequence. Moreover,

Ext $p^{-1}BP_*$ and Ext $v_1^{-1}BP_*/p$ can be computed by the theory of Morava stabilizer algebras discussed above. The latter Ext group is closely related to Ext $v_1^{-1}BP_*/p^\infty$ since there is a short exact sequence

$$(8.21) \quad 0 \to v_1^{-1}BP_*/p \to v_1^{-1}BP_*/p^\infty \xrightarrow{p} v_1^{-1}BP_*/p^\infty \longrightarrow 0 .$$

Hence the Ext groups for the two middle terms of 8.20 are known, and we are left with computing Ext $BP_*/(p^\infty, v_1^\infty)$. Unfortunately, this seems to be just as difficult as computing Ext BP_* itself, so we have gained very little unless we iterate the procedure as follows.

Define BP_*BP-comodules M^n and N^n inductively as follows. $N^o = BP_*$, $M^n = v_n^{-1}N^n$ (where $v_o = p$) and N^{n+1} is the quotient in the short exact sequence

$$(8.22) \qquad 0 \to N^n \to M^n \to N^{n+1} \to 0$$

For $n = 0, 1$ this sequence is 8.17 and 8.19 respectively, and one could write $N^n = BP_*/(p^\infty, v_1^\infty \cdots v_{n-1}^\infty)$ and $M^n = v_n^{-1}BP_*/(p^\infty, v_1^\infty \cdots v_{n-1}^\infty)$. We can splice together the short exact sequences 8.21 to get a long exact sequence

$$(8.23) \qquad 0 \to BP_* \to M^o \to M^1 \to M^2 \cdots$$

Theorem 8.24. The long exact sequence 8.23 leads to a first quadrant cohomology spectral sequence converging to Ext BP_* with $E_1^{s,t} = \mathrm{Ext}^t M^s$. □

We call this the chromatic spectral sequence. We should warn the reader that it is not at all suited for computing the Novikov E_2-term through a given range of dimensions. We have other devices for that [54], [55]. Its purpose rather is to highlight certain structural patterns in Ext BP_*, as will be explained below.

Very few of the groups $E_1^{s,t}$ have actually been computed. In [39], we compute $E_1^{0,t}$ (which is just one copy of Q in $E_1^{0,0}$ in dimension 0) and $E_1^{1,t}$ for all primes, and $E_1^{2,0}$ for $p > 2$. (We also found the corresponding groups $E_\infty^{s,t}$.) It would be interesting to know all of $E_1^{2,t}$ (especially $E_1^{2,0}$ for $p = 2$) and $E_1^{3,0}$. It is certainly possible (but not easy) to compute these groups with existing techniques. Our knowledge of $\mathrm{Ext}\, v_2^{-1}BP_*/I_2$, which originally motivated the whole program, has hardly been exploited.

To relate $\mathrm{Ext}\, M^n$ to $\mathrm{Ext}\, v_n^{-1}BP_*/I_n$, we need to define some more comodules M_i^{n-i}, which we do by induction on i by setting $M_0^n = M^n$ and M_{i+1}^{n-i-1} is the kernel of the short exact sequence

$$(8.25) \qquad 0 \to M_{i+1}^{n-i-1} \to M_i^{n-i} \xrightarrow{v_i} M_i^{n-i} \to 0 .$$

For $n = 1$ and $i = 0$, this is the sequence 8.21, and one could write

$$M_i^{n-i} = v_n^{-1}BP_*/(p, v_1, \cdots v_{i-1}, v_i^\infty, v_{i+1}^\infty \cdots v_{n-1}^\infty).$$

In particular, $M_n^0 = v_n^{-1}BP_*/I_n$. Each sequence 8.25 gives a long exact sequence of Ext groups and a Bockstein type spectral sequence going from $\mathrm{Ext}\, M_{i+1}^{n-i-1}$ to $\mathrm{Ext}\, M_i^{n-i}$. Hence, once can in principle compute $\mathrm{Ext}\, M^n$ in terms of $\mathrm{Ext}\, v_n^{-1}BP_*/I_n$, which is accessible through the theory described earlier in this section. In particular, Theorem 8.15 gives a vanishing parabola (instead of a vanishing line), i.e.

<u>Corollary 8.26</u> In the chromatic spectral sequence, $E_1^{s,t} = 0$ if $(p-1)\nmid s$ and $t > s^2$. □

We will now explain how one can use this apparatus to prove Theorems 5.10 and 5.14(a). One can set up chromatic spectral sequences converging to Ext BP_*/I_n by making a long exact sequence

(8.26) $$0 \to BP/I_n \to M_n^0 \to M_n^1 \to M_n^2 \quad \cdots ,$$

where the M_n^i are defined by 8.25. One gets

Theorem 8.27 The long exact sequence 8.26 leads to a first quadrant cohomology spectral sequence converging to Ext BP_*/I_n with $E_1^{s,t} = \mathrm{Ext}^t M_n^s$. □

In the case $n = 1$, we know $E_1^{0,t} = \mathrm{Ext}^t v_1^{-1} BP_*/p$ (Examples 8.10 and 8.11). The image of Ext BP_*/p in this group is simply the subgroup of elements which are permanent cycles in the chromatic spectral sequence. The differentials originating in $E_r^{0,t}$ are easily computed in this case and one finds $E_2^{0,t} = E_\infty^{0,t}$.

Finally, we will explain our use of the word 'chromatic' and define nth order phenomena in the Novikov E_2-term. Both terms refer to various types of periodicity. Ext $v_n^{-1} BP_*/I_n$ is v_n-periodic, i.e. multiplication by v_n induces an isomorphism between $\mathrm{Ext}^{s,k} v_n^{-1} BP_*/I_n$ and $\mathrm{Ext}^{s,k+2(p^n-1)} v_n^{-1} BP_*/I_n$. Moreover, M^n can be shown to be a direct limit of comodules in which increasinly large powers of v_n give similar isomorphisms. Specifically,

Proposition 8.28 Let $M^n(i) = v_n^{-1} BP_*/(p^{1+i}, v_1^{p^i}, v_2^{p^{2i}}, \cdots v_{n-1}^{p^{(n-1)i}})$. Then multiplication by $p(v_1 v_2^2 \cdots v_{n-1}^{n-1})^{(p-1)p^i}$ gives a comodule map $M^n(i) \to M^n(i+1)$ and $M^n = \varinjlim M^n(i)$. Moreover, $v_n^{p^{ni}} \in \mathrm{Ext}^0 M^n(i)$ and multiplication by it gives an isomorphism $\mathrm{Ext}^{s,k} M^n(i) \xrightarrow{\approx} \mathrm{Ext}^{s,k+2p^{ni}(p^n-1)} M^n(i)$. □

Since $\mathrm{Ext}\, M^n = \varinjlim \mathrm{Ext}\, M^n(i)$, the former is a direct limit of periodic groups under periodic maps, or weakly periodic. Each element of Ext $M^n(i)$ can be multiplied nontrivially by $v_n^{p^{ni}}$ and we call this property <u>nth order periodicity</u>.

Hence by <u>nth order phenomena in the Novikov spectal sequence</u> we mean the subquotient of Ext BP_* isomorphic to $E_\infty^{n,*}$ (the nth column) of the chromatic spectral sequence, and related homotopy elements.

We see then that the filtration of Ext BP_* for which the chromatic E_∞-term is the associated trigraded group, is the filtration by order of periodicity. The chromatic spectral sequence is like a spectrum in the astronomical sense that it resolves the Novikov E_2-term Ext BP_* into various 'wavelengths' or orders of periodicity. Hence the adjective 'chromatic'.

References

1. J. F. Adams, On the groups J(X), IV, Topology 5(1966), 21-71.

2. J. F. Adams, Lectures on generalized cohomology, Lecture Notes in Math., Vol. 99 (Springer-Verlag, 1969).

3. J. F. Adams, Localization and completion with an addendum on the use of Brown-Peterson homology in stable homotopy, University of Chicago Lecture Notes in Mathematics, 1975.

4. J. F. Adams, On the nonexistence of elements of Hopf invariant one, Ann. of Math. 72(1960), 20-103.

5. J. F. Adams, A periodicity theorem in homological algebra, Proc. Cambridge Phil. Soc. 62(1966), 365-377.

6. J. F. Adams, Stable homotopy and generalized homology, University of Chicago Press, 1974.

7. J. F. Adams, Stable homotopy theory, Lecture Notes in Math., Vol. 3 (Springer-Verlag, 1966).

8. J. F. Adams, On the structure and applications of the Steenrod algebra, Comm. Math. Helv. 32(1958), 180-214.

9. S. Araki, Typical formal groups in complex cobordism and K-theory, Kinokumiya Book-Store, Kyoto, 1974.

10. M. G. Barratt, M. E. Mahowald, and M. C. Tangora, Some differentials in the Adams spectral sequence-II, Topology, 9(1970), 309-316.

11. A. K. Bousfield, Types of acyclicity, J. Pure Appl. Algebra 4(1974), 293-298.

12. A. K. Bousfield and D. M. Kan, Homotopy limits, completions and localizations, Lecture Notes in Math., Vol.304 (Springer-Verlag, 1972).

13. Browder, The Kervaire invariant of framed manifolds and its generalizations, Ann. of Math. 90(1969), 157-186.

14. V. M. Buhstaber and S. P. Novikov, Formal groups, power systems and Adams operators, Math. USSR Sbornik 13(1971), 70-116.

15. H. Cartan and S. Eilenberg, Homological Algebra, Princeton University Press, 1956.

16. P. Cartier, Modules associés à un groupe formel commutatif. Courbes typiques, C. R. Acad. Sci. Paris, 265(1967), A129-132.

17. A. Fröhlich, Formal groups, Lecture Notes in Math., Vol. 74 (Springer-Verlag, 1968).

18. M. Hazewinkel, Formal groups and applications, Academic Press (to appear).

19. M. Hazewinkel, A universal formal group and complex cobordism, Bull. A.M.S. 81(1975), 930-933.

20. D. C. Johnson, H. R. Miller, W. S. Wilson, and R. S. Zahler, Boundary homomorphisms in the generalized Adams spectral sequence and the non-triviality of infinitely many γ_t in stable homotopy Reunion sobre teoria de homotopia, Northwestern Univ. 1974, Soc. Mat. Mexicana, 1975, 47-59.

21. P. S. Landweber, Annihilator ideals and primitive elements in complex bordism, Ill. J. Math 17(1973); 273-283.

22. P. S. Landweber, $BP_*(BP)$ and typical formal groups, Osaka J. Math. 12(1975), 357-369.

23. M. P. Lazard, Commutative formal groups, Lecture Notes in Math., Vol. 443 (Springer-Verlag, 1975).

24. M. P. Lazard, Groupes analytiques p-adiques, IHES Pub. Math. No. 26(1965).

25. M. P. Lazard, Sur les groupes formels a un parametre, Bull. Soc. Math. France, 83(1955) 251-274.

26. A. Liulevicius, The factorization of cyclic reduced powers by secondary cohomology operations, Mem. Amer. Math. Soc. 42(1962).

27. M. E. Mahowald, The metastable homotopy of S^n, Memoirs A.M.S. 72, 1967.

28. M. E. Mahowald, A new infinite family in ${}_2\pi_*^S$, Topology 16 (1977), 249-256.

29. M. E. Mahowald, Some remarks on the Arf invariant problem from the homotopy point of view, Proc. Symp. Pure Math. A.M.S. Vol. 22.

30. M. E. Mahowald and M. C. Tangora, On secondary operations which detect homotopy classes, Bol. Soc. Math. Mexicana (2) 12(1967), 71-75.

31. M. E. Mahowald and M. C. Tangora, Some differentials in the Adams spectral sequence, Topology 6(1967), 349-369.

32. J. P. May, The cohomology of restricted Lie algebras and of Hopf algebras, J. Alg. 3(1966), 123-146.

33. J. P. May, The cohomology of restricted Lie algebras and of Hopf algebras; application to the Steenrod algebra, Thesis, Princeton University 1964.

34. J. P. May, Matric Massey products, J. Alg. 12(1969), 533-568.

35. R. J. Milgram, The Steenrod algebra and its dual for connective K-theory, Reunion sobre teoria de homotopia, Northwestern Univ. 1974, Soc. Mat. Mexicana, 1975, 127-158.

36. H. R. Miller, Some algebraic aspects of the Adams-Novikov spectral sequence, Thesis, Princeton University, 1974.

37. H. R. Miller and D. C. Ravenel, Morava stabilizer algebras and the localization of Novikov's E_2-term, Duke Math. Journal 44 (1977) 433-446.

38. H. R. Miller, D. C. Ravenel, and W. S. Wilson, Novikov's Ext^2 and the nontriviality of the gamma family, Bull. Amer. Math. Soc., 81(1975), 1073-1075.

39. H. R. Miller, D. C. Ravenel, and W. S. Wilson, Periodic phenomena in the Adams-Novikov spectral sequence, Ann. of Math. (to appear).

40. H. R. Miller and W. S. Wilson, On Novikov's Ext^1 modulo an invariant prime ideal, Topology, 5(1976), 131-141.

41. J. Morava, Extensions of cobordism comodules, (to appear).

42. J. Morava, Structure theorems for cobordism comodules, (to appear somewhere).

43. New York Times, editorial page, June 2, 1976.

44. S. P. Novikov, The methods of algebraic topology from the viewpoint of cobordism theories, Math. U.S.S.R.-Izvestiia 1 (1967), 827-913.

45. S. Oka, A new family in the stable homotopy groups of spheres, Hiroshima J. Math., 5(1975), 87-114.

46. S. Oka, A new family in the stable homotopy groups of spheres II, Hiroshima J. Math. 6(1976), 331-342.

47. S. Oka, Realizing some cyclic BP_*-modules and applications to homotopy groups of spheres, Hiroshima Math: J. 7(1977), 427-447.

48. S. Oka and H. Toda, Nontriviality of an element in the stable homotopy groups of spheres, Hiroshima Math. J. 5(1975), 115-125.

49. D. G. Quillen, The Adams conjecture, Topology 10(1971), 1-10.

50. D. G. Quillen, On the formal group laws of unoriented and complex cobordism, Bull. A.M.S. 75(1969), 115-125.

51. D. C. Ravenel, The cohomology of the Morava stabilizer algebras, Math. Z. 152(1977), 287-297.

52. D. C. Ravenel, Computations with the Adams-Novikov spectral sequence at the prime 3 (to appear).

53. D. C. Ravenel, Localization with respect to certain periodic homology theories, to appear.

54. D. C. Ravenel, A May spectral sequence converging to the Adams-Novikov E_2-term, (to appear).

55. D. C. Ravenel, A new method for computing the Adams-Novikov E_2-term, (to appear).

56. D. C. Ravenel, The nonexistence of odd primary Arf invariant elements in stable homotopy, Math. Proc. Cambridge Phil. Soc. (to appear).

57. D. C. Ravenel, The structure of BP_*BP modulo an invariant prime ideal, Topology 15(1976), 149-153.

58. D. C. Ravenel, The structure of Morava stabilizer algebras, Inv. Math. 37(1976), 109-120.

59. D. C. Ravenel and W. S. Wilson, The Hopf ring for complex cobordism, J. of Pure and Applied Algebra (to appear).

60. Science. June 7, 1976.

61. N. Shimada and T. Yamamoshita, On the triviality of the mod p Hopf invariant, Jap. J. Math. 31(1961), 1-24.

62. C. L. Siegel, Topics in Complex Function Theory, Vol I. Wiley-Interscience, 1969.

63. L. Smith, On realizing complex bordism modules, Amer. J. Math. 92(1970) 793-856.

64. L. Smith, On realizing complex bordism modules IV, Amer. J. Math. 99(1971), 418-436.

65. V. P. Snaith Cobordism and the stable homotopy of classifying spaces, (to appear).

66. N. E. Steenrod and D. B. A. Epstein, Cohomology operations, Ann. of Math. Studies, 50.

67. M. C. Tangora, On the cohomology of the Steenrod algebra, Math. Z. 116(1970), 18-64.

68. E. Thomas and R. S. Zahler, Nontriviality of the stable homotopy element γ_1, J. Pure Appl. Algebra 4(1974), 189-203.

69. H. Toda, Composition methods in homotopy groups of spheres, Ann. of Math. Studies 49.

70. H. Toda, Extended p-th powers of complexes and applications to homotopy theory, Proc. Japan Acad. 44(1968), 198-203.

71. H. Toda, An important relation in homotopy groups of spheres, Proc. Japan Acad. 43(1967), 893-942.

72. H. Toda, p-primary components of homotopy groups, IV, Mem. Coll. Sci., Kyoto, Series A 32(1959), 297-332.

73. H. Toda, On spectra realizing exterior parts of the Steenrod algebra, Topology 10(1971), 53-65.

74. J. S. P. Wang, On the cohomology of the mod-2 Steenrod algebra and the non-existence of elements of Hopf invariant one, Ill. J. Math 11(1967), 480-490.

75. R. S. Zahler, The Adams-Novikov spectral sequence for the spheres, Ann. of Math 96(1972), 480-504.

76. R. S. Zahler, Fringe families in stable homotopy, Trans. Amer. Math. Soc., 224(1976), 243-253.

77. M. E. Mahowald, The construction of small ring spectra, (to appear).

78. S. Oka, Ring spectra with few cells, (to appear).

Some remarks on the lambda algebra

by Martin C. Tangora

Recently there has been an increased interest in the lambda algebra, as in connection with the Brown-Gitler spectra or in Mahowald's recent work on the double suspension and the η_j family. The purpose of the present note is to allay unreasonable fears about the difficulty of working in this algebra and to discuss its utility in connection with its original raison d'être, the Adams spectral sequence.

In the first part we will assemble in one place the basic defining formulas of the algebra for all primes, and show how to generate tables of differentials and relations with ease, by means of some tricks with binomial coefficients.

In the second part we will comment on what we call the Curtis table of the Adams spectral sequence and give two examples, one in which the information in the table is very useful and another in which it is not.

The references I recommend are Whitehead's notes [10] for a brief introduction and a convenient table, Curtis's notes [3] for the full theoretical development, and Wang's paper [9] for some formal properties. All these emphasize the 2-primary case. The odd-primary case is not well documented, to my knowledge, but see Bousfield-Kan [2]. The original reference is the paper by six authors [1] but it now has chiefly historical value.

There is a lambda algebra for each prime p. In each case this is a bi-graded differential algebra over the field with p elements, which provides an E_1 term for the Adams spectral sequence. Certain sub-algebras provide E_1 terms for the unstable Adams spectral sequences for each S^n (and each p).

The lambda algebra is a relatively small model for the cobar construction and seems to contain implicitly all the information one could wish for about the cohomology of the Steenrod algebra and its unstable analogues, including chain-level constructions of all possible Massey products. Its structure is given completely by two or three formulas. Calculation of its homology in a certain bi-grading is possible, at least in theory, without prior knowledge of any other bi-grading. The algebra is Koszul in the sense of Priddy [7].

The bad news is that the algebra is not commutative, and is too large to be used for high-dimensional computations except in certain cases or with considerable support from large high-speed electronic computers. These defects are perhaps inevitable in any system which contains so much information. (Certainly non-commutativity is to be expected in any algebra mod 2 which can produce symmetric Massey products such as $\langle h_0, h_1, h_0\rangle = h_1^2$.)

1. Formulas and tables

In this section we will assemble the basic formulas for all primes and show how to make light work of the construction of the basic tables of differential and relations.

We consider first the prime 2. Here Λ is a differential graded associative algebra over the field of two elements, multiplicatively generated by elements $\lambda_0, \lambda_1, \lambda_2, \ldots$ with the natural grading $|\lambda_i| = i$. Products are subject to the generating relations

$$(1) \qquad \lambda_i \lambda_{2i+1+n} = \sum_{j\geq 0} \binom{n-j-1}{j} \lambda_{i+n-j}\lambda_{2i+1+j} \qquad (i\geq 0,\ n\geq 0)$$

and the formula for the differential is

$$(2) \qquad d(\lambda_n) = \sum_{j\geq 1} \binom{n-j}{j} \lambda_{n-j}\lambda_{j-1} \qquad (n\geq 0)$$

The relations lead us to call a monomial product "admissible" (or "allowable") if in the subscript sequence $i_1,\dots,i_s$ we have $2i_r \geq i_{r+1}$ at each place. The admissible monomials then form an additive basis. (Observe that this is a more "liberal" condition than the corresponding condition in the Steenrod algebra.)

It is convenient to suppress the lambdas from the notation and to write, for example, 11 3 3 in place of $\lambda_{11}\lambda_3\lambda_3$.

For each $n \geq 1$, denote by $\Lambda(n)$ the subspace generated by admissible monomials of which the first factor is λ_i with $i < n$. Then, for each such n, $\Lambda(n)$ is an E_1 term for the unstable Adams spectral sequence converging to the homotopy of S^n. The element $i_1\dots i_s$ has Adams filtration s and homotopy dimension Σi_r.

Exercise 1 is to calculate H^1, which is to say, determine which generators are cycles.

Anyone who is serious about calculating in this algebra will want to know how to make up tables without unreasonable effort. I would like to point out that some well known properties of binomial coefficients can make this task completely painless, at least for $p = 2$. The following lemma can be found in any good book on cohomology operations.

Old lemma: If the integers m and n are written in the number base p (p prime) as $m_j\dots m_2m_1$ and $n_k\dots n_2n_1$, then

$$\binom{m}{n} = \prod \binom{m_i}{n_i} \pmod p$$

This allows quick calculation of any binomial coefficient mod p, but to make a table one should write out the Pascal triangle mod p, using the Pascal recursion relation

$$(3)\qquad \binom{m}{n} = \binom{m-1}{n-1} + \binom{m-1}{n}$$

which holds (over the integers) for all (m,n) except (0,0).

Exercises 2, 3, and 5 are to write out the Pascal triangle

for the corresponding prime, as far as $m = 16$, 27, and 25 respectively. There is of course no Exercise 4.

With the mod 2 lambda algebra in mind, we write

$$A(k,j) = \binom{k-j-1}{j} \tag{4}$$

We need a table of $A(k,j)$. Such a table may be obtained from the mod 2 Pascal triangle by shearing it, i.e., dropping the jth column by $j-1$ rows. (I trust that you would have noticed this sooner than I did.) So the coefficient table is no more trouble to write out than the Pascal triangle. You will also observe that the jth column is periodic with period 2^e where e is the number of digits in the integer j written base p. (This follows from the Old Lemma.) The Pascal recursion relation (3) becomes

$$A(k,j) = A(k-1,j) + A(k-2,j-1) \tag{5}$$

which can be used as a check on the table.

One needs a table of the relation for $\lambda_i \lambda_j$ for i fixed and j variable, one table for each i, but all of these tables are isomorphic, as they all have the same coefficients. Even the table for the differential embeds in this same standard table, because the formula for d should have been written

$$d(\lambda_{k-1}) = \sum_{j\geq 1} \binom{k-j-1}{j} \lambda_{k-j-1}\lambda_{j-1} \qquad (k\geq 1) \tag{6}$$

which shows that d behaves just like λ_{-1}, whatever that means, and its table is the same table but with the first column deleted.

I would like to mention that the Adem relations can and should be treated in this spirit, but I will give the details elsewhere.

We turn now to the case where p is an odd prime, which is more complicated. For each odd prime p we have an algebra Λ with two kinds of multiplicative generators, $\mu_0, \mu_1, \mu_2, \ldots$ and $\lambda_1, \lambda_2, \lambda_3, \ldots$, with the grading $|\mu_i| = iq$ and $|\lambda_i| = iq-1$, where as always $q = 2(p-1)$.

For the coefficients we introduce the notation

$$(7)\qquad A(k,j) = (-1)^{j+1}\left(\binom{(k-j)(p-1)}{j} - 1\right)$$

$$(8)\qquad B(k,j) = (-1)^{j}\left(\binom{(k-j)(p-1)}{j}\right)$$

Then the generating relations are of four kinds:

$$(9)\qquad \lambda_i\lambda_{pi+k} = \sum_{j\geq 0} A(k,j)\lambda_{i+k-j}\lambda_{pi+j}$$

$$(10)\qquad \lambda_i\mu_{pi+k} = \sum_{j\geq 0}\left(A(k,j)\lambda_{i+k-j}\mu_{pi+j} + B(k,j)\mu_{i+k-j}\lambda_{pi+j}\right)$$

$$(11)\qquad \mu_i\nu_{pi+k+1} = \sum_{j\geq 0} A(k,j)\mu_{i+k-j}\nu_{pi+j+1}$$

where ν may stand for either λ or μ. There is such a set of formulas for each $k \geq 0$ with $i \geq 1$ for (9) and (10) and $i \geq 0$ for (11). The differential is given by

$$(12)\qquad d(\lambda_k) = \sum_{j\geq 1} A(k,j)\lambda_{k-j}\lambda_j \qquad (k\geq 1)$$

$$(13)\qquad d(\mu_k) = \sum_{j\geq 0} A(k,j)\lambda_{k-j}\mu_j + \sum_{j\geq 1} B(k,j)\mu_{k-j}\lambda_j \qquad (k\geq 0)$$

Thus a monomial product is admissible if at each place

$$(14)\qquad pi_k \geq i_{k+1} \quad \text{if} \quad \nu_k = \mu_k$$

$$(15)\qquad pi_k - 1 \geq i_{k+1} \quad \text{if} \quad \nu_k = \lambda_k$$

These formulas were set down by Bousfield and Kan, incidentally correcting a mistake in sign in the original six-author paper (where the odd-primary case was relegated to an appendix).

Once you have seen the trick of shearing the Pascal triangle, you want to use it everywhere, but now the signs seem to interfere.

One option is to generate the tables of the coefficients $A(k,j)$ and $B(k,j)$ intrinsically by means of the recursion formula

$$(16)\qquad A(k,j) \equiv \sum_{i=0}^{p-1} A(k-i-1, j-i) \pmod p$$

which is derived from the rth-order Pascal recursion formula

$$(17)\qquad \binom{m}{n} = \sum_{i=0}^{r}\binom{r}{i}\binom{m-r}{n-r+i}$$

with $r = p-1$. The exceptional values are $m < r$ for (16) and therefore $k = j+1$ ($\leq p-2$) for (17).

Formula (16) holds good with B replacing A throughout, the only exceptional value being $k = j = 0$.

Alternatively we try to write out the entire coefficient triangle at once from the Pascal triangle. The definition (7) indicates that we are to write the Pascal triangle mod p, change certain signs, discard all the rows not congruent to -1 mod $p-1$, and shear. Similar steps are required for the B triangle.

However, both triangles can be generated at one stroke, as follows: Write the Pascal triangle; change all the signs in checkerboard fashion, i.e., multiply each entry by $(-1)^{m+n}$ where m and n are the row and column indices; extract the rows of interest into two separate triangles, and shear each. It may seem inefficient to change the signs throughout the entire Pascal triangle when we will only be using two rows out of every $p-1$ rows (if $p=3$ this objection already loses force), but on the contrary, the checkerboard sign change is easier to accomplish than any other, because of the following proposition.

<u>Reflection Lemma</u>: Let m and n be non-negative integers and let p^e be any power of the prime p which is greater than m and n. Then

$$(18) \qquad \binom{m}{n} \equiv (-1)^{m+n} \binom{x-n}{x-m} \pmod p \quad \text{where } x = p^e - 1.$$

Thus the checkerboard sign change can be accomplished by reflecting part of the Pascal triangle in a median -- or better yet, simply by relabelling the triangle.

I do not know whether (18) is a new identity. It can be proved by reducing to the special case $e = 1$ (i.e., $m,n < p$) using the Old Lemma, and then proving the special case by induction on m.

To illustrate this process, which may sound harder than it really is, we show how the first five rows of the B table for p = 3 are obtained from the first nine rows of the Pascal triangle mod 3. We write '+' for +1 and '-' for -1.

Here are the first nine rows before and after the checkerboard or reflection sign change:

```
+                    +
+ +                  - +
+ - +                + + +
+ 0 0 +              - 0 0 +
+ + 0 + +            + - 0 - +
+ - + + - +          - - - + + +
+ 0 0 - 0 0 +        + 0 0 + 0 0 +
+ + 0 - - 0 + +      - + 0 - + 0 - +
+ - + - + - + - +    + + + + + + + + +
```

These rows are indexed 0 through 8. The B triangle is based on the rows divisible by p-1 = 2, viz.,

```
+
+ + +
+ - 0 - +
+ 0 0 + 0 0 +
+ + + + + + + + +
```

It only remains to shear, so that B(k,j) is the entry in the kth row and j^{th} column of

```
+
+
+ +
+ - +
+ 0 0
```

where the rows and columns are indexed beginning with 0, and where we have only written five rows because the sixth row requires the tenth row of the Pascal triangle which is out of range.

2. The Curtis table

I am using the term "Curtis table" for the kind of table of the Adams spectral sequence which appears in Curtis's notes [3] and in Whitehead's book [10]. I hope the term is not an injustice to Salomonsen or to Whitehead.

The table is a display chart for the stable and unstable Adams spectral sequence -- for p = 2 in all the references. It lists in a very abbreviated form all the "interesting" elements in the homology of the lambda algebra.

What information do we seek from the algebra? We want the structure of the E_2 term of the Adams spectral sequence, stable and unstable, first of all the additive structure, and eventually the multiplicative structure; and as much as possible of the structure of the homotopy groups, including sphere of origin and Hopf invariant.

The Curtis table is designed to give the stable and unstable additive structure for all S^n, and also Hopf invariant and sphere of origin. This is a wealth of information in a very compact and readable form. However, if we want to investigate products, especially higher products, the table is only a point of departure. We will look at two examples to show that for such purposes the table is sometimes helpful and sometimes not.

This is not meant to be an argument against the table. Like the famous Toda tables, this table should be on the desk of evryone who wants information about homotopy groups. Moreover, the computation of the table can be done by machine. In fact the computation of the table is an interesting problem in its own right, which I plan to discuss elsewhere. But for our present purpose we will take the table as a given.

The examples are stable and 2-primary. The coordinates of the table are $t-s$ and s, and we will say that an element is "at $(t-s,s)$".

<u>First example:</u> $t - s = 23$, $s = 7$.

At (23,7) the May spectral sequence ([6],[8]) told us that there is a single generator for Ext which has been denoted i. In the Curtis table this element is denoted 13 1 2 4 1 1 1. This means that the element in question first appears in the unstable Adams spectral sequence for the 14-sphere (since it begins with 13) and that its Hopf invariant is the element 1 2 4 1 1 1 at (11,6), i.e., $P^1h_1^{\ 2}$. Finally, it means that the minimal representative for this (co-)homology class in Λ has 13 1 2 4 1 1 1 as its leading (maximal) term, but we do not know what other terms we may have to add to get a (co-)cycle.

Suppose we want to investigate the role played by i in the multiplicative structure of Ext -- information which may be needed in determining Adams differentials. From the May spectral sequence we can easily learn such relations as $h_0h_2i \neq 0$, which is enough to show that $d_2(i) \neq 0$ in the stable Adams spectral sequence, and such as i^2 and i^3 being non-zero. We can also deduce from the May spectral sequence alone that i is represented by the Massey products

(19) $\langle h_0, c_0, d_0\rangle \doteq i = \langle h_0^{\ 2}h_3, h_3, h_0^{\ 4}, h_3\rangle$

Other important relations are difficult or impossible to arrive at by such means. The important fact that $h_3i \neq 0$ has been proved by a sequence of Massey product manipulations starting from the relation $h_0^{\ 4}x = h_0e_0g$ which was obtained from homotopy information [4]. The fact that all powers of i are non-zero has been proved by entirely different methods [5]. The question whether $iy = 0$ at (61,13) is still open.

So we may be motivated to seek a cycle representative for i in Λ. We must complete to a cycle the leading term given by the Curtis table. There is an algorithm for this process which we will not discuss now, but I have carried out this calculation and it has the disturbing dénouement that the minimal representative has 254 terms.

Imagine trying to check whether this 254-term polynomial multiplied by some other polynomial is indeed homologous at (61,13) to such-and-such. Or trying to verify the Massey products (19) using this polynomial. Surely this cycle is of little utility -- although for theoretical purposes it is important, and its leading term has told us important information.

I can show that there is another representative with only a single term, but that would take us off on a long sidetrack.

The conclusion seems to be that in some instances the Curtis table, though helpful for sphere of origin and Hopf invariant, is not helpful for multiplicative information.

Second example: $t - s = 52$, $s = 5$.

At (52,5) the May spectral sequence gave us a generator D_1 of Ext but this element does not seem to be related to its neighbors. The question naturally arises whether $h_0D_1 = 0$ at (52,6), since there is an element at (52,6) of lower May filtration. The answer to this question eluded me for years but I have recently found it with the help of the Curtis table.

You may well object that the Curtis table does not go out to the 52-stem, but we are not considering that problem here.

Suppose, though, that you want to find a representative of D in the lambda algebra. Without the Curtis table, you would have to calculate the homology of Λ at the bi-grading (52,5). Even if you had a cycle which you believed to be the right class, you would have to study the boundaries from (53,4) to be sure that your candidate was not a boundary. Now at (53,4) there are about two thousand six hundred elements in the vector-space basis, so you would have to look at linear combinations of the boundaries of well over two thousand monomials. With the Curtis table, you learn that only about a dozen of these monomials are really significant.

The Curtis table, then, tells us that the minimal representative for D_1 has leading term 4 7 11 15 15. Thus the element D_1 is already present in the Adams spectral sequence for S^5 and it has h_3c_2 at (48,4) as its Hopf invariant.

If we try to complete this to a cycle, we find that this monomial is already itself a cycle.

This is a real windfall, and there is another windfall in the fact, almost immediate from the relations (1), that left multiplication of this cycle by λ_0 gives zero.

Proposition: $h_0D_1 = 0$ at (52,6).

You lose some, you win some.

University of Illinois at Chicago Circle
Chicago, Illinois 60680
Research partially supported by the Science Research Council through Oxford and by the National Science Foundation through UICC. Computing services were provided by the UICC Computer Center.

References

[1] Bousfield, Curtis, Kan, Quillen, Rector, and Schlesinger, Topology 5 (1966), 331-342. MR 33 #8002.

[2] Bousfield, A.K., and D.M. Kan, The homotopy spectral sequence etc., Topology 11 (1972), 79-106, especially pp. 101-102. MR 44 #1031.

[3] Curtis, E.B. Simplicial homotopy theory. Lecture Notes, Aarhus Universitet, 1967. MR 42 #3785. Reprinted, slightly revised and enlarged, in Advances in Math. 6 (1971) 107-209. MR 43 #5529. The table is on p. 104 of the Aarhus notes (to the 23-stem) and p. 190 of Advances (to the 16-stem).

[4] Mahowald, M.E., and M.C. Tangora, Some differentials in the Adams spectral sequence, Topology 6 (1967) 349-369. MR 35 #4924. (See Proposition 5.1.3.)

[5] Margolis, H.R., S.B. Priddy and M.C. Tangora, Topology 10 (1971) 43-46. MR 45 #9318.

[6] May, J.P. Dissertation, Princeton, 1964.

[7] Priddy, S.B., Koszul resolutions, Trans.A.M.S. 152 (1970) 39-60. MR 42 #346.

[8] Tangora, M.C., Math.Z. 116 (1970), 18-64. MR 42 #1112.

[9] Wang, J.S.P., On the cohomology of the mod-2 Steenrod algebra etc., Illinois J. Math. 11 (1967), 480-490. MR 35 #4917.

[10] Whitehead, G.W. Recent advances in homotopy theory. Regional Conference Series (A.M.S. - Conference Board), 1970. MR 46 #8208. The table is on pp. 71-73 (to the 22-stem).